T0402113

Springer Theses

Recognizing Outstanding Ph.D. Research

For further volumes:
http://www.springer.com/series/8790

Aims and Scope

The series "Springer Theses" brings together a selection of the very best Ph.D. theses from around the world and across the physical sciences. Nominated and endorsed by two recognized specialists, each published volume has been selected for its scientific excellence and the high impact of its contents for the pertinent field of research. For greater accessibility to non-specialists, the published versions include an extended introduction, as well as a foreword by the student's supervisor explaining the special relevance of the work for the field. As a whole, the series will provide a valuable resource both for newcomers to the research fields described, and for other scientists seeking detailed background information on special questions. Finally, it provides an accredited documentation of the valuable contributions made by today's younger generation of scientists.

Theses are accepted into the series by invited nomination only and must fulfill all of the following criteria

- They must be written in good English.
- The topic should fall within the confines of Chemistry, Physics, Earth Sciences, Engineering and related interdisciplinary fields such as Materials, Nanoscience, Chemical Engineering, Complex Systems and Biophysics.
- The work reported in the thesis must represent a significant scientific advance.
- If the thesis includes previously published material, permission to reproduce this must be gained from the respective copyright holder.
- They must have been examined and passed during the 12 months prior to nomination.
- Each thesis should include a foreword by the supervisor outlining the significance of its content.
- The theses should have a clearly defined structure including an introduction accessible to scientists not expert in that particular field.

Peter A. Byrne

Investigation of Reactions Involving Pentacoordinate Intermediates

The Mechanism of the Wittig Reaction

Doctoral Thesis accepted by
University College Dublin, Ireland

Author
Dr. Peter A. Byrne
School of Chemistry and Chemical Biology
University College Dublin, Belfield
Dublin 4
Ireland

Supervisor
Prof. Dr. Declan Gilheany
School of Chemistry and Chemical Biology
University College Dublin, Belfield
Dublin 4
Ireland

ISSN 2190-5053 ISSN 2190-5061 (electronic)
ISBN 978-3-642-32044-6 ISBN 978-3-642-32045-3 (eBook)
DOI 10.1007/978-3-642-32045-3
Springer Heidelberg New York Dordrecht London

Library of Congress Control Number: 2012943970

Printed on acid-free paper

Springer is part of Springer Science+Business Media (www.springer.com)

Parts of this thesis have been published in the following journal articles:

1. "Unequivocal experimental evidence for a unified Li salt-free Wittig reaction mechanism for all phosphonium ylide types: Reactions with β-heteroatom substituted aldehydes are consistently selective for *cis*-oxaphosphetane derived products"—P. A. Byrne, D. G. Gilheany, *Journal of the American Chemical Society*, **2012**, DOI: 10.1021/ja300943z

2. "A Convenient & Mild Chromatography-Free Method for the Purification of the Products of Wittig and Appel Reactions"—P. A. Byrne, K. V. Rajendran, J. Muldoon, D. G. Gilheany, *Organic and Biomolecular Chemistry*, **2012**, 10, 3531–3537. DOI: 10.1039/C2OB07074J

3. "*erythro* β-Hydroxyphosphonium Salts"—P. A. Byrne, H. Müller-Bunz, D. G. Gilheany, in preparation, to be submitted to *Phosphorus, Sulfur, Silicon Relat. Elem.*

4. "Anomalous *Z*-selectivity in the reactions of keto-stabilised phosphonium ylides with *ortho*-heteroatom substituted benzaldehydes"—P. A. Byrne, L. J. Higham, P. McGovern, D. G. Gilheany, in preparation, accepted for publication in *Tetrahedron Lett.*, manuscript ID TETL-D-12-01594 *Tetrahedron Letters*.

5. "Observation of Wittig reaction products, intermediates, and their quench products by variable temperature NMR"—P. A. Byrne, Y. Ortin, J. Muldoon, D. G. Gilheany, in preparation to be submitted to *Eur. J. Org. Chem.*

6. "The modern mechanistic interpretation of the mechanism of the Wittig reaction"—review paper to be submitted to *Acc. Chem. Res.*

Supervisor's Foreword

The Wittig reaction involves the formation of an alkene and a phosphine oxide from a phosphonium ylide and an aldehyde or ketone. It is perhaps the most widely used C=C bond forming reaction in chemistry.[1,2] In general, the selectivity for *E* or *Z* alkene depends on the nature of the phosphonium ylide.[3] "Reactive" ylides (R_2 = alkyl) consistently reacts with all types of aldehyde to give predominantly the *thermodynamically disfavoured* *Z*-alkene, while "stabilised" ylides (e.g. R^2 = CO_2Me) consistently gives mostly *E*-alkene in its reactions. "Semi-stabilised" (e.g. R^2 = phenyl) ylides gives mixtures containing similar proportions of *Z* and *E* alkene, with the relative quantity of each depending on the exact structure of the ylide and the reaction conditions.

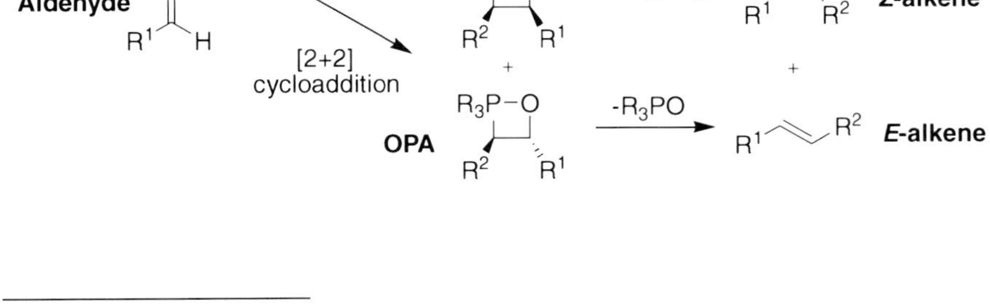

[1] Maryanoff BE, Reitz AB (1989) Chem Rev 89:863.

[2] Vedejs E, Peterson MJ (1994) In: Eliel EL, Wilen SH (eds) Topics in stereochemistry, vol 21. Wiley, New York, p 1.

[3] Johnson AW (1993) Ylides and imines of phosphorus. New York, Wiley, pp 221–305 (Chaps. 8 and 9).

Since the discovery of the Wittig reaction in 1953, its reaction mechanism has never been definitively settled, with many different variants proposed and disproved. The initially favoured mechanism that involved initial formation of a *betaine* followed by ring closure to oxaphosphetane (OPA) and subsequent OPA cycloreversion to give the products has since fallen into disfavour. In its place has risen the [2+2] cycloaddition mechanism, whereby OPA is formed directly and irreversibly from ylide + aldehyde, with the stereochemistry of the alkene being decided in the OPA-forming step.[4] Several of the developments that led to this consensus on the mechanism occurred many years apart from each other. This fact, along with the poorly understood effects of lithium cation on the reaction, and the differing selectivities observed in the reactions of different types of ylide have led to confusion on the issue in the wider world of chemistry, which is reflected in the content of many undergraduate textbooks and lectures on the Wittig reaction. The viewpoint persists that different mechanisms operate in different Wittig reactions, with some involving reversible formation of OPA or even involving betaine.

Peter's thesis describes the discovery of an effect that operates in representative examples of *all* types of phosphonium ylide under Li salt-free conditions. Reactions of both aromatic and aliphatic aldehydes bearing a heteroatom substituent in the β-position (two carbons away) relative to the carbonyl group uniformly show dramatically increased selectivity for Z-alkene or its precursor, *cis*-OPA, compared with analogous aldehydes lacking such a substituent. An example of this remarkable swing in selectivity is shown below. Ylide (**A**) reacts with benzaldehyde (**B**) to give alkene with *Z*/*E* ratio of 7:93, whereas its reaction with the β-heteroatom substituted 2-chlorobenzaldehyde (**C**) gives an alkene *Z*/*E* ratio of 94:6! The origin of the effect lies in the existence of a bonding interaction between the phosphorus and the β-heteroatom in the cycloaddition transition state leading to the OPA intermediate.

This is the first time that an effect that is common to all Wittig reactions has been discovered. Since it is consistent across all types of ylide and aldehyde, there must be a single mechanism in operation in these and therefore all other Li salt-free Wittig reactions. The effect can only be rationalised in the context of the direct cycloaddition mechanism, and so the work in this thesis shows unequivocally, and for the first time, that all Wittig reactions occur by the same [2+2] cycloaddition mechanism under Li salt-free conditions. The issue of the mechanism of one of the most important reactions in organic chemistry is thus settled.

---

[4] Vedejs E, Marth CF (1988) J Am Chem Soc 110:3948.

As a remarkable added bonus, the thesis also describes the development of a new chromatography-free method for the removal of phosphine oxide from the alkene product of the Wittig reaction, thus addressing the main operational difficulty facing an experimentalist employing this reaction. The technique involves the conversion of phosphine oxide to chlorophosphonium salt by addition of oxalyl chloride to the crude product. Cyclohexane is then added, and the alkene product is isolated by filtration of the solution to remove phosphorus and subjected to aqueous work-up. This technique can also be applied to the removal of phosphine oxide from the products of many other reactions.

Dublin, August 2012 Prof. Dr. Declan Gilheany

Contents

Symbols and Abbreviations

app d	Apparent doublet (spectral)
app t	Apparent triplet (spectral)
Ar	Aryl group
AcOH	Acetic acid
β-HPS	Beta-hydroxyphosphonium salt
CPS	Chlorophosphonium salt
c-C_6H_{11}	Cyclohexyl
COSY	Correlation NMR spectroscopy
d	Doublet (spectral)
dd	Doublet of doublets (spectral)
ddd	Doublet of doublet of doublets (spectral)
dq	Doublet of quartets (spectral)
DBP	Dibenzophosphole
DBU	1,8-Diazabicyclo[5,4,0]undec-7-ene
DMF	*N,N*-dimethylformamide
DMSO	Dimethyl sulfoxide
δ	Delta (spectral—chemical shift in ppm)
∇^2	Del squared (Laplacian derivative)
EWG	Electron withdrawing group
HSQC	Heteronuclear single quantum correlation NMR spectroscopy
HMBC	Heteronuclear multiple bond correlation NMR spectroscopy
HPLC	High pressure liquid chromatography
HRMS	High resolution mass spectrometry
Hz	Hertz
J	NMR coupling constant, in Hz
k	Rate constant
KHMDS	Potassium hexamethyldisilazide (potassium bis(trimethylsilyl)amide)
LiHMDS	Lithium hexamethyldisilazide (lithium bis(trimethylsilyl)amide)
lit.	Literature reference
m	Meta
m	Multiplet (spectral)

$[M]^+$	Mass of molecular ion (mass spectrometry)
MHz	Megahertz
MP	Melting point
NOESY	Nuclear overhauser effect NMR spectroscopy
o	Ortho
OPA	Oxaphosphetane
p	Para
π	Pi (orbital)
q	Quartet (spectral)
ρ	Rho (Hammett plot slope)
$\rho(\mathbf{r})$	Charge density at position **r**
SCOOPY	α-substitution plus carbonyl olefination via β-oxido phosphonium ylides
s	Singlet (spectral)
σ	Sigma (orbital)
σ^*	Sigma star (antibonding orbital)
TLC	Thin layer chromatography
TOCSY	Total correlation NMR spectroscopy
TS	Transition state
UV	Ultraviolet (spectroscopy)

Chapter 1
Introduction to the Wittig Reaction and Discussion of the Mechanism

1.1 The Wittig Reaction: Introduction, Utility and Recent Developments

The Wittig reaction [1] is perhaps the most commonly used method for the synthesis of alkenes. Several excellent reviews on the topic have previously been written [2–5]. The reaction (see Scheme 1.1) occurs between a carbonyl compound (aldehyde or ketone in general, **2**) and a phosphonium ylide (**1**). The latter species is a carbanion stabilised by an adjacent phosphorus substituted with three carbons, giving alkene (**3**) and phosphine oxide (**4**) as the by-product. The ylide can be represented by resonance structures **1a** (fully ionic ylide form) and **1b** (ylene form), which show between them the ionic character of the P–C bond and the contribution to the stabilisation of the carbanion by phosphorus.

There has been much discussion in the literature over the exact nature of the ylide P–C_α bond [6–8], and the nature of similar bonds in related structures such as the P–C bond in phosphorus stabilised carbanions [9], the P=O bond in phosphine oxides, [8, 10, 11] and both the P–N and the P–C bonds of iminophosphoranes [12]. Studies based on topological analysis of the Laplacian derivative of the charge density, $\nabla^2\rho(\mathbf{r})$ [7, 10, 12], indicate that the P–O bond of phosphine oxide and the P–C and P–N bonds of iminophosphorane are composed of a polar covalent σ bond and an electrostatic interaction (essentially an ionic bond) between the positively charged phosphorus atom and the lone pair(s) on atom Z (where Z = C, N or O, the atom bound to phosphorus). On this basis, there is no π-bonding interaction of the lone pairs of Z with the σ^* orbitals of the bonds of the other substituents to phosphorus, so phosphorus is not formally hyperconjugated. Similar topological analysis of $\nabla^2\rho(\mathbf{r})$ for phosphonium ylides suggests that there may be "negative hyperconjugation" or back-bonding of the lone pair on the ylide α-carbon into the σ^* orbitals of the bonds of phosphorus to its other substituents, although it indicates that the P–C bond is still heavily polarised towards carbon [6, 7]. Leyssens and Peeters [9] showed in a recent publication that negative hyperconjugation is in

P. A. Byrne, *Investigation of Reactions Involving Pentacoordinate Intermediates*, Springer Theses, DOI: 10.1007/978-3-642-32045-3_1,

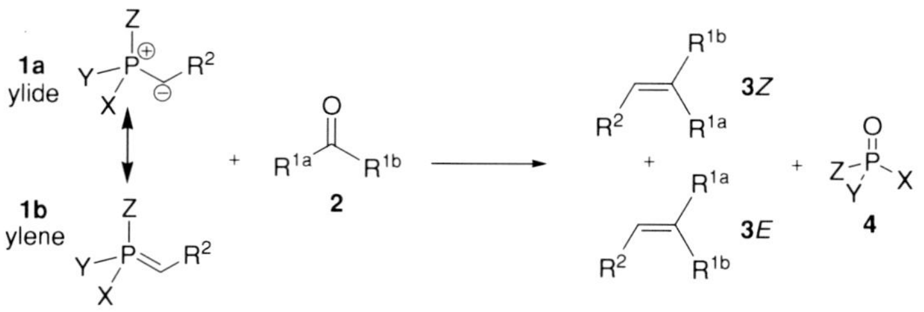

Scheme 1.1 The Wittig Reaction. X, Y and Z may each be alkyl, aryl or alkoxy, and needn't necessarily be the same. R^2 may be alkyl, aryl, vinyl, or an electron withdrawing group (e.g. an ester). The carbonyl reactant (**2**) may be formaldehyde ($R^{1a} = R^{1b} = H$), an aldehyde (R^{1a} = alkyl or aryl, $R^{1b} = H$), or a ketone (R^{1a} = alkyl or aryl, R^{1b} = alkyl or aryl)

operation in the bonding of phosphorus-stabilised carbanions through calculations using the "finite difference Fukui function". This involved the determination of the change in the electron density distribution for the carbanion upon ionisation (removal of an electron from the carbanion lone pair). Electron density was found to decrease in regions corresponding to the σ^* acceptor orbitals of the bonds of phosphorus to its other substituents, thus showing that these regions contain some of the electron density of the carbanion lone pair. They also showed that certain physical characteristics of the carbanion that differ from those of the neutral parent compound are as a consequence of negative hyperconjugation in the carbanion, such as a shorter P–C bond and longer bonds and smaller bond angles for the bonds of phosphorus to its other substituents. Ylides will be represented in the ylene form (**1b**) throughout this thesis.

There are a number of features of the Wittig reaction that help to make its use so widespread[1]:

- The reaction is regiospecific—the alkene double bond is invariably formed between the ylide α-carbon and the carbonyl carbon of the aldehyde or ketone.
- Reaction conditions are usually mild, even in comparison to the other extremely useful regiospecific phosphorus-based olefination, the Wadsworth–Emmons reaction [2, 13].
- The starting materials are frequently easily obtainable. Many stabilised ylides (which have a conjugating group such as an ester or sulfone as the ylide α-carbon substituent R^2) and phosphonium salts are now commercially available. The same is true for aldehydes and ketones. Ylides are easily accessible by in situ deprotonation of the parent phosphonium salt, which may itself be obtained in a straightforward manner from the reaction of a phosphine with an alkylating agent, particularly if the air stable and readily available triphenylphosphine is used. Many syntheses of carbonyl reagents are also available.

[1] See Ref. [3], Chap. 8, pp. 221–273.

- Ylides are often tolerant of functional groups, meaning that the reaction can be very suitable for use in syntheses of complex molecules.
- The phosphonium group that activates the ylidic carbon in the reaction (and permits relatively facile deprotonation of the parent phosphonium salt) is eliminated as phosphine oxide in the course of alkene formation, so no separate step is required to remove it. The phosphine oxide can usually be removed—most frequently by chromatography—although in some cases this does present some difficulty. It is pertinent to briefly mention at this point that a method has been developed in the course of this project that facilitates the separation of the alkene product from phosphine oxide without the need for modification of starting materials, and which allows the regeneration of phosphine from phosphine oxide. Thus phosphine oxide removal may henceforth no longer be a major problem.
- The stereoselectivity of the reaction depends very much on the nature of the reactants being used, and frequently it is possible to direct the stereoselectivity of the reaction towards the desired alkene isomer by expedient choice of these reactants.

Despite the fact that Wittig first reported on the reaction almost 60 years ago, it still remains the subject of many modern publications. Amongst the recent developments reported in the literature are:

- New methods for generation of the ylide [14–20, 32]
- Methods for reactions showing atypical stereoselectivity [21–25]
- New reaction conditions [26–29]
- One pot reactions starting either from phosphine [15, 16, 28–34] or from alcohol [35]
- Synthetic studies [31, 36–42]
- Catalytic variants [43]
- Investigations on reaction kinetics and the relative nucleophilicity of phosphonium ylides [44]
- Mechanistic studies on reactions of ketones [45]
- Reactions in which unusual alkene *Z/E* selectivity is induced e.g. by the use of phosphonium ylides with modified "spectator substituents" on phosphorus [21, 23, 46], by the use of trialkylgallium base to generate the ylide [47], and by the addition of methanol at low temperature to the oxaphosphetane adducts from reactions of non-stabilised ylides [22]
- Computational studies on the reaction mechanism [48–50] and on the stability of heteroatom-stabilised ylides including phosphonium ylides [51]
- Selected recent examples of publications in which use had been made of Wittig reactions include the synthesis of photochromic dithienylethenes [52], the enantioselective synthesis of highly substituted cycloalkanes by organocatalysed domino Michael-Wittig reactions [53], the one-step synthesis of *Z*-allylic esters

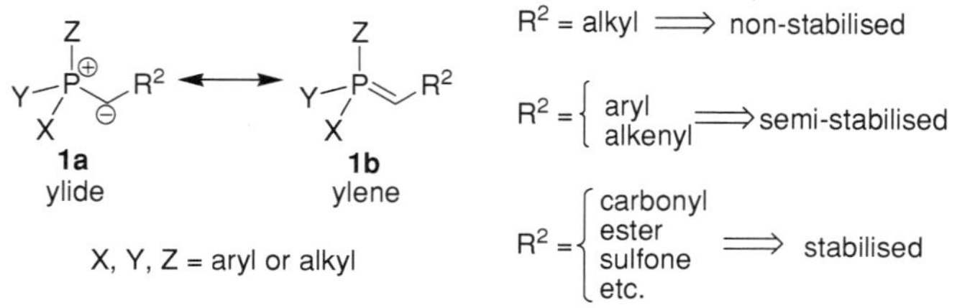

Fig. 1.1 Classification of ylides

and alcohols by a SCOOPY-modified Wittig reaction [54],[2] and the stereoselective total synthesis of (-)-spirofungin A (in which Wittig reactions are employed for several steps) [55].

The scope of the related reaction of phosphonium ylides with *N*-sulfonyl imines has also been very effectively broadened, with the development of reactions of non-stabilised, [56] semi-stabilised [57] and stabilised [58] ylides that can be tuned to give complete *E or* Z selectivity by choice of the appropriate *N*-sulfonyl alkyl group.

1.2 Classification and Reactivity Trends of Phosphonium Ylides

Phosphonium ylides are broadly categorised according to the nature of the substituent(s) attached to the α-carbon (R^2 in Fig. 1.1). If R^2 is an alkyl group, the ylide is referred to as being *non-stabilised* or *reactive*, as ylides of this type readily react with ambient moisture, and as such are not stable in air. If R^2 is a phenyl or alkenyl group, the ylide is somewhat less prone to hydrolysis owing to the conjugative stabilisation by the unsaturated group, and so these ylides are called *semi-stabilised* ylides. If R^2 is a carbonyl, ester, nitrile, sulfone or other such conjugating group, the ylide is stable in air and is thus referred to as a *stabilised* ylide.

A very significant aspect of the Wittig reaction is that, broadly speaking, the nature of the phosphonium moiety used in a Wittig reaction dictates the stereoselectivity of the reaction. Certainly, in the case of triphenylphosphine-derived ylides it is true that in general the *E* or *Z* selectivity can be predicted based on the degree of anion stabilisation conferred on the ylide by its α-substituent (R^2). These trends are set out below. For reactions of ylides derived from other phosphines, prediction of the stereochemical outcome is not so straightforward and may be further complicated by the presence or absence of dissolved salts (especially Li salts, *vide infra*).

Alkylidenetriphenylphosphoranes (Fig. 1.2a) are the most frequently used non-stabilised ylides. These show very consistently high *Z*-selectivity in Li-salt free

[2] For details on the SCOOPY-modification of the Wittig reaction, see Ref. [3], Chap. 8, pp. 241, 242.

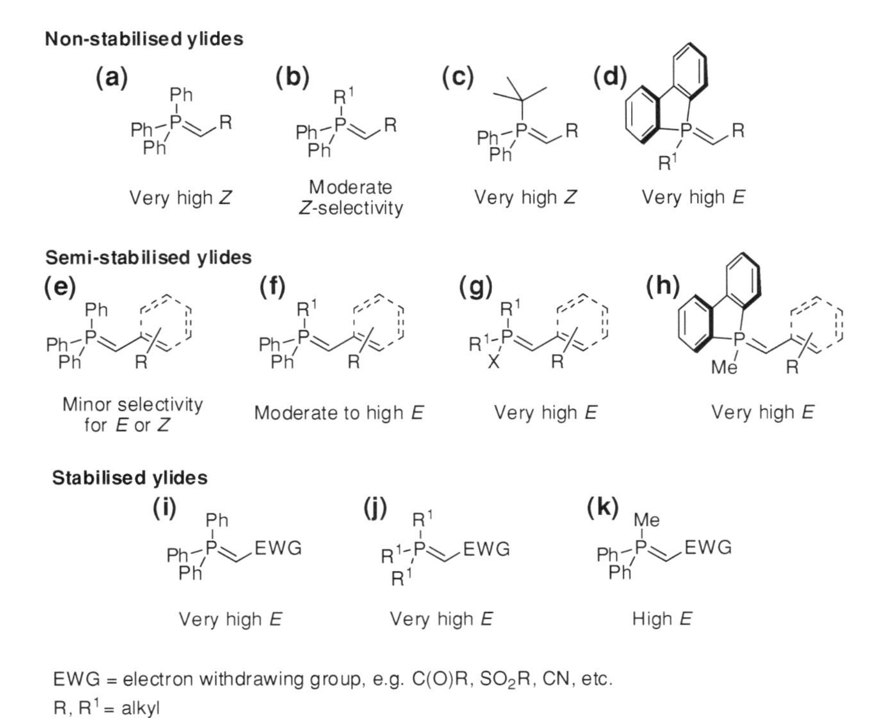

Fig. 1.2 Summary of observed selectivity trends for reactions of representative ylides

reactions with a broad range of aldehydes [4]. The only known exceptions involve reactions of ethylidenetriphenylphosphorane with aromatic aldehydes, and these reactions have been shown not to proceed under kinetic control (see Sect. 1.4.1). Selectivity is generally highest in reactions with tertiary aldehydes, although in most cases even reactions of primary aldehydes still show overwhelming selectivity for the *Z*-alkene. By comparison, (alkylidene)alkyldiphenylphosphoranes (Fig. 1.2b) typically show very significantly reduced *Z*-selectivity, with some reactions with primary aldehydes even showing moderate *E*-selectivity [59]. Alkylidene-*t*-butyldiphenylphosphoranes (Fig. 1.2c) show comparable *Z*-selectivity to alkylidenetriphenylphosphoranes. (Alkylidene)alkyldibenzophospholanes (Fig. 1.2d) show extremely high *E*-selectivity [59]. A generally applicable observation on the reactions of all of these non-stabilised ylides is that selectivity for *Z*-alkene is highest for a particular ylide in reactions with tertiary aldehydes, while reactions with primary aldehydes show the highest *E*-selectivity.

Wittig reactions of the most commonly encountered semi-stabilised and stabilised ylides, derived from triphenylphosphine, are generally not hugely selective for one isomer or another. Allylidene or benzylidene triphenylphosphoranes (Fig. 1.2e) generally react to give approximately equal proportions of *E* and

Z alkene, while semi-stabilised alkyldiphenylphoshine-derived ylides (Fig. 1.2f) show moderate and in some cases very high *E*-selectivity [4]. Semi-stabilised ylides with two or more of the *P*-phenyl groups replaced by alkyl groups (Fig. 1.2g) [4] and also those derived from methyldibenzophosphole (Fig. 1.2h) [60] show exceptionally high *E*-selectivity. As with non-stabilised ylides, the general observation applies that for a given semi-stabilised ylide *Z*-selectivity is highest for reactions with tertiary aldehydes, and *E*-selectivity is highest for reactions with primary aldehydes. Also, for a given aldehyde, *Z*-selectivity is highest with triarylphosphine-derived semi-stabilised ylides bearing bulky aryl groups, and *E*-selectivity is highest with ylides bearing two or more *P*-alkyl spectator ligands, or with methyldibenzophosphole-derived ylides.

Many stabilised ylides are air stable and can be isolated and purified, so reactions of such ylides are often carried out using the pre-formed ylide, without the need for in situ generation. Wittig reactions of stabilised ylides derived from both triphenylphosphine (Fig. 1.2i) and trialkylphosphine (Fig. 1.2j) generally show very high selectivity for *E*-alkene in polar aprotic solvents [4]. Methyldiphenylphosphine-derived ester stabilised ylides (Fig. 1.2k) show decreased although still predominant *E*-selectivity. In many cases diminished *E*–selectivity and even predominant *Z*-selectivity is observed in reactions of stabilised ylides in alcohol solvents.

1.3 Proposed Mechanisms: Description and Evaluation in the Light of Experimental Evidence

A large number of mechanisms have been proposed for this reaction throughout its history. Johnson was able to enumerate a total of eight in his review of the Wittig reaction [3]. Each of these is discussed below.

1.3.1 The Betaine Mechanism

This mechanism was originally proposed by Wittig and Schöllkopf, and initially suggested that a betaine (dipolar species **5** in Scheme 1.2) was the sole intermediate in the reaction [61].[3] It was later revised to what would now be widely regarded as the "betaine mechanism"—nucleophilic addition of ylide to carbonyl to form betaine (similar to an Aldol reaction), followed by ring closure to oxaphosphetane (OPA, species **6**) and decomposition of OPA to alkene and phosphine oxide (see Scheme 1.2) [63].

[3] This, somewhat bizarrely given the experimental evidence now available, has recently been suggested by a computational paper to be the only intermediate in the reaction of methylidenetrimethylphosphorane and ethanal in DMSO [62].

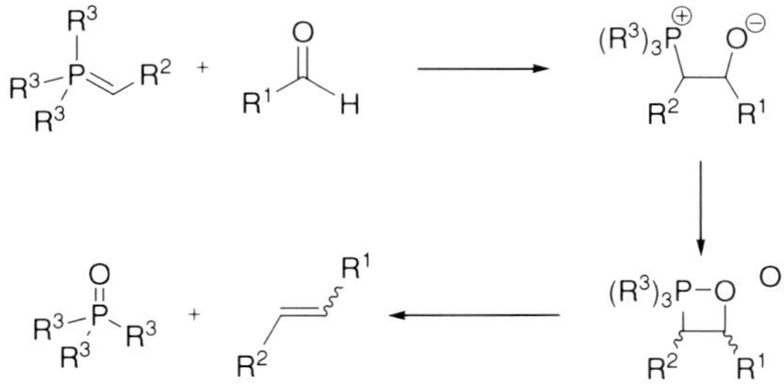

Scheme 1.2 The betaine mechanism for the Wittig reaction

At the time of their proposal, there was a significant amount of experimental evidence strongly suggesting the involvement of betaine intermediates in the Wittig reaction. There was also an absence of evidence to the contrary and the alternative explanations for the observed phenomena that have since been amassed had not yet been recognised. The most significant evidence suggesting the involvement of betaines came from the Wittig reaction of several ylides (generated using phenyllithium from the parent phosphonium bromide salt) with carbonyl compounds at low temperature, which gave a precipitate that on warming yielded alkene and phosphine oxide but if treated with HBr gave β-hydroxyphosphonium salt (β-HPS). Subsequently the precipitate was isolated and shown to be a betaine-LiBr complex [64] (which can also be detected spectroscopically) [65]. Betaines (and hence OPAs) were generated by means independent of the Wittig reaction, either by nucleophilic cleavage of epoxide (of defined stereochemistry) with phosphine [66, 67] or by deprotonation of β-HPS (of defined stereochemistry) [61, 64, 68], and were shown to decompose to alkene and phosphine oxide. Based on this evidence, it was reasonable to conclude that betaine was an intermediate in the Wittig reaction.

Some further specifics from these studies were used in explanation of the experimentally observed *Z*/*E* ratios from Wittig reactions, and are salient to the discussion:

In reactions of ylides derived from triphenylphosphine, non-stabilised ylides were known to react with high *Z*-selectivity in the absence of Li salts, and to show much decreased *Z*-selectivity in the presence of Li, while semi-stabilised ylides showed little or no selectivity, and stabilised ylides generally reacted with high *E*-selectivity. Deprotonation of *erythro*-(1-hydroxy-1-phenylprop-2-yl)triphenylphosphonium bromide (see Scheme 1.3b for definition of *erythro* and *threo* β-HPSs) with ^{t}BuOK in diethyl ether gave the *Z*-alkene product almost exclusively, indicating high stereospecificity (and thus irreversibility) of conversion of the *erythro*-betaine, and *cis*-OPA necessarily produced, to *Z*-alkene under Li-salt free conditions, while similar experiments in the presence of Li showed considerable conversion to the *E*-alkene (see Scheme 1.3) [64]. These observations are in keeping with the observation of high *Z*-selectivity in Wittig reactions of non-

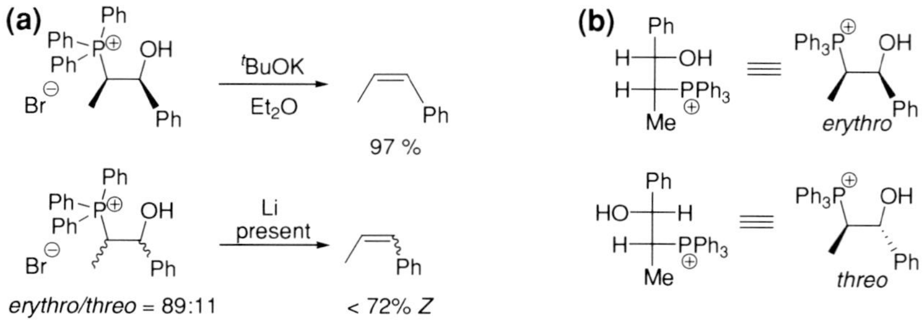

Scheme 1.3 **a** Experiments of Schlosser and Christmann on deprotonation of β-hydroxyphosphonium salts (β-HPSs) derived from non-stabilised ylide [64]. **b** Fischer projections of Schlosser's β-HPSs, showing that the β-HPS derived from *cis*-OPA is *erythro*, and that derived from *trans*-OPA is *threo*

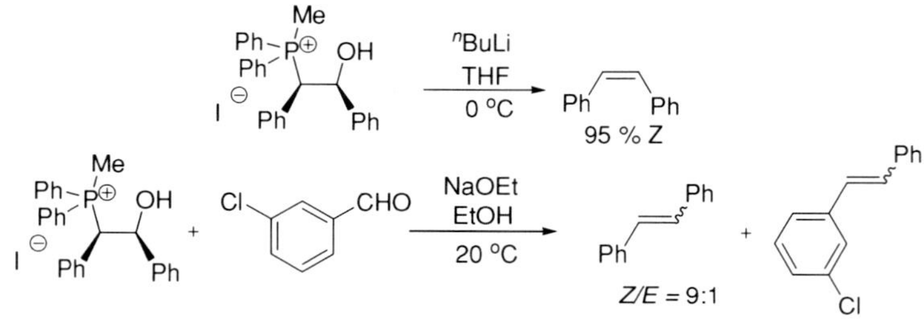

Scheme 1.4 Deprotonation experiments on β-HPSs derived from semi-stabilised ylides [68]

stabilised ylides such as ethylidenetriphenylphosphorane in the absence of Li, and of lower selectivity in the presence of Li.

Deprotonation of *erythro*-(2-hydroxy-1,2-diphenyleth-1-yl)methyldiphenylphosphonium by ^{n}BuLi (Li present) in THF showed almost completely stereospecific conversion to *Z*-alkene, indicating that this reaction too might be under kinetic control (see Scheme 1.4) [68]. Deprotonation experiments using the same β-HPS in ethanol with sodium ethoxide base in the presence of reactive aldehyde *m*-chlorobenzaldehyde gave a large amount of crossover product, while the stilbene that did form showed a *Z/E* ratio of around 90:10, indicating significant reversal of either *erythro*-betaine or *cis*-OPA to ylide and aldehyde under the reaction conditions [68]. It has since been shown that, at least in certain circumstances, the addition of methanol *at low temperature* (i.e. before OPA decomposition occurs) to the Wittig reactions of non-stabilised ylides causes very high *E*-selectivity in the reactions, which in the absence of methanol (or if it is added after warming to room temperature) would show high *Z*-selectivity [48].

The reaction of triphenylphosphine with epoxide ethyl-*trans*-2-phenylglycidate in refluxing ethanol in the presence of *m*-chlorobenzaldehyde gave a large amount of *m*-chlorocinnamate (crossover product) as well as the expected ethyl *Z*-cinnamate by the mechanism shown in Scheme 1.5 [67]. Subjecting the *cis*-epoxide to the

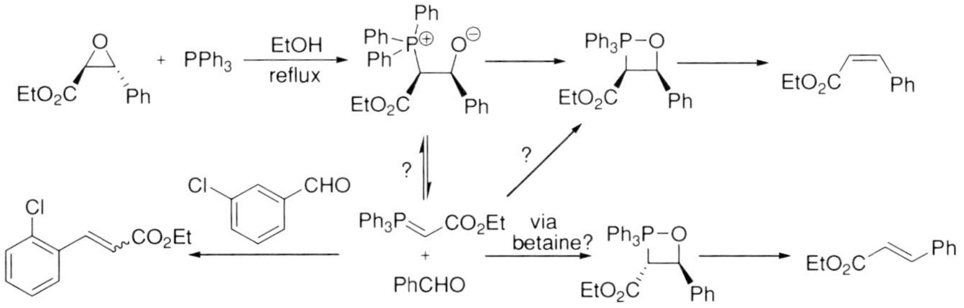

Scheme 1.5 Independent betaine generation by nucleophilic cleavage of epoxide with phosphine [67]. Reversal of OPAs derived from stabilised ylides to Wittig starting materials had not been unequivocally ruled out [67] at the time of the formulation of the betaine mechanism, so would have been considered in principle to be a possibility, as indicated in this scheme

Fig. 1.3 Two diagrams of the transition state leading to the *anti*-conformer of *erythro*-betaine, which has been proposed as the kinetically favoured path for the Wittig reaction

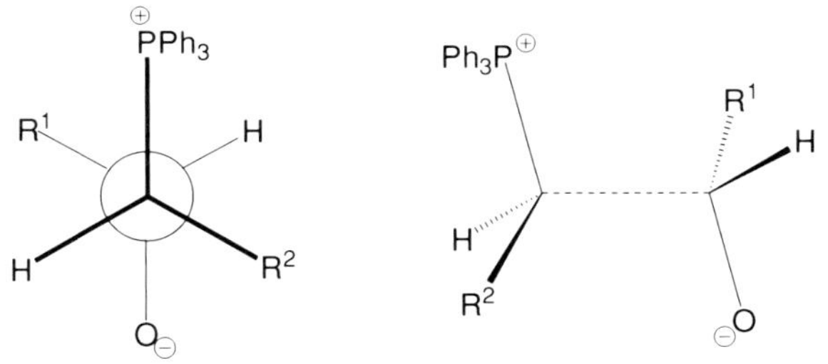

same reaction conditions gave *E*-cinnamate and a smaller proportion of crossover product.

The above experimental results were used to inform the rationalisation of the observed selectivity in Wittig reactions in the context of the betaine mechanism. The reactions of non-stabilised alkylidenetriphenylphosphoranes with aldehydes typically show high *Z*-selectivity. This was rationalised within the context of the betaine mechanism as being due to addition of the ylide to the carbonyl in such a way as to give *anti*-betaine (with a P–C–C–O dihedral angle of 180°, as shown in Fig. 1.3). The minimisation of steric repulsion in the transition state (TS) leading to an *anti*-betaine would dictate that there should also be a 180° dihedral angle between the vicinal R^1 and R^2 groups, thus giving an *anti-erythro*-betaine, which could undergo bond rotation to the *syn*-conformation and ring-close to *cis*-OPA, which ultimately leads to *Z*-alkene. So the high *Z*-selectivity was proposed to be due to kinetically favoured formation of *erythro*-betaine.

Reactions of semi-stabilised triphenylphosphine-derived ylides typically show little or no diastereoselectivity, while stabilised ylides derived from triphenylphosphine usually react with high *E*-selectivity. These observations were rationalised by postulating an increased propensity to reversal of the kinetically favoured *erythro*-betaine and/or *cis*-OPA in reactions of these ylides (especially stabilised ylides), so that there would be equilibration of the intermediates by reversal to

Wittig reactants. The increased reversal was supposed to be as a result of the greater stability, and thus longer lifetime, of the intermediates in reactions of stabilised ylides. If this scenario were true, greater *E*-selectivity should be observed if the irreversible decomposition to alkene and phosphine oxide were faster for *trans*-OPA than for *cis*-OPA, or if OPA formation were irreversible and ring-closure was faster for the *threo*-betaine than for the *erythro* isomer.

If it is shown experimentally that both isomers of a particular OPA form irreversibly and decompose stereospecifically, then this means there is no equilibration of OPAs. If individual isomers of OPA (or of alkene) can be stereospecifically generated via a betaine from a precursor of known, defined stereochemistry (e.g. *cis*-OPA from *cis*-epoxide or *erythro*-β-HPS), or indeed if a mixture of OPA isomers of known diastereomeric ratio can be generated from a mixture of diastereomers of precursors of the same known diastereomeric ratio, then this shows that *E*-selectivity is not the result of the existence of an equilibrium between the betaines and faster ring-closing of the *threo*-betaine to *trans*-OPA. This would necessarily mean that *E*-selective Wittig reactions that proceed through these intermediates are under kinetic control, and therefore that the observed high *E*-selectivity is as the result of a kinetic preference for the formation of *trans*-OPA. The betaine mechanism is therefore unable to account for kinetic *E*-selectivity.

The betaine mechanism has now been all but refuted. The experimental evidence supporting this assertion will be addressed in much greater detail in the next section (Sect. 1.4), but it is appropriate briefly to discuss why at this point. *Uncomplexed* betaines have never been observed spectroscopically in a Wittig reaction mixture, or in an independent betaine generation experiment. OPAs are also necessarily produced in each of the independent generations of betaine, and only they can be detected by NMR in solution in such experiments, or indeed in Wittig reactions [59, 60, 69–71] It has also been conclusively shown that solutions containing only OPA (as confirmed by NMR monitoring of the solution) undergo acid quenching reactions to give β-HPS [71] and react with LiBr to give a betaine-LiBr complex [65]. So, the formation of β-HPSs by acid quenching of Wittig reaction mixtures, or the formation of betaine-LiBr complexes do not require an uncomplexed betaine as the precursor. Betaines have also been shown conclusively not to be involved in the Wittig reactions of dibenzophosphole-derived non-stabilised ylides (see Sect. 1.4.2) [72]. Kinetic control has been shown to operate in all but a few exceptional Wittig reactions, so equilibration of Wittig intermediates cannot be responsible for *E*–selectivity in reactions of stabilised and semi-stabilised ylides—and thus the betaine mechanism is not in operation.

1.3.2 Bergelson's "C–P–O–C" Betaine Mechanism

Bergelson et al. proposed an alternative betaine mechanism involving initial attack of the carbonyl oxygen at phosphorus to form a "C–P–O–C" betaine (with charges on the carbon atoms), followed by ring closure to OPA and decomposition of OPA

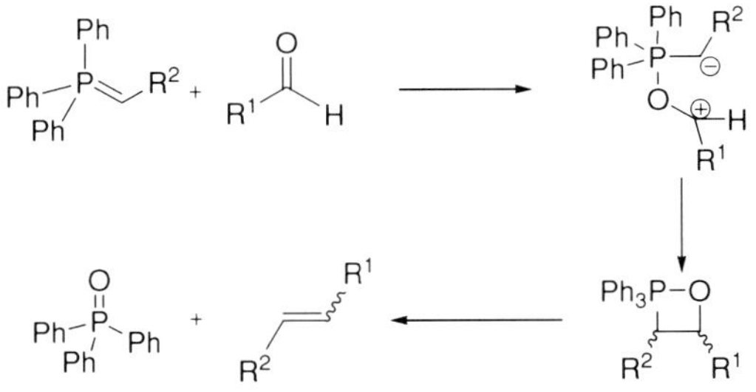

Scheme 1.6 Alternative betaine mechanism

to alkene, as shown in Scheme 1.6 [73]. This proposal was refined by Schneider to account for the observation of *Z*-selectivity in the reactions of non-stabilised ylides by considering the trigonal bipyramidal geometry of the substituent about phosphorus in the transition state (TS) that would result from initial attack of oxygen at phosphorus and in particular the steric effects of the *P*-phenyl groups in this geometry [74]. It can be assumed that the bimolecular step of this mechanism would be the rate-determining step, and thus this mechanism would predict the observation of a negative ρ value for carbonyl reactants since a positive charge is developed at the carbonyl carbon. However, carbonyl compounds have been experimentally found to show positive ρ for the Wittig reaction in reactions of all ylide types. For this reason, and because no dipolar intermediate can be observed by NMR in Wittig reactions, this mechanism has been ruled out.[4]

1.3.3 Schweizer Mechanism

On the basis of the observation that certain semi-stabilised and stabilised phosphonium ylides and conjugated carbonyl compounds reacted in alcohol solvents to give vinylphosphine oxides in addition to the expected alkene and phosphine oxide products, Schweizer et al. proposed that in this medium, these Wittig reactants initially form a betaine, which becomes protonated by the alcohol, and then undergo net elimination of water to give a vinylphosphonium salt. This could undergo attack nucleophilic attack at phosphorus by either ethoxide or hydroxide. Attack by ethoxide would give an alkoxyphosphonium salt, which could itself undergo nucleophilic attack at carbon with the concomitant formation of a P=O double bond by elimination of either the vinyl group as alkene, giving triphenylphosphine oxide (the expected Wittig product), or of a benzene, giving vinyl phosphine oxide. Hydrolysis (attack by hydroxide) would give similar results. See Scheme 1.7 [75].

P-chiral phosphonium salts are configurationally stable (i.e. the stereochemistry at phosphorus is invariant). Since the above mechanism invokes the involvement

[4] See, for example, Ref. [67], and also Sect. 1.4.3 later, in which the evidence for the currently accepted mechanism is presented.

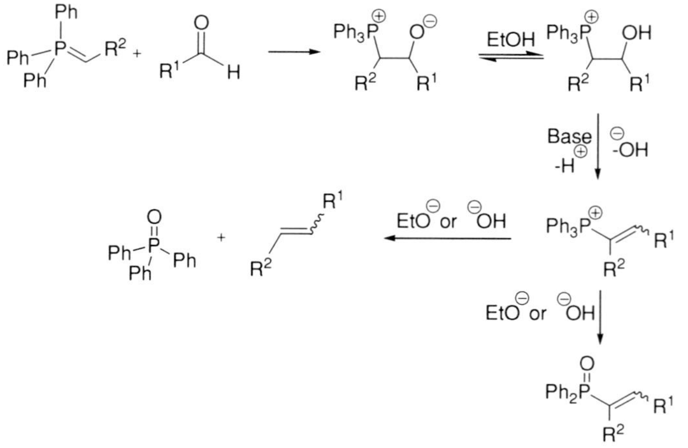

Scheme 1.7 Proposed mechanism for Wittig reactions and vinylphosphine oxide formation in alcohol solvent

of a nucleophilic attack at phosphorus, if it is in operation then Wittig reactions involving *P*-chiral phosphonium ylides should undergo at least partial inversion of configuration at phosphorus. It has been shown that Wittig reactions of (benzylidene)ethylmethylphenylphospohorane with benzaldehyde give phosphine oxide with retention of configuration at phosphorus in ether solvent [76], while the same reaction (in which the ylide was generated from the parent phosphonium salt using ethoxide base) in ethanol was shown subsequent to the publication of Schweizer's paper to proceed with retention of configuration at phosphorus [77]. It was thus proposed that formation of Wittig products by OPA formation (from betaine) and the formation of vinylphosphine oxide occur by separate pathways. It has since been shown that, at least in certain circumstances, the addition of methanol *at low temperature* (i.e. before OPA decomposition occurs) to the Wittig reactions of non-stabilised ylides causes very high *E*-selectivity in the reactions, which in the absence of methanol (or if it is added after warming to room temperature) show high *Z*-selectivity [48]. This was proposed to be as a result of β-HPS formation from OPA and methanol, and deprotonation of the β-HPS by methoxide to give β-hydroxy ylide, which can re-form OPA (non-stereospecifically) by proton transfer from the hydroxyl group to the ylidic carbon. Thus the vinylphosphonium salts produced in the study of Schweizer et al. are likely to have been produced from reaction of OPA, not betaine, with ethanol.

1.3.4 Olah's Single Electron Transfer Mechanism

The Wittig reaction has been proposed by Olah and Krishnamurthy to proceed to proceed by a one electron transfer mechanism [78], i.e. the transfer of an electron

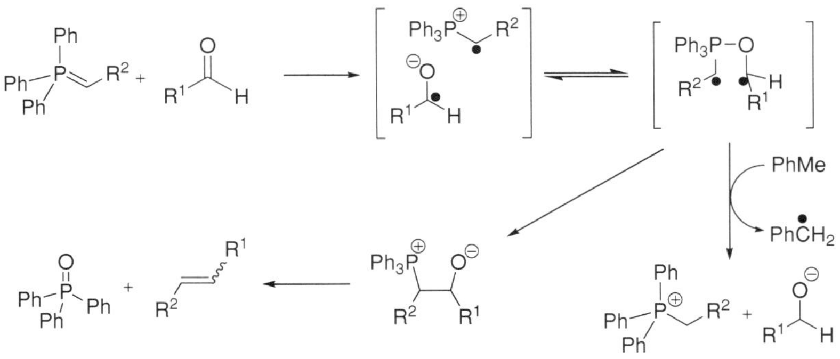

Scheme 1.8 Proposed mechanism for the Wittig mechanism involving single electron transfer from ylide to carbonyl species

from the ylide to the carbonyl compound to initially give a tight radical ion pair, supposed to be in equilibrium with a P-O bonded diradical species (see Scheme 1.8), which was thought to go on to form betaine and then alkene. This was based on the fact that attempted reactions in refluxing solvents of non-stabilised ylides with adamantanone, 4-hydroxyadamantan-2-one, bicyclo[1,3]nonan-9-one or benzophenone gave the carbonyl reduction product (alcohol) and phosphonium salt as the only products after work-up, or in addition to the Wittig products. The alcohol formation was particularly favoured in the reactions of sterically bulky ylides isopropylidenetriphenylphosphorane and (diphenylmethylidene)triphenylphosporane. The hydrogen source was the reaction solvent, as reactions in toluene gave benzylated toluenes. The one electron transfer mechanism was also advocated by Yamataka and co-workers on the basis of their observation that there is no significant kinetic isotope effect for the reaction of isopropylidenetriphenylphosphorane with benzaldehyde having a ^{14}C-labelled carbonyl group [79].

A number of tests for radical involvement in Wittig reactions are described in the review of Vedejs and Peterson [4]. One of these is presented here. The reaction of (2,3-diphenylcyclopropyl)methylidenetriphenylphosporane with 2,3-diphenylcyclopropane carboxaldehyde gives the expected Wittig product in high yield with the expected high *Z*-selectivity (see Scheme 1.9a, path A) [80]. The formation of radical ions (path B) should result in exceptionally fast ring cleavage and thus not give Wittig products, by analogy with the cyclopropylcarbinyl radicals, whose ring opening reactions show extremely large unimolecular rate constants (Scheme 1.9b) [80]. The absence of ring-opened products in the above Wittig reaction rules out the involvement of radical ions or diradicaloids such as the those in Scheme 1.9a path B.

This strongly implies that electron transfer is not an intrinsic part of the Wittig reaction mechanism, although electron transfer may be possible between suitable Wittig reactants under the right experimental conditions. In cases where it is possible, this probably leads to the formation of side-products. In one of their publications, Vedejs and Marth list a series of examples of reactions which, if radical species were involved in the manner that is proposed in this mechanism,

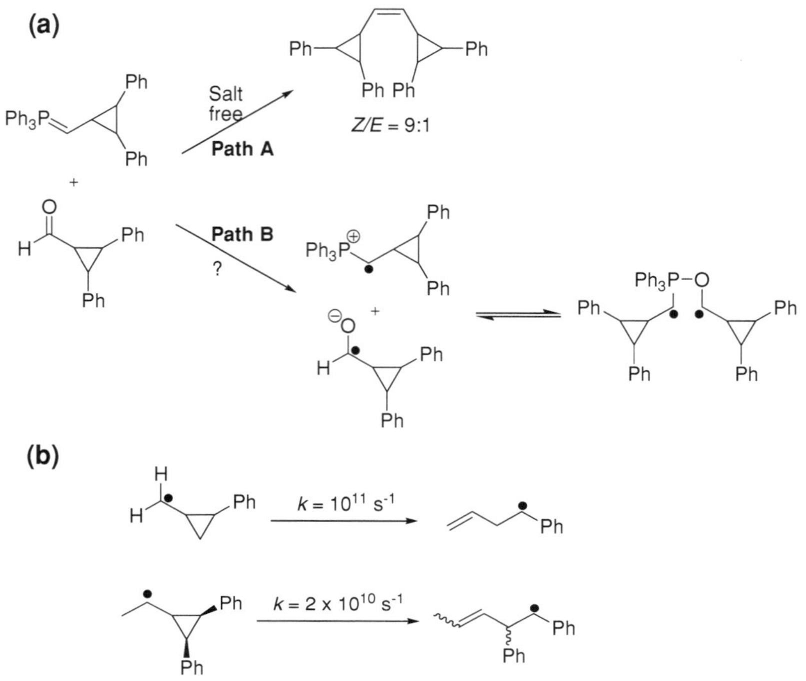

(a)
Ph$_3$P
Ph
Ph
+
O
H
Ph
Ph
Salt free
Path A
Ph
Ph
Ph Ph
Z/E = 9:1
Path B
?
Ph$_3$P
O
H
Ph$_3$P–O
(b)
H
H
Ph
k = 10^{11} s^{-1}
k = 2 x 10^{10} s^{-1}

Scheme 1.9 **a** Test for involvement of radical species in the Wittig reaction, **b** Rates of ring opening for cyclopropylcarbinyl radicals

should give products derived from radical-type reactions, but instead react to give normal Wittig products [72].

1.3.5 Bestmann's "P–O–C–C" Betaine Mechanism

Bestmann proposed a mechanism involving direct *cis*-selective cycloaddition of ylide and carbonyl to give OPA with oxygen in an axial position in the phosphorus-centred trigonal bipryramid (although no rationale for the *cis*-selectivity was presented), followed by pseudorotation about phosphorus to place oxygen in an equatorial site, subsequent cleavage of the P–C bond to give a "P–O–C–C" betaine and finally scission of the C–O bond to give alkene and phosphine oxide, as shown in Scheme 1.10 [81]. Betaines derived from non-stabilised ylides were supposed to be very short-lived and thus to quickly decompose to Wittig products, whereas betaines with a stabilising group on the carbanion carbon (i.e. derived from a stabilised ylide) were thought to be longer lived due to the greater stabilisation of the negative charge, so that rotation could occur about the C–C bond before C–O bond breakage occurred, thus giving rise to *E*-alkene.

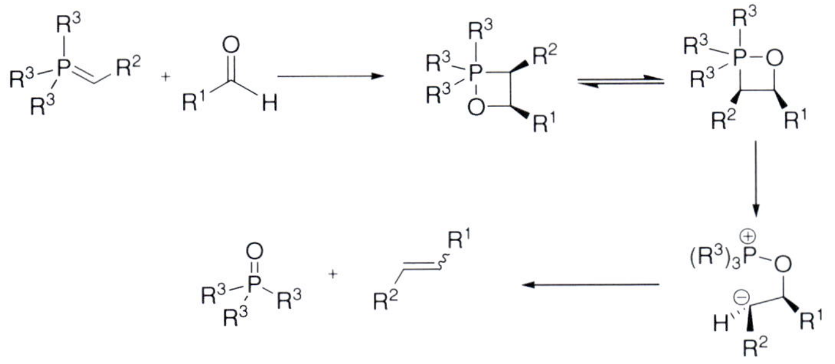

Scheme 1.10 Bestmann mechanism for the Wittig reaction

OPA formation has been shown to be irreversible and stereospecific for Wittig reactions of all ylide types by stereospecific decomposition of β-hydroxyphosphonium salts (β-HPSs) to alkene [60, 69]. Thus there can be no equilibration of intermediates after OPA has been formed, and intermediates such as the betaine shown in Scheme 1.10 can play no part in the Wittig reaction.

1.3.6 McEwen's Spin-Paired Diradical Mechanism

McEwen and co-workers have proposed two mechanisms involving the initial formation of spin-paired diradical intermediates that subsequently ring-closed to OPA (see Scheme 1.11). The first of these involved an entity with a C–C bond and an unpaired electron on each of phosphorus and oxygen, presumably formed by transfer of one electron from each of the C = O and P = C bonds into the new C–C bond, and movement of the remaining electron from each bond on to oxygen and phosphorus, respectively [82]. The second involved a P–O bond with two carbon-

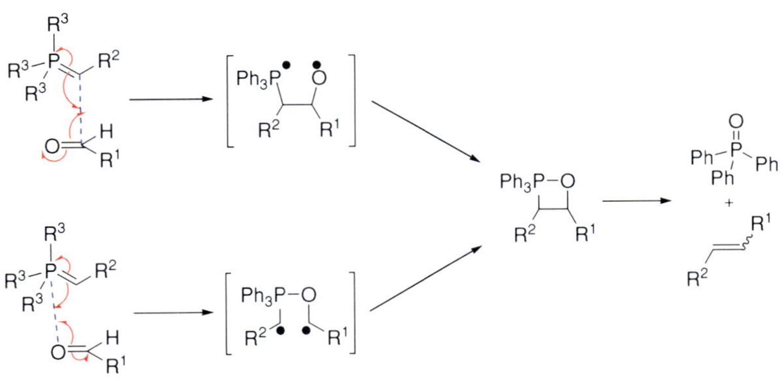

Scheme 1.11 Mechanisms involving spin-paired diradical intermediates prior to OPA formation

centred radicaloids formed by an analogous process [83] (this differs from the mechanism of Olah, in which electron transfer from ylide to carbonyl was suggested to occur before the occurrence of P–O bonding). The kinetically favoured pathway was postulated to favour *cis*-OPA formation due to orthogonal approach of the ylide and aldehyde in the diradical *intermediate*, which was proposed for non-stabilised ylides to be short lived, while it would be longer lived for stabilised ylides and could thus undergo bond rotation to give *trans*-OPA and *E*-alkene.

The first case, involving the C–C bonded diradical, can be ruled out by the same experiment that quashed the involvement of betaines in reactions of non-stabilised ylides [72]. The second, involving the P–O bonded diradical, can be ruled out for the same reasons as were invoked for the Bergelson–Schneider mechanism [see Sect. 1.3.2], and in particular by the fact that this mechanism would predict a negative ρ value for carbonyl compounds in the Wittig reaction where a positive value is observed.

1.3.7 Schlosser Mechanism

Schlosser and Schaub proposed the formation of OPA by cycloaddition of ylide and aldehyde [84]. The kinetically favoured OPA-forming TS was postulated to be planar, with trigonal bipyramidal geometry about phosphorus (i.e. complete re-organisation had—occurred of the substituents about phosphorus to the new geometry). The ylide α-substituent was supposed to cause the *P*-phenyl group on the same side of the forming ring to be oriented ca. 50° out of the plane of the bipyramid (which contains the phosphorus and ylide α-carbon atoms, as well as two *P*-phenyl *ipso* carbons), while the *P*-phenyl group on the opposite side of the ring was only oriented ca. 10° out of the same plane, since it should have no great steric interactions with the hydrogen substituent at the ylide α-carbon (see Fig. 1.4). This orientation of the *P*-phenyl groups was proposed to mean that placement of the aldehyde substituent on the same side of the forming ring as the ylide α-substituent would result in smaller 1–3 steric interactions than if the aldehyde substituent side was on the opposite side, where a *P*-phenyl *ortho*

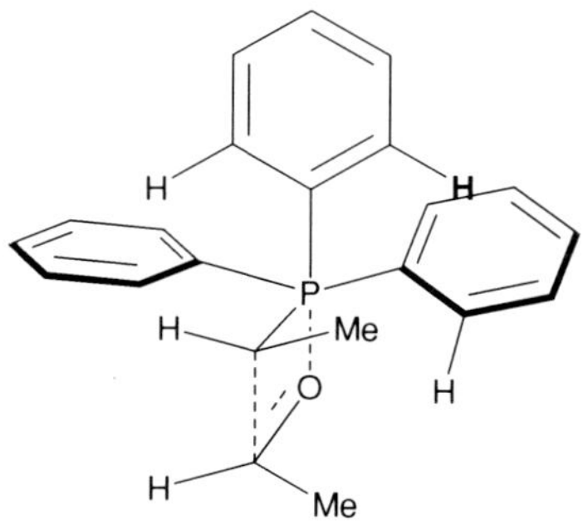

Fig. 1.4 Transition state for OPA formation according to the Schlosser model for the Wittig reaction

hydrogen would be pointing straight at it. This TS therefore leads to *cis*-OPA, and is also consistent with the drop in *Z*-selectivity observed when the (non-stabilised) ylide bears less sterically bulky phosphorus substituents.

Based on the above rationale for the cycloaddition TS, it can be concluded that *cis*-OPA should be the most thermodynamically stable OPA isomer. However, there exists no known example of isomerisation of *trans*-OPA to *cis*-OPA (whether by reversal to Wittig reactants or otherwise). On this basis it can be concluded that *trans*-OPA is thermodynamically favoured over *cis*-OPA, which is intuitively sensible, as the former should have similar 1–3 steric interactions but significantly decreased 1–2 interactions compared to the latter. Therefore the cycloaddition TS cannot be product-like (i.e. it must be *early*), as otherwise it would be *trans*-selective, and so could not have a trigonal bipyramidal arrangement of the substituents about phosphorus (an arrangement that is intrinsic to the explanation of selectivity in the Schlosser mechanism). Also, it is difficult to rationalise *E*–selectivity in reactions of stabilised ylides in the context of this mechanism, involving as it does direct OPA formation, given that OPA formation has been shown to be irreversible for reactions of all ylide types [60, 69]. Although the rationale put forth to explain the observed selectivity in the context of this mechanism is not consistent with experimental observations, it is close to the currently accepted mechanism (see Sect. 1.4) in that it involves OPA formation by direct [2 + 2] cycloaddition of ylide and aldehyde.

1.3.8 Vedejs Cycloaddition Mechanism

Vedejs advanced the proposal of direct irreversible cycloaddition of the ylide and aldehyde to give OPA, followed by irreversible and stereospecific cycloreversion of the OPA to give phosphine oxide and alkene [60, 86]. Reactions of aldehydes with non-stabilised alkylidenetriphenylphosphoranes were suggested to proceed through an early puckered TS (meaning that the carbonyl C=O bond and ylide P–C bond approach each other at a relatively large angle) in which both bond formation and rehybridisation about the atoms in the forming ring are at an early stage (see Fig. 1.5a). The fact that the four substituents bound to phosphorus in the ylide remain in a nearly tetrahedral arrangement about phosphorus in the TS has particular consequences for the shape of the TS. This disposition of the phosphorus substituents means that planar approach of the ylide and aldehyde is disfavoured for two reasons—firstly one of the *P*-phenyl groups must necessarily project in the direction of the forming P–O bond, and secondly there exists the possibility of large 1–3 steric interactions between the phosphorus substituents and the aldehyde substituent. Puckering relieves both of these unfavourable interactions, and minimises 1–2 steric interactions (between aldehyde and ylide substituents) if the large substituent on the carbonyl is placed in a pseudo-equatorial position and the ylide α-substituent is pseudo-axial in the forming ring. This energetically favoured TS leads to *cis*-OPA. The lowest energy *trans*-selective TS would necessarily be

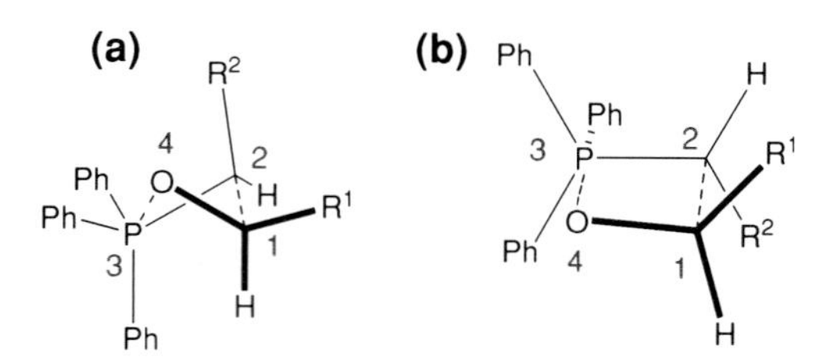

Fig. 1.5 **a** Proposed *cis*-selective cycloaddition transition state in reactions of alkylidenetriphenylphosphoranes. **b** Proposed *trans*-selective cycloaddition transition state in reactions of stabilised ylides in Vedejs mechanism [60, 86]

significantly higher in energy whether puckered (in which case it would suffer from either increased 1–2 interactions by placing R^2 pseudo-equatorial in Fig. 1.5a or from increased 1–3 interactions by placing R^1 pseudo-axial) or planar (due to increased 1–3 interactions, and "blocked approach" of the carbonyl oxygen to phosphorus). Thus, *cis*-OPA is thought to be formed selectively under conditions of kinetic control, and to decompose irreversibly and stereospecifically to *Z*-alkene and phosphine oxide by *syn*-cycloreversion

Reactions of stabilised ylides were proposed by Vedejs and co-workers to proceed through a later transition state in which bond formation and rehybridisation about the reactive centres are advanced (see Fig. 1.5b). That the geometry of the substituents about phosphorus (including the carbonyl oxygen) is close to trigonal bipyramidal in such a TS was postulated to mean that 1–3 interactions would be much less significant, so the driver for TS puckering is removed. Planar approach was thus thought to be eminently possible and probably favourable over puckering given the more advanced nature of bond formation. 1–2 steric interactions would naturally be lower in a planar *trans*-selective TS than in a planar *cis*-selective TS, which would suffer significantly from torsional strain between R^1 and R^2. *trans*-OPA is formed selectively and irreversibly, and hence high selectivity for *E*-alkene is seen after cycloreversion of the OPA.

The cycloaddition mechanisms proposed by Vedejs for different ylides could also be used to account for the selectivity observed in reactions of ylides with different substitution patterns at phosphorus. Computational results by Aggarwal, Harvey and co-workers largely confirm the predictions of this cycloaddition mechanism, although the nature of the *trans*-selective cycloaddition TS in reactions of stabilised ylides is proposed by them on the basis of their computational results to be puckered (but still energetically favoured) to take into account the effect of the interaction of the dipoles along the ylide C–R^2 (where R^2 is a stabilising group such as an ester) and aldehyde C–O bonds [48, 49].

The finer details of the mechanism and especially the experimental and computational evidence that has been amassed in its favour are described fully in the next section on the currently accepted mechanism for the Wittig reaction (Sect. 1.4), so further discussion of the cycloaddition is deferred until then.

1.4 Modern Mechanistic Interpretation of the Wittig Reaction

There are many questions to be answered in considering how the reaction of a phosphonium ylide and aldehyde to give an alkene and phosphine oxide progresses. These include:

1. Whether the reaction proceeds reversibly or irreversibly, i.e. is it under kinetic or thermodynamic control?
2. What is the nature of the first-formed intermediate in the reaction? Many intermediates have been proposed, but OPAs and betaines are the only ones for which sufficient experimental evidence been presented to enable them to gain widespread acceptance.
3. How does the observed selectivity for *Z* or *E* alkene arise? The answer to this question is intrinsically linked to the first two points.

1.4.1 The Operation of Kinetic Control in the Wittig Reaction

The substantial evidence available on this issue strongly suggests that, in all but a few exceptional cases, the Li-salt free Wittig reaction proceeds irreversibly. Below is presented the evidence that has been put forth implying the operation of kinetic control in the reactions of each of the three major classes of ylide. The exceptional cases in which thermodynamic control has been shown to operate in Li-salt free reactions will also be described.

1.4.1.1 Non-stabilised Ylides

It is generally accepted that OPA must occur at some point along the reaction coordinate from phosphonium ylide and aldehyde to alkene and phosphine oxide—either as an intermediate or TS. OPAs have been shown to be the only observable intermediates many times in reactions of non-stabilised ylides [65, 68, 70, 86]. It has also been demonstrated in such reactions that the final step of the Wittig reaction is decomposition of OPA to alkene and phosphine oxide, since the rate of OPA decomposition equals the rate of alkene formation [60].

The operation of kinetic control in a Wittig reaction is established if the kinetic OPA *cis/trans* ratio (usually determined by low temperature ^{31}P NMR) is identical to the *Z/E* ratio of the alkene produced by warming of the OPA. There are surprisingly few examples in which kinetic control has been absolutely proven in this way. Many of the examples of the application of low temperature ^{31}P NMR

Table 1.1 Wittig reactions for which kinetic control (and stereospecific decomposition of OPA) have been demonstrated by direct comparison of the OPA *cis/trans* and alkene *Z/E* ratios.[a] X = halide counter-ion

(i) Base (ii) $R^{1a}C(O)R^{1b}$, -78 °C; $-ZY_2P{=}O$, Decomposition temperature

Entry	Y	Z	R^2	R^{1a}	R^{1b}	Base	OPA *cis/ trans* ratio	Alkene *Z/E* ratio	Ref.
1	Ph	Ph	*n*-Pr	*n*-C_5H_{11}	H	LiHMDS	5.8:1	5.8:1	[70]
2	Ph	Ph	*n*-Pr	Ph	H	NaHMDS	100:0[b]	96:4	[71]
3	Ph	Ph	*n*-Pr	Ph	H	NaHMDS	ca.95:5	95:5[c]	[87]
4	Ph	Ph	*n*-Pr	Ph	H	NaHMDS	≥98:2	96:4	[70]
5[d]	DBP	Ph	Me	$(CH_2)_2Ph$	H	$NaNH_2$	6:94	5:95	[59]
6	Ph	Ph	Me	*c*-C_6H_{11}	Me	KHMDS	95:5	–	[85]
7[e]	Et	Ph	Me	Me	*c*-C_6H_{11}	KHMDS	4:96	10:90	[88]
8[f,g]	Ph	2-fur	Me	Ph	H	NaHMDS	94:6	94:6	[23]
9[f,g]	2-fur	Ph	Me	Ph	H	NaHMDS	96:4	96:4	[23]
10[f,g]	2-fur	2-fur	Me	Ph	H	NaHMDS	98:2	98:2	[23]

[a] Phosphonium salt counter-ion is bromide in all cases except entries 5–7, for which it is iodide. The decomposition temperature is the temperature at which OPA decomposition to alkene was effected experimentally. Unless otherwise specified, decomposition was carried out at "room temperature" (15–20 °C)

[b] Only one OPA signal observed in ^{31}P NMR spectrum obtained at −20 °C

[c] The paper in question states that the reaction "did not exhibit stereochemical drift", but does not quote the alkene *Z/E* ratio

[d] The specific ylide used in this reaction, incorporating the dibenzophosphole (DBP) system, is shown in Fig. 1.6a [59]. The experimental OPA decomposition temperature for this reaction was 110 °C

[e] OPA decomposition temperature not specified

[f] "2-fur" indicates a *P*-fur-2-yl group. OPA decomposition in this reaction was carried out in refluxing THF, quoted as 70 °C

[g] In a separate reaction, *n*-BuLi was used to generate the ylide. Stereospecific OPA decomposition was also noted from the OPA in this case, but with the exception of entry 8 the yield of alkene was much lower than when NaHMDS was employed

monitoring of the OPA intermediate of a Wittig reaction have been used to demonstrate the operation of OPA equilibration in reactions involving aromatic or tertiary aldehydes and/or trialkylphosphonium alkylides (see discussion later). However, stereospecific conversion of the OPA diastereomers produced in the Wittig reactions shown in Table 1.1 to alkene has been demonstrated by low temperature ^{31}P NMR, implying the operation of kinetic control in these reactions [23, 60, 71, 72, 88, 89].

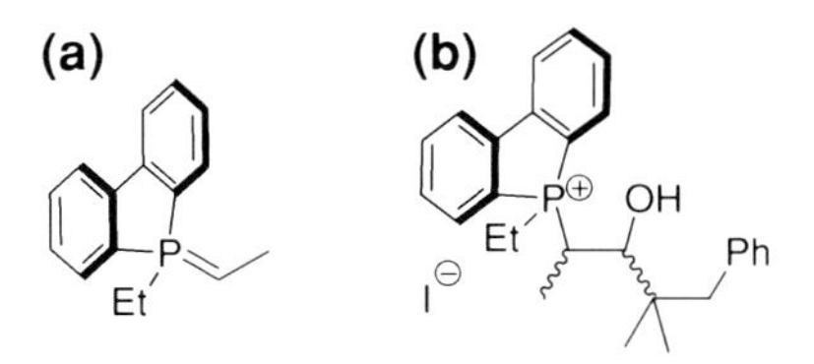

Fig. 1.6 **a** Ethylidene ethyldibenzophospholane [59], **b** *erythro* or *threo* (3-hydroxy-4,4-dimethyl-5-phenylpent-2-yl)ethyldibenzophospholium iodide, incorporating the dibenzophosphole (DBP) system [69]

As shown in Table 1.1 entry 1, stereospecific conversion of the OPA (*cis/trans* ratio of 5.8:1) produced in the reaction of hexanal with *n*-butylidenetriphenylphosphorane (generated using LiHMDS) to a 5.8:1 mixture of *Z* and *E* dec-4-ene has been reported [70, 71]. This is consistent with the fact that LiBr has been shown to exert an effect on the stereoselectivity of OPA formation in the reaction of $Ph_3P{=}CHCH_3$ and $PhCH_2CH_2CHO$, but not to affect the stereochemical ratio of the OPA formed from these reactants if added to a pre-formed solution of the OPA to give betaine-LiBr complex [65]. These observations suggest that OPA formation is irreversible in reactions of aliphatic aldehydes, even in the presence of Li^+, and that the diminished *Z*-selectivity in these reactions arises from the effect of Li^+ on the initial formation of OPA.

OPAs are generally air and temperature sensitive. The decomposition products from exposure to (relatively) high temperatures are alkene and phosphine oxide, so at low temperature the OPAs derived from non-stabilised ylides are kinetically stable. If the phosphorus of the OPA is constrained to being in a five-membered ring, as in dibenzophosphole (DBP) derived OPAs, then the OPA is kinetically stable at room temperature, and requires heating to induce alkene formation [59]. Such OPAs can conveniently be monitored spectroscopically at room temperature, and as a consequence they have been used in a number of studies to elucidate details of the Wittig reaction mechanism [59, 69]. DBP-derived OPAs have been generated from Wittig reactions (e.g. from the ylide shown in Fig. 1.6a) and by β-HPS deprotonation (e.g. of the β-HPS shown in Fig. 1.6b). They have also been used in similar investigations on reactions of semi-stabilised ylides [see Sect. 1.4.1.2].

Individual OPA isomers generated by a process independent of the Wittig reaction (by deprotonation of β-hydroxyphosphonium salt using sodium or potassium base, which transiently gives betaine and then OPA) have been demonstrated by low temperature ^{1}H and ^{31}P NMR to undergo stereospecific decomposition to alkene, i.e. *erythro* β-HPS gave only *cis*-OPA and hence only *Z*-alkene, and likewise *threo* β-HPS gave *trans*-OPA and hence *E*-alkene [69, 70, 88]. This is shown in Table 1.2. There are also some other examples where kinetic control in Wittig reactions of non-stabilised ylides has been proven in this manner.[5]

[5] See Ref. [4], p. 24, Table 6.

Table 1.2 β-Hydroxyphosphonium salts for which stereospecific decomposition to alkene has been demonstrated for either both isomers or just the *erythro* isomer

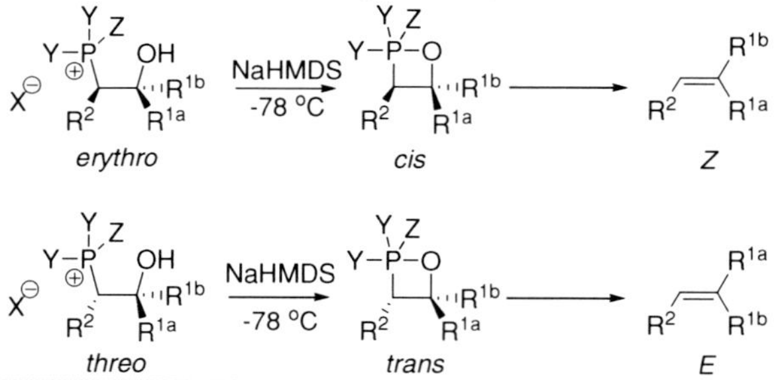

Entry	Y	Z	R^2	R^{1a}	R^{1b}	OPA decomposition temperature[a] (°C)	Ref.
1	Ph	Et	Me	$C(Me)_2CH_2Ph$	H	20	[69]
2	DBP[b]	Et	Me	$C(Me)_2CH_2Ph$	H	110	[69]
3	Ph	Ph	Me	$C(Me)_2CH_2Ph$	H	20	[69]
4	DBP[b]	Et	Me	$C(Me)_2CH_2Ph$	H	110	[59][c]
5	Ph	Ph	*n*-Pr	Ph	H	20	[70, 71]
6	Ph	Et	Me	*c*-C_6H_{11}	Me	>−50	[89]

[a] Temperature at which OPA decomposition was effected experimentally
[b] The specific β-HPS used in this reaction, incorporating the dibenzophosphole (*DBP*) system, is shown in Fig. 1.6b
[c] Deprotonation carried out using $NaNH_2$

Another method for OPA generation independent of a Wittig reaction involves S_N2 epoxide ring-opening by lithium diphenylphosphide to give β-oxidophosphine, which is quaternised with methyl iodide to give betaine and hence OPA. *cis*-Epoxide was shown to give *E*-alkene stereospecifically by this method via *trans*-OPA, while *trans*-epoxide stereospecifically gave *Z*-alkene via *cis*-OPA [89, 90]. The latter is shown in Scheme 1.12.

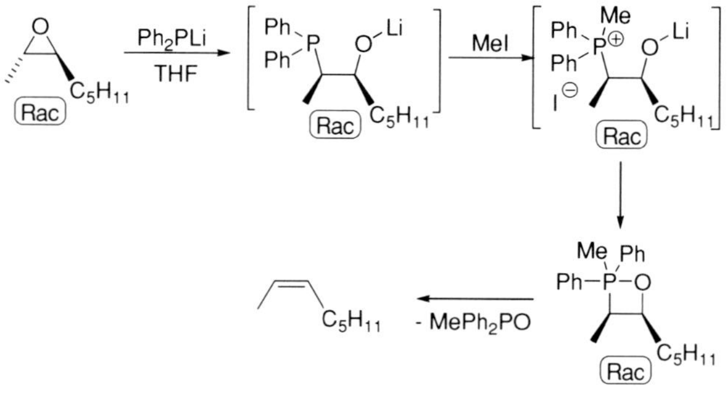

Scheme 1.12 Reaction system for which OPA is produced by nucleophilic cleavage of an epoxide by diphenylphosphide and quaternisation of the resulting phosphine and shown to decompose stereospecifically to alkene [89, 90]

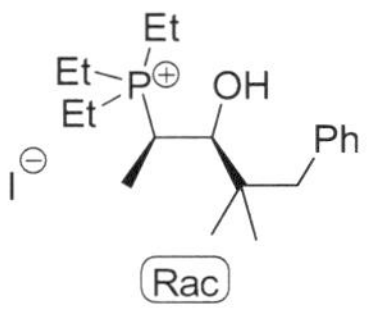

Fig. 1.7 *erythro*-(3-hydroxy-4,4-dimethyl-5-phenylpent-2-yl)triethylphosphonium iodide

It can be concluded from these results that the formation of OPA is irreversible, as *cis* and *trans* isomers do not interconvert under Li-salt free conditions, and that OPA decomposition to alkene and phosphine oxide is stereospecific, irreversible, and occurs through a *syn*-cycloreversion. Therefore the alkene *Z/E* ratio is identical to the kinetic OPA *cis/trans* ratio, and the stereochemistry of the alkene product is decided in a TS in the step(s) leading from starting materials to OPA.

There are some exceptional cases where equilibration of OPA diastereomers has been shown to occur under Li-salt free conditions, all involving non-stabilised ylides. This process has been termed "stereochemical drift", and has been observed only for the OPAs produced in the reactions of trialkylphosphonium alkylides with tertiary or aromatic aldehydes [69, 70, 87], and in reactions of triphenylphosphonium alkylides with aromatic aldehydes [65]. In the latter case stereochemical drift was only evident at temperatures at or above those at which alkene formation could also occur.

Deprotonation of *erythro*-*β*-hydroxyphosphonium salt at −78 °C, as shown in Fig. 1.7, would be expected to give only the *cis*-OPA that would be produced in the Wittig reaction of ethylidenetriethylphosphorane and 2,2-dimethyl-3-phenylpropanal. However, the reaction mixture upon warming to −40 °C was observed to consist of a 80:20 mixture of *cis* and *trans* OPA [69]. On warming to −5 °C, the amount of *trans*-OPA was observed to increase at the expense of the *cis*-OPA, while a small amount of alkene formation also occurred. Decomposition of OPA to alkene and phosphine oxide was effected at >30 °C giving a final *Z/E* ratio of 17:83. Some 2,2-dimethyl-3-phenylpropanal was also isolated, indicating reversal to ylide and aldehyde. Reversal was confirmed by the observation of crossover product in an experiment involving deprotonation of the same *β*-HPS in the presence of *m*-chlorobenzaldehyde. At least some of this reversal is likely to be due to reversal of the betaine transiently produced upon *β*-HPS deprotonation, but OPA reversal must be involved in view of the fact that the OPA *cis/trans* ratio was observed to change after the passage of some time with an increase in temperature. The *threo*-diastereomer of this *β*-HPS decomposed stereospecifically to *E*-alkene [69].

Stereochemical drift was shown to operate in the reaction of ethylidenetriphenylphosphorane with benzaldehyde by a series of crossover experiments [65]. This was observed only above −30 °C. When spectroscopic monitoring of the reactions was carried out below this temperature, no crossover products could be observed. The mechanism of equilibration of the OPA isomers was shown conclusively in this case to involve cycloreversion to Wittig starting materials. Li-salt free OPA equilibration was also demonstrated for these reactants by an experiment in which a 3:1 mixture of *erythro* and *threo* (1-hydroxy-1-phenylprop-2-yl)

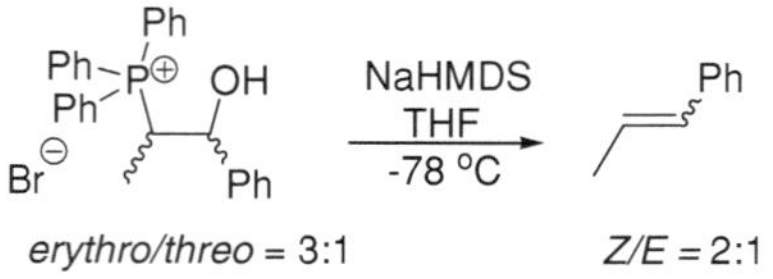

NaHMDS THF -78 °C

erythro/threo = 3:1 *Z/E* = 2:1

Scheme 1.13 Deprotonation of a mixture of *erythro* and *threo* (1-hydroxy-1-phenyl-*n*-pent-2-yl)triphenylphosphonium bromide

triphenylphosphonium bromide gave a 2:1 mixture of *Z* and *E*-1-phenylprop-1-ene [70], as shown in Scheme 1.13.

The closely related β-HPS *erythro*-(1-hydroxy-1-phenyl-*n*-pent-2-yl)triphenylphosphonium bromide (R = phenyl, R^1 = *n*-propyl; see Table 1.2 entry 5) was shown to undergo stereospecific conversion to *Z*-1-phenylpent-1-ene in the same study [70], and as mentioned above, the Li-salt free Wittig reaction that gives the same OPA (benzaldehyde + *n*-butylidenetriphenylphosphorane) proceeded stereospecifically [87], or with negligible stereochemical drift [70]. Thus it appears that the operation of kinetic control or otherwise in reactions of alkylidenetriphenylphosphoranes with benzaldehyde is heavily dependent on the nature of the alkylidene moiety.

It is noteworthy that all instances of stereochemical drift observed thus far involve the production of increased amounts of *trans*-OPA at the expense of the *cis*-OPA, or the production of a greater proportion of *E*-alkene than had been present of the *trans*-OPA. In other words, stereochemical drift always results in an increased proportion of *trans*-OPA, since *E*-alkene must be produced from *trans*-OPA, and alkene formation is irreversible. An obvious conclusion is that the *trans*-OPA is thermodynamically favoured; this is corroborated by the observation that *trans*-OPAs typically require higher temperatures to effect their decomposition to alkene and phosphine oxide than do their *cis* counterparts [69, 70, 88]. Thus the proportion of *Z*-alkene produced in a Wittig reaction always represents a lower bound to the kinetic selectivity for the *cis*-OPA; in other words high selectivity for *Z*-alkene is directly indicative of kinetic control, and cannot occur "by mistake" (i.e. due to equilibration) as long as the *cis*-OPA is indeed thermodynamically disfavoured.

1.4.1.2 Semi-stabilised Ylides

It is not possible, in general, to detect OPAs derived from semi-stabilised ylides by low temperature NMR. Consequently, a kinetic diastereomeric OPA ratio cannot be established to compare with the final alkene *Z/E* ratio. However, in one case an OPA produced in a Wittig reaction of a semi-stabilised ylide has been reported. As shown in Scheme 1.14, the reaction of cyclohexanecarboxaldehyde with the phenyldibenzophosphole-derived semi-stabilised ylide allylidenephenyldibenzophospholane in THF-*d8* at -78 °C gave an OPA (^{31}P NMR $\delta = -72$ ppm) that decomposed above -50 °C to 1-cyclohexyl-1,3-butadiene with *Z/E* = 5:95 [59].

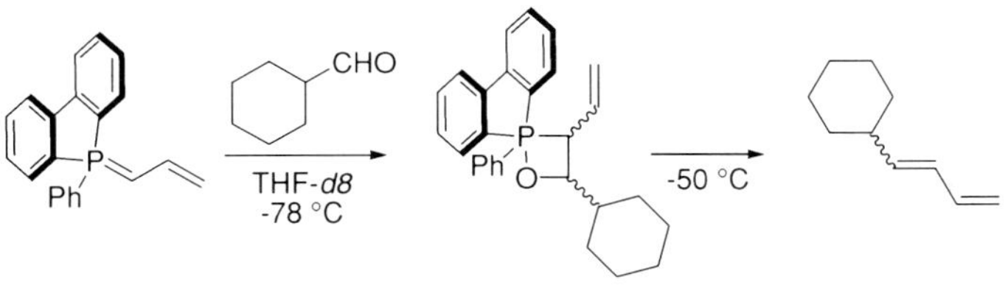

Scheme 1.14 Wittig reaction of allylidenephenyldibenzophospholane and cyclohexanecarboxal dehyde

Such a "constrained" OPA (in which the phosphorus atom is a spiro centre between four and five-membered rings) is kinetically relatively stable (compared to other semi-stabilised ylide-derived OPAs) [90] because increasing bond angle strain is induced in the phospholane ring on going from OPA to phosphine oxide. This effectively raises the barrier to OPA decomposition to alkene and phosphine oxide, and may even lower the barrier to OPA formation.

In addition to the impossibility of spectroscopic observation of OPAs in Wittig reactions of semi-stabilised ylides, with the above singular known exception, crossover experiments with such ylides[6] do not give meaningful results because the reaction of the semi-stabilised ylide proceeds so quickly that the reaction is already complete by the time the crossover ylide or aldehyde is added. Thus, proof for the operation of kinetic control in Wittig reactions of semi-stabilised ylides requires OPA generation from a source other than a Wittig reaction, and demonstration of stereospecific conversion either (a) of the precursor of *cis* or *trans*-OPA to *Z*- or *E*-alkene, respectively, or (b) of a mixture of the precursors of known diastereomeric ratio to alkene of corresponding *Z/E* ratio. This has been done for reactions of semi-stabilised ylides by forming alkene from either (1) β-HPS by deprotonation or (2) epoxide by nucleophilic ring cleavage by phosphide and subsequent quaternisation of phosphorus.[7] Both methods presumably form betaine and OPA transiently *en route* to alkene, and have both been described in connection with reactions of non-stabilised ylides in Sect. 1.4.1.1.

Each of these methods suffers from the fact that betaine generation is inherent in the process of OPA formation. The formation of OPA in Wittig reactions does not necessarily require the intermediacy of a betaine. There may exist processes in experiments involving β-HPS deprotonation or nucleophilic cleavage of epoxide that do not necessarily operate in Wittig reactions which result in the OPA *cis/trans* ratio being different to the diastereomeric ratio of the starting material. For example, dissociation of the transiently formed betaine into ylide and aldehyde ("betaine

[6] In crossover experiments for Wittig reactions, a more reactive ylide or aldehyde is added to the reaction mixture after the intermediate has formed but not completely decomposed in order to "mop up" any starting material that re-forms by Wittig reversal.

[7] The β-HPSs in this case were actually obtained by a route involving nucleophilic ring opening of an epoxide. Phosphide cleavage of an epoxide of defined stereochemistry, addition of acid to give β-hydroxyphosphine and quaternisation of phosphorus with methyl iodide or triflate gives β-HPS.

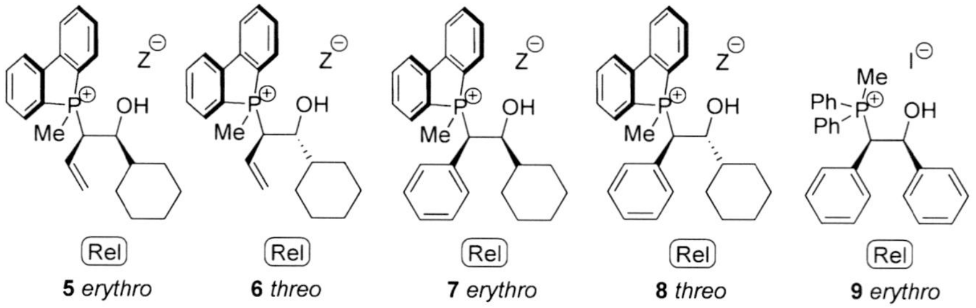

Chart 1.1 β-hydroxyphosphonium salts used for the independent OPA generation experiments described in references 47 and 71. Counter-ion Z is iodide or triflate (trifluoromethanesulphonate)

reversal"), and non-stereospecific recombination of these Wittig starting materials to give OPA may occur in either type of experiment, whereas Wittig reactions do not necessarily form betaines. Also in β-HPS deprotonation experiments, there exists the possibility of the formation of β-hydroxy ylide (by deprotonation of the α-carbon, not the β-hydroxy group), which then undergoes non-stereospecific proton transfer to form betaine and hence OPA. This latter process has been demonstrated to occur in corresponding experiments involving β-HPSs derived from stabilised ylides when strong bases are used [see Sect. 1.4.1.3] [59]. In each case, stereospecific decomposition of the OPA mixture of changed diastereomeric ratio to alkene product with a *Z/E* ratio that is not identical to the initial β-HPS *erythro/threo* ratio or epoxide diastereomeric ratio could be erroneously interpreted as indicating the operation of some equilibration of OPA diastereomers in the corresponding Wittig reaction. The same problem of the apparent operation of stereochemical drift applies in crossover experiments involving β-HPS deprotonation. Despite such limitations, these methods have been successfully used to demonstrate the irreversibility of OPA formation in reactions of semi-stabilised ylides.

The *cis* and *trans*-OPAs that would be produced in the reaction of allylidenemethyldibenzophospolane with cyclohexanecarboxaldehyde were independently synthesised in separate experiments by deprotonation of *erythro* and *threo* β-HPSs **5** and **6** respectively (see Chart 1.1) [60]. Similarly, the *cis* and *trans*-OPAs from the reaction of benzylidenemethyldibenzophospolane with cyclohexanecarboxaldehyde were obtained by deprotonation of β-HPSs **7** and **8** respectively (see Chart 1.1). Each deprotonation experiment gave a single OPA (as evaluated by low temperature ^{31}P NMR), which decomposed stereospecifically to the expected alkene isomer—so *erythro* β-HPS gave Z-alkene, and *threo*-β-HPS gave *E*-alkene. These experiments prove that there is no equilibration of either the betaine or of the OPA which are each necessarily formed in the path from β-HPS to alkene. The corresponding Wittig reaction must occur under kinetic control, since the formation of each of the possible intermediates is irreversible.

Kinetic control has also indirectly been shown to be in operation in the reactions of *unconstrained* semi-stabilised ylides (i.e. where the phosphorus atom of the ylide is not part of a ring system), although as alluded to above, observation of

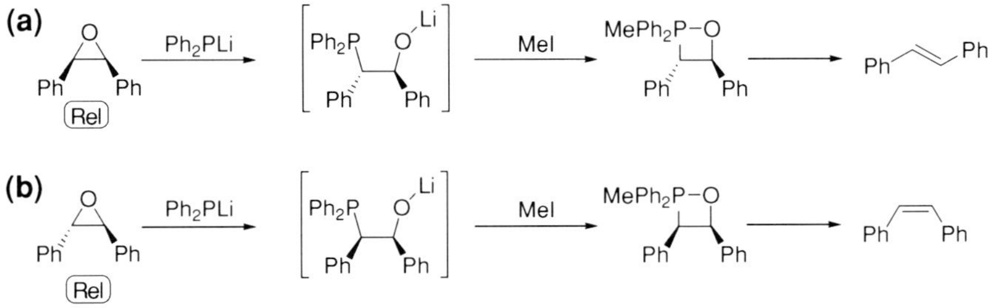

Scheme 1.15 Stereospecific conversion of **a** *cis*-stilbene oxide to *E*-stilbene and **b** *trans*-stilbene oxide to *Z*-stilbene by nucleophilic ring cleavage and quaternisation of the resulting β-oxidophosphine

the transient OPA intermediate by NMR is not possible in these cases. Deprotonation of (2-hydroxy-1-phenyl-2-phenyleth-1-yl)methyldiphenylphosphonium iodide **9** with each of *n*-BuLi, NaHMDS and KHMDS gave *Z*-alkene stereospecifically [83]. Treatment of stilbene oxide with lithium diphenylphosphide gave a β-oxidophosphine intermediate, which was quaternised with methyl iodide, resulting in the formation of alkene, presumably by transient formation of a betaine which then underwent ring-closure to OPA as shown in Scheme 1.15 [90]. The process was shown to proceed with inversion—*trans*-stilbene oxide gave *Z*-stilbene, and *cis*-stilbene oxide gave *E*-stilbene

These results show that there is no inter-conversion of OPA intermediates derived from semi-stabilised ylides that have been generated by non-Wittig processes (and also that there is no interconversion of the associated betaines, whether they are involved in the Wittig mechanism or not). This confirms that both OPA formation and OPA decomposition to alkene and phosphine oxide are irreversible, at least in the case of Wittig reactions of *P*-phenyl-5*H*–dibenzophosphole-derived semi-stabilised ylides, and also most significantly in the case of Wittig reactions of unconstrained methyldiphenylphosphine-derived ylides with benzaldehyde.

1.4.1.3 Ester-Stabilised Ylides

No OPAs derived from stabilised ylides have ever been detected by NMR, not even dibenzophosphole derived OPAs. Consequently, proof for the operation of kinetic control in reactions of these ylides rests on the demonstration of stereospecific conversion of *erythro* or *threo* β-HPS to *Z* or *E* alkene, respectively. *erythro*-β-HPSs derived from primary, secondary, tertiary, and aromatic aldehydes and ester stabilised ylide (ethoxycarbonylmethylidene)methyldiphenylphosphorane were obtained by addition of phosphino enolate [$Ph_2PCHCOOEt$]Li to aldehyde followed by acid quenching to give β-hydroxyphosphine, as shown in Scheme 1.16. The *erythro*-β-hydroxyphosphine was isolated by column chromatography and then quaternised with methyl triflate to give *erythro*-β-HPS [60].

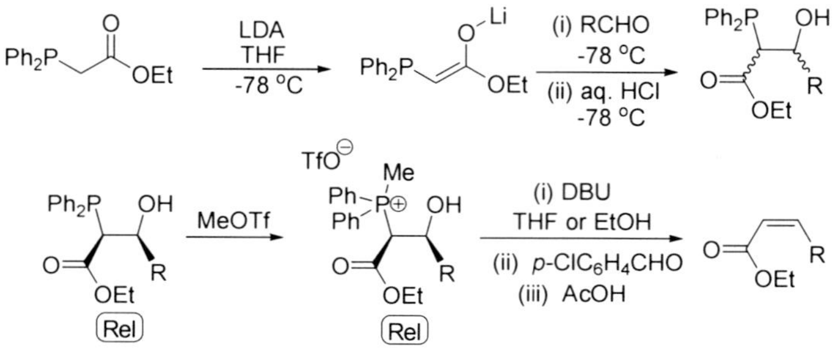

Scheme 1.16 Synthesis of β-hydroxyphosphine, and hence β-HPS corresponding to the acid quench product of a Wittig reaction of a stabilised ylide, and stereospecific conversion of the *erythro*-β-HPS to *Z*-alkene. R = Ph, CH_2CH_2Ph, *c*-C_6H_{11}, or CMe_2CH_2Ph [60]

Deprotonation experiments using the β-HPSs thus obtained were carried out in THF and ethanol. DBU (1,8-diazabicyclo[5,4,0]undec-7-ene), mesityllithium and KHMDS bases were used in separate experiments. Two equivalents of *p*-chlorobenzaldehyde were added to the reaction mixture 30 seconds after the addition of base. This aldehyde reacts with any ylide produced by potential reversal of OPA or betaine to Wittig starting materials, thus preventing recombination of the reactants derived from reversal. Delayed addition of *p*-chlorobenzaldehyde prevented it from exerting any effect on the betaine that is necessarily transiently produced in β-HPS deprotonation. High yield and stereospecificity and negligible crossover were observed in the conversion of the *erythro*-β-HPSs to *Z*-alkene (*Z/E* $\geq$ 98:2) using DBU as base for the cases with R = CH_2CH_2Ph, *c*-C_6H_{11}, or CMe_2CH_2Ph. Greater production of each of crossover alkene and *E*-cinnamate was observed in the case with R = Ph. In deprotonation experiments at 20 °C using KHMDS in THF, the production of crossover product was observed to have increased relative to the experiments with DBU—8 % for the case with R = CH_2CH_2Ph, 5 % for R = *c*-C_6H_{11}, 23 % for R = CMe_2CH_2Ph and 48 % for R = Ph. The production of crossover alkene showed that ylide derived from some form of reversal to Wittig starting materials had to have been present. *E*-Cinnamate could not have resulted from any OPA or betaine reversal process since ylide would have been intercepted as crossover product, and it was shown not to result from isomerisation of the *Z*-cinnamate under the reaction conditions.

The pathway by which crossover product and *E*-alkene were formed was elucidated using α-deuterated *erythro*-β-HPSs (see 1–17). Under "strong base conditions", deprotonation of α-deuterated *erythro*-β-HPS at low temperature (−78 °C) yielded a very high proportion of *Z*-alkene with high deuterium content, while at higher temperatures a much lower proportion of *Z*-alkene was produced, *and* its deuterium content was also much lower. Most significantly, no deuterium could be detected under any circumstances in the *E*-alkene or crossover alkene. None of these observations are consistent with reversal of the betaine or OPA produced by β-HPS deprotonation, because if this was undergoing reversal then the *E*-alkene or crossover alkene should show some deuterium content.

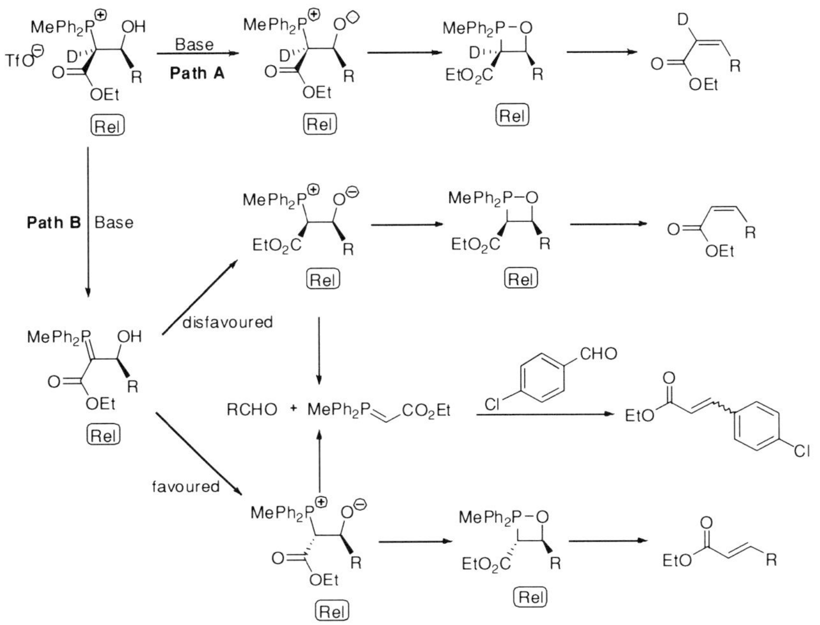

Scheme 1.17 Experiments on α-deuterated *erythro*-β-HPSs. R = Ph, CH_2CH_2Ph, *c*-C_6H_{11}, or CMe_2CH_2Ph [60]

In explanation of the experimental observations (see Scheme 1.17), it was proposed that the base can remove either the hydroxyl proton (**path A**) or the α-deuterium (**path B**). In the former case, the betaine formed reacts to give *cis*-OPA and hence *Z*-cinnamate with full deuterium content. In the latter case, a β-hydroxy ylide (that has lost its deuterium) results, which can undergo proton transfer (by an unknown mechanism) to either face of the β-hydroxy ylide yielding *erythro* or *threo* betaine (the latter being favoured). This can either undergo ring closure to give *cis* or *trans*-OPA and hence deuterium-free *Z* or *E*-alkene respectively, or can revert to ylide and aldehyde and hence to crossover alkene. Dedeuteration is more likely at higher temperature (due to poorer discrimination of acidic sites by the base), so that more of each of the crossover product and the *E*-alkene are observed. There is also a lower proportion of deuterium in the *Z*-alkene that is produced, as some of it is derived from non-deuterated *erythro*-betaine, which is itself derived from β-hydroxy ylide. The fact that the *Z*-alkene alone retains the deuterium label in these experiments confirms that *cis*-OPA is formed irreversibly and that it undergoes stereospecific decomposition to *Z*-alkene. This means that OPA is necessarily the final intermediate in the Wittig reaction of stabilised ylides, regardless of how it is formed.

That crossover alkene is produced in these deprotonation experiments (especially the ones involving strong base) indicates that it is highly likely that some ylide is present after deprotonation has occurred. It has been explained above how it was shown that this ylide *cannot* be derived from OPA. If formed, it must thus result from C–C bond scission in either the betaine or the β-hydroxy ylide. There is no definitive evidence to prove which entity undergoes reversal to ylide and aldehyde. Betaine reversal is not an issue for the *erythro*-betaine produced by hydroxyl deprotonation of the initial *erythro*-β-HPS, as proved by the complete absence of deuterium labelled *E*-enoate in all cases, but based on the above results, it is in principle possible. If betaine reversal does occur, a possible explanation given by Vedejs and Fleck [60] is that β-HPS deprotonation by weak base (e.g. DBU) produces a *syn*-betaine rotamer that can easily cyclise to OPA, whereas proton transfer involving a β-hydroxy ylide or β-HPS deprotonation by strong base (e.g. KHMDS) may produce rotamers that are more prone to reversal to Wittig starting materials.

It may be assumed that betaine (whether produced as *syn* or *anti* rotamer) is energetically uphill from Wittig starting materials based on its apparent readiness to decompose to ylide and aldehyde or to alkene and phosphine oxide via OPA. Certainly OPAs produced in Wittig reactions of non-stabilised and semi-stabilised ylides have been demonstrated to be the only observable intermediates in those reactions, so in these cases, OPA is more stable than betaine. Since *cis*-OPA is shown by the above experiments to be formed irreversibly, and to decompose very quickly to alkene and phosphine oxide, *E*-selectivity in reactions of stabilised ylides cannot be the result of the decomposition of *trans*-OPA being more rapid than that of *cis*-OPA [see Sect. 1.3.1]. The remaining possibility for involvement of betaine in reactions of stabilised ylides requires that they equilibrate via ylide and aldehyde, and that ring closure for *threo*-betaine is kinetically preferred compared to that for the *erythro* isomer. This would require there to be a large energy barrier between betaine and OPA. In light of the rapidity of alkene formation in β-HPS deprotonation experiments, and of the fact that the first step of Wittig reactions of stabilised ylides (the bimolecular step) is rate-limiting, this seems highly unlikely. All signs indicate the operation of kinetic control in reactions of stabilised ylides. Since the betaine mechanism cannot account for high *E*-selectivity under kinetic control, it can be concluded that it is not in operation in reactions of stabilised ylides.

Another possibility for the formation of crossover product is that the β-hydroxy ylide itself perhaps might undergo a Wittig reaction with the *p*-chlorobenzaldehyde, giving the OPA shown in Scheme 1.18. This OPA can give crossover alkene

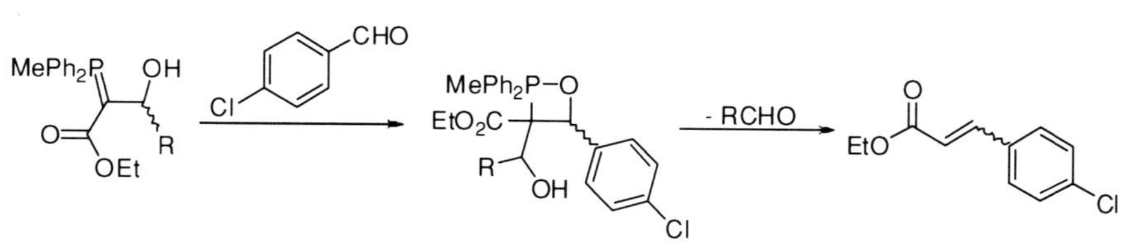

MePh$_2$P OH O OEt R CHO Cl MePh$_2$P–O EtO$_2$C R OH Cl - RCHO O EtO Cl

Scheme 1.18 Postulated mechanism for the formation of crossover alkene by a Wittig reaction of β'-hydroxy ylide

either directly by elimination of the non-crossover aldehyde, or by forming β'-hydroxy enoate, which itself can eliminate aldehyde.

Since the β-hydroxy ylide pathway can not be involved in a normal Wittig reaction, and since it plays at most a minor role in stabilised ylide-derived β-HPS deprotonation experiments in the absence of strong base, it may be concluded that *erythro*-β-HPSs derived from stabilised ylides decompose stereospecifically to *Z*-alkenes upon deprotonation. Thus, the typically observed high *E*-selectivity in Wittig reactions of ester-stabilised ylides results from a kinetic preference for the formation of *trans*-OPA (i.e. the pathway leading to *E*-alkene), since OPA decomposition to alkene and phosphine oxide is shown by these results to be stereospecific, and is in general irreversible. Irreversible formation of *trans*-OPA in reactions of stabilised ylides can be inferred from the fact that *cis*-OPA is formed irreversibly, since *trans*-OPAs are generally thermodynamically favoured over *cis*-OPA and decompose more slowly than *cis*-OPAs (with few exceptions) [84], indicating a greater barrier to cycloreversion for *trans*-OPAs.

1.4.2 The Nature of the First Formed Intermediate in the Wittig Reaction

The non-involvement of betaine intermediates (or spin-paired diradical C–C bonded species proposed by McEwen and co-workers) [83] in Wittig reactions of non-stabilised ylides has been conclusively demonstrated using a *P*-phenyl-dibenzophosphole-derived ylide, isopropylidenephenyldibenzophospholane (see Scheme 1.19) [72]. The OPA derived from this ylide can be detected by NMR. It is pseudorotationally restricted because of the high ring strain that would be induced if the five-membered ring were forced to span two equatorial sites. The two possible OPA pseudorotamers that have the five-membered ring spanning axial and equatorial sites interconvert sufficiently slowly at low temperature that they can be distinguished by ^{31}P NMR. Their equilibrium pseudorotameric ratio was established by observation of the ^{31}P NMR of the Wittig reaction of isopropylidenephenyldibenzophospholane and 3-phenylpropanal at −78 °C (see Scheme 1.19). Interconversion of the OPA pseudorotamers was relatively slow at this temperature, and the equilibrium ratio was established to be 1.8:1. The kinetic OPA pseudorotameric ratio (initial ratio) for this Wittig reaction in Et_2O was

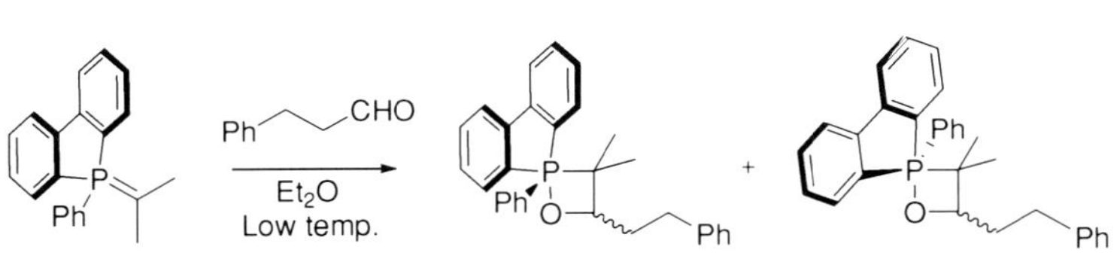

Ph⌒⌒CHO; Et_2O, Low temp.

Scheme 1.19 Wittig reaction of isopropylidenephenyldibenzophospholane with 3-phenylpropanal to give OPA pseudorotamers that are distinguishable by NMR

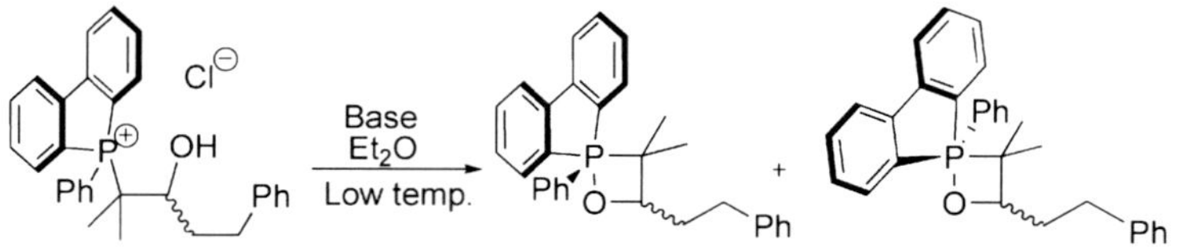

Scheme 1.20 β-HPS deprotonation to give OPA pseudorotamers that are distinguishable by NMR

observed to be 6.5:1 by the addition of the aldehyde to the ylide at −109 °C, and subsequent ^{31}P NMR observation of the reaction mixture at −78 °C.

The kinetic OPA pseudorotameric ratio produced in the deprotonation of the corresponding β-HPS in Et_2O at −109 °C (see Scheme 1.20), which must necessarily proceed through a betaine, was also observed by ^{31}P NMR at −78 °C, and was found to be 1:4.2—different from the kinetic pseudorotameric ratio produced in the Wittig reaction, and also from the equilibrium ratio of the pseudorotamers at this temperature.

The kinetic OPA pseudorotameric ratios were also determined in a similar manner for this reaction at low temperature in THF and CH_2Cl_2. In each case, the ratio was found to be different for the OPA formed by Wittig reaction compared to that formed by β-HPS deprotonation. That a different kinetic pseudorotameric ratio is produced in each instance shows that the formation of OPA by Wittig reaction does not occur through a betaine intermediate, at least in the case of non-stabilised ylides.

In addition to this very conclusive direct evidence that OPAs are the first-formed and only intermediates in Wittig reactions, the very substantial evidence presented above for the operation of kinetic control in Wittig reactions can itself be interpreted as evidence for this same thing. Experiments that necessarily produce betaines and hence OPAs by processes independent of the Wittig reaction have been shown to result in stereospecific conversion to alkene. Since high *E*-selectivity (or high selectivity for *trans*-OPA) in a kinetically controlled process cannot be accounted for within the context of the betaine mechanism, it must be discounted as a possible mechanism for the Wittig reaction.

As a final point, it should also be noted that there is no easy way to account for how changing the substituents at phosphorus changes the selectivity in reactions of each ylide type using the betaine mechanism.

1.4.3 Oxaphosphetane Structure and Pseudorotation

Westheimer's rule on apical entry and departure requires that any bond made to phosphorus to form a trigonal bipyramidal compound must initially place the new substituent in an axial site, and that upon departure of a substituent from a trigonal bipyramidal compound the bond broken must be to a substituent in an apical site [92]. Trigonal bipyramidal compounds are stabilised to the greatest extent possible if the most electronegative atom(s) occupy the apical sites. This arrangement of the substituents results in the HOMO being as low in energy as is possible [93]. Based on these concepts, the most stable OPA pseudorotamers should have oxygen in an axial site, and formation of OPA in the Wittig reaction should initially give a pseudorotamer of OPA with apical oxygen. Furthermore, the OPA pseudorotamer that undergoes decomposition to alkene and phosphine oxide must have ring carbon-3 (formerly the ylide α-carbon) in an axial site and therefore, necessarily, oxygen in an equatorial site. There must be a pseudorotation process to bring about the formation of the "C-apical" OPA from the "O-apical". Although in one of their publications Vedejs and Marth question whether the formation of an OPA with the ring carbon-3 apical is necessary for OPA decomposition to alkene and phosphine oxide to occur [91], the available computational evidence suggests that it *is* necessary [49]. Furthermore, if P–C bond breakage is possible for any OPA pseudorotamer, then this would constitute a notable exception to Westheimer's rule. Below is presented the experimental data that has been amassed pertaining to the structure of OPAs produced in Wittig reactions and to OPA pseudorotation.

Vedejs and Marth were able to resolve signals due to two different pseudorotamers of a dibenzophosphole-derived OPA (see Chart 1.2 structures **7–9**) by low temperature ^{1}H NMR, and to prove that these pseudorotamers had oxygen in the apical position [91]. This was established by an evaluation of the one bond coupling ^{13}C-^{31}P constants $^{1}J_{PC}$ in the ^{13}C NMR for the phosphorus-bound carbons in the dibenzophosphole unit. In such a system there must necessarily be one carbon apical and one equatorial. Indeed, by low temperature ^{13}C NMR, there appear three signals for quaternary aromatic carbons with small coupling constants to phosphorus (less than 20 Hz)—diagnostic of aromatic carbons in an apical position [91] or not directly joined to phosphorus and thus not in the trigonal bipyramid—and a fourth quaternary aromatic carbon showing coupling to phosphorus of 132 Hz. The latter is characteristic of an sp^{2}-hybridised equatorial carbon in a phosphorus-centred trigonal bipyramid [94]. The OPA ring carbon (δ 54.5) decisively shows $^{1}J_{PC} = 82$ Hz, which is indicative of an equatorial aliphatic (sp^{3}) carbon in a phosphorus-centred trigonal bipyramid. This means that the most stable OPA in solution has either the structure **7** or **8** in Chart 1.2, but not structure **9**. In a separate publication, Vedejs and Marth disclosed the resolution by low temperature ^{1}H and ^{31}P NMR of signals due to two different pseudorotamers of another DBP-derived OPA (shown above in Scheme 1.19) [72], for which the presence of an apical oxygen was established in a similar manner to that described above in this paragraph. This data was used to establish that other dibenzophosphole-derived OPAs (**10–13** in Chart 1.2) as well as some examples of unconstrained OPAs are also likely to have oxygen in the apical position in their favoured solution pseudorotamer(s). Other spectra collected on unconstrained OPAs with the ylide α-carbon ^{13}C labelled show $^{1}J_{PC}$ values of very similar magnitude for carbon-3 in the OPA ring [71].

Bangerter et al. resolved signals due to pseudorotameric OPAs **14a** and **14b** (see Chart 1.2) and in a separate experiment did the same for OPA pseudorotamers **15a** and **15b** by low temperature ^{31}P NMR [88]. They established a $^{1}J_{PC}$ value of

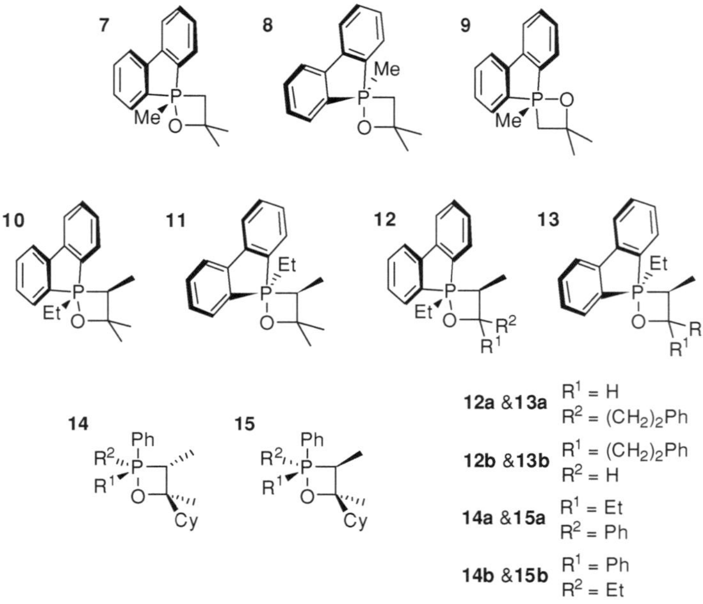

7 8 9

10 11 12 13

14 15

12a &13a R^1 = H, R^2 = $(CH_2)_2Ph$

12b &13b R^1 = $(CH_2)_2Ph$, R^2 = H

14a &15a R^1 = Et, R^2 = Ph

14b &15b R^1 = Ph, R^2 = Et

Chart 1.2 Dibenzophosphole-derived OPAs **7**, **8** & **10–13** with oxygen in apical position and **9** with oxygen in equatorial position [91], and ethyldiphenylphosphine-derived OPAs **14** and **15** [88]

86.7 Hz for ring carbon-3 in OPAs **15a** and **15b**, thus showing that the dominant solution structure for these OPAs has oxygen in an axial site [94]. The $^1J_{PC}$ values for the *P*-ethyl group of **15a** and **15b** labelled with ^{13}C at the methylene group were determined to be 92.4 and 106.8 Hz respectively. These values are characteristic of an equatorial sp^3-hybridised group [94], which leads to the conclusion that the alkyl group is preferentially placed in an equatorial position in both pseudorotamers [88].

In the studies of Vedejs and Marth [72, 91] and Bangerter et al. [87], the rate of pseudorotation of each OPA was established by line shape analysis of variable temperature NMR spectra of the resolved OPA pseudorotamers. It was concluded that in each case the rate of OPA pseudorotation is too fast relative to the rate of OPA decomposition to have any effect on the decomposition rate.

The ^{13}C NMR spectrum of each of the OPAs whose two possible O-apical pseudorotamers are shown as structures **10** and **11,** as **12a** and **13a**, and as **12b** and **13b** respectively (see Chart 1.2) contains only one set of signals [91]. This can mean either that the two O-apical pseudorotamers are undergoing rapid interconversion, or that one pseudorotamer is heavily favoured in solution. Since the pseudorotamers of OPAs very similar to these can be resolved by low temperature NMR, Bangerter et al. [88] concluded that one of the O-apical pseudorotamers is predominant in solution in the case of each of the OPAs **10/11**, **12a/13a** and **12b/13b**, so that the pseudorotamers do not interconvert rapidly at low temperature.

Further evidence indicating the presence of a strongly favoured OPA pseudorotamer in solution from the present work is described later, in Sect. 2.4.5.

1.4.4 How Does the Observed Selectivity for Z or E *Alkene Arise? An Explanation of the Currently Accepted Mechanism and Source of Stereoselectivity*

1.4.4.1 Summary of Experimental Evidence on the Wittig Reaction Mechanism

a. OPAs are the only observable intermediates in reactions of non-stabilised ylides.
b. Betaines and diradical species have been proven not to play a part in the reactions of these ylides.
c. Stereospecific conversion of *erythro* and *threo* betaines that would be derived from semi-stabilised ylides (generated either from β-HPSs or epoxides) to *Z* and *E* alkene respectively have been demonstrated. The OPAs (generated by β-HPS deprotonation) that would be produced in the Wittig reactions of dibenzophosphole-derived ylides have been observed by low temperature NMR. The *cis*-OPAs produced have been shown to decompose stereospecifically to Z-alkene and the *trans*-OPAs have been shown to decompose stereospecifically to *E*-alkene. A DBP-derived OPA has also been observed in the case of one Wittig reaction, and was demonstrated to undergo stereospecific decomposition to alkene and phosphine oxide.
d. Stereospecific conversion of *erythro*-betaines derived from stabilised ylides (and generated from *erythro*-β-HPSs) to *Z*-alkenes has also been observed.
e. It seems safe to assume that conversion of *trans*-OPA derived from stabilised ylides to *E*-alkene is irreversible and stereospecific, since *E*-alkene formation is generally very favoured in such reactions, and also *trans*-OPAs are typically more stable than their *cis* counterparts.
f. OPAs are obligatory at some point along the Wittig reaction coordinate (either as intermediates or TSs). Decomposition of OPA derived from non-stabilised ylides has been shown to occur at the same rate as alkene formation [69–71]. The stereospecific decomposition of α-deuterated *erythro*-β-HPSs derived from stabilised ylides to *Z*-alkene with retention of the deuterium label shows that the intermediate *cis*-OPA is formed irreversibly and decomposes stereospecifically. Thus it has been shown that OPA is in fact the final intermediate in the Wittig reaction (regardless of how it is formed).
g. The betaine mechanism cannot account for kinetic *E*-selectivity (particularly important for reactions of stabilised ylides).
h. These facts, all together, strongly suggest that OPAs are the first-formed and only intermediates formed in Li-salt free Wittig reactions.

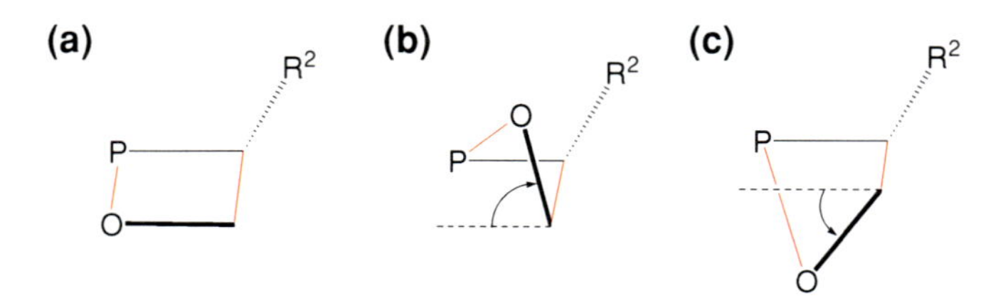

Fig. 1.8 **a** Planar TS with puckering angle of 0°, **b** TS with positive puckering angle, **c** TS with negative puckering angle

1.4.4.2 Non-stabilised Ylides

By the convention employed in the paper of Aggarwal, Harvey and coworkers [49], when the ylide P=C bond and the aldehyde C=O bond are parallel (i.e. a planar TS), the puckering angle of the TS is defined as 0° (see Fig. 1.8a). If the dihedral angle between the carbonyl C=O bond and the ylide P–C bond is smaller than the corresponding angle in the planar TS (see Fig. 1.8b), then the TS puckering angle is positive. If the dihedral angle is greater than in the planar TS, then the TS puckering angle is negative (see Fig. 1.8c).

The following is Vedejs's rationale for the mechanism of the Wittig reaction of non-stabilised ylides [85]. Alkylidenetriphenylphosphoranes (non-stabilised ylides) react preferentially through an early, puckered, transition state in which the carbonyl substituent occupies a *pseudo*-equatorial position (see Fig. 1.9a). The forming P–O and C–C bonds are long, and rehybridisation about the ring atoms is not particularly advanced. In this lowest energy TS, the puckering angle is *positive*. This arrangement minimises steric interactions between the carbonyl substituent and the *P*-phenyl groups (referred to as "1–3 interactions", with the numbering of the ring positions as shown in Fig. 1.9), which are still in a pseudo-tetrahedral arrangement about phosphorus. It also results in the minimisation of steric interactions between the carbonyl substituent and the ylide substituent ("1–2 interactions"), and allows the forming P–O bond to avoid the *P*-phenyl group that is necessarily projecting in the direction of carbonyl approach to the ylide. The steric interactions of the ylidic substituent R^2 with the carbonyl substituent ("1–2 interactions") and with the *P*-phenyl groups ("2–3 interactions") are minimised if it is in a pseudo-axial site, as shown in Fig. 1.9a, and hence this TS leads to *cis*-OPA and *Z*-alkene. Calculations on the Wittig reaction of Ph_3P=CHMe and MeCHO at the B3LYP/6-31G*(THF) level of theory (employs a continuum solvent with the dielectric properties of THF) confirm this to be the lowest energy transition state, and indicate that there may be a stabilising hydrogen bonding interaction between the aldehyde oxygen and one of the *P*-phenyl C-H bonds in *both* diastereomeric TSs [49].

There is no clear picture of what the exact geometry of the lowest energy *trans*-selective TS in reactions of alkylidenetriphenylphosphoranes might be. A possible

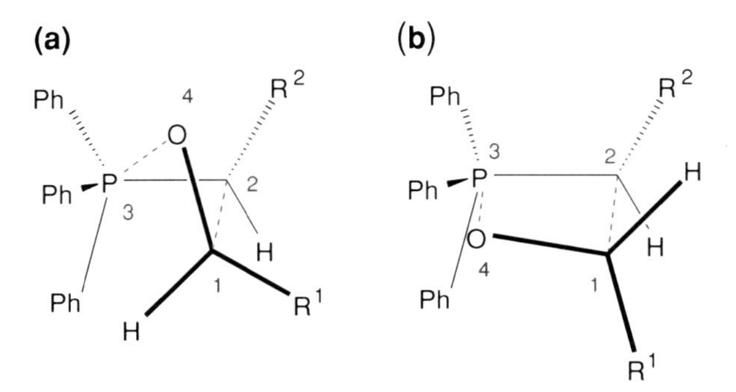

Fig. 1.9 Cycloaddition transition states leading to OPA in Wittig reactions of alkylidenetriphenylphosphoranes: **a** kinetically favoured *cis*-selective cycloaddition transition state showing a *positive* angle of puckering, with R^2 in a pseudo-axial site and R^1 in a pseudo-equatorial site; **b** lowest energy *trans*-selective transition state. R^1 is the large carbonyl substituent, and R^2 is the large ylide α-substituent

puckered *trans*-selective TS with the ylide α-carbon substituent in a pseudo-equatorial site (Fig. 1.9a with R^2 and H swapped at carbon-2) would suffer from large 1–2 interactions and hence would be disfavoured relative to the *cis*-selective TS. A possible planar *trans*-selective TS is shown in Fig. 1.9b. Like the *cis*-selective TS described above, bond formation and rehybridisation about the ring atoms are each not particularly advanced—so the TS is early. 1–2 steric interactions between the carbonyl substituent and the ylide α-carbon substituent are avoided in this TS due to the *trans* arrangement of these substituents. However, it is less stable than the *cis*-selective TS because one of the *P*-phenyl groups encumbers the approach of the carbonyl oxygen to phosphorus, and because of large 1–3 steric interactions between the carbonyl substituent R^1 and one of the *P*-phenyl groups. As with the *cis*-selective TS there may exist hydrogen bonding between the carbonyl oxygen and a *P*-phenyl C-H in this TS [49].

It is postulated that the OPAs form irreversibly (as has been very comprehensively shown experimentally), and decompose stereospecifically to alkene and phosphine oxide by [2 + 2] cycloreversion, so stereoselectivity is decided in the cycloaddition step. The computational results of Aggarwal, Harvey and coworkers [49] on the reaction of $Ph_3P{=}CHMe$ + MeCHO indicate that *cis*-OPA is less thermodynamically stable than the *trans*-OPA due to increased 1–2 steric interactions in the former, and that OPA formation is exothermic. The barrier to OPA cycloreversion to alkene and phosphine oxide is found to be higher for *cis*-OPA than for *trans*-OPA. Based on this, it can be surmised that in some circumstances the barrier to cycloreversion to alkene and phosphine oxide for *cis*-OPA is similar to the barrier to reversal to ylide and aldehyde. This may provide an explanation for the observation of stereochemical drift in certain reactions of alkylidenetriphenylphosphoranes—in particular those with aromatic aldehydes [65, 87].

Non-stabilised ylides for which one or more of the *P*-phenyl groups are replaced by alkyl group(s) show much lower *Z*-selectivity than alkylidenetriphenylphosphoranes [4]; in the context of the cycloaddition mechanism this is explained by decreased 1–3 interactions and thus a lower propensity to cycloaddition TS puckering. Computational results on the reaction of Me_3P=CHMe + MeCHO at the B3LYP/6-31G* (THF) level of theory indicate that both the *cis* and *trans*-selective cycloaddition TSs are planar, and very similar in energy—the role of 1–2 steric interactions does not particularly militate against the *cis*-selective TS in this case due to the very long forming bonds in an early TS [49]. In reactions of alkylidenetrialkylphosphoranes with tertiary or aromatic aldehydes, an extra factor is known to be at play in the experimentally observed *E*-selectivity. The *cis*-OPA undergoes reversal to ylide and aldehydes, while the *trans*-OPA does not [69, 87]. and thus *trans*-OPA accumulates and gives rise to *E*-alkene. The computational study, referred to earlier, found that the kinetic barriers to *cis*-OPA cycloaddition and reversal to ylide and aldehyde were similar, which is consistent with the experimentally observed depletion of this intermediate. The lower *Z*-selectivity in reactions of alkyldiphenylphosphine-derived ylides compared to alkylidenetriphenylphosphoranes is, however, truly dependent on the energetic discrimination of the TSs in the OPA forming cycloaddition step, as irreversible OPA formation has been proven for such ylides [69]. Thus in those reactions, the higher *E*-selectivity reflects an increased kinetic advantage (or decreased kinetic disadvantage) for *trans*-OPA formation.

A further aspect of the currently accepted mechanism for Wittig reaction of non-stabilised ylides has to do with pseudorotation of the OPA intermediate. The OPA formed by the cycloaddition step is postulated to have the oxygen in the apical position, in accordance with Westheimer's rule on apical entry and departure for a phosphorus-centred trigonal bipyramid [92]. This OPA, (which could be any one of a number of pseudorotamers with apical oxygen) is proposed to undergo rapid pseudorotation to a less stable pseudorotamer with ring carbon-3 in an apical position, and the ring oxygen in an equatorial site. This pseudorotamer undergoes [2 + 2] cycloreversion to alkene and phosphine oxide. The computational work of Aggarwal, Harvey and co-workers confirms the operation of this process [49].

Calculations on the reactions of MeCHO with Ph_3P=CHMe and Me_3P=CHMe respectively show that betaines are significantly higher in energy than OPAs and likewise TSs involving *anti* addition of the reactants (in the style of an aldol reaction to form an *anti*-betaine) are higher in energy than cycloaddition TSs [49].

The cycloaddition mechanism of Vedejs for the Li-salt free Wittig reaction of non-stabilised ylides—that being direct, irreversible cycloaddition of ylide and aldehyde to give OPA, followed by facile pseudorotation of the OPA and then by irreversible, stereospecific [2 + 2] cycloreversion of OPA to alkene and phosphine oxide—is consistent with the following facts:

- OPA formation has been shown to be irreversible [69, 70, 88], and the OPA *cis/trans* ratio corresponds to the *Z/E* ratio of the alkene ultimately obtained from the reaction [70, 71, 88]. This implies that the stereoselectivity is determined in the OPA forming step. OPA decomposition has been shown to occur at the same rate as alkene formation for reactions of alkylidenetriphenylphosphoranes [69–71, 88]. Thus OPA is the final intermediate formed in the Wittig reaction, and phosphine oxide is formed by *syn*-elimination.

- Tests for the involvement of radicaloids or ionic species (betaines) indicate that neither type of species play a part in the Wittig reaction.
- OPA accumulates during the reaction, and OPA decomposition is the final step of the reaction. Therefore this step is rate-determining. It has been observed that reactions of non-stabilised ylides with benzaldehydes have ρ values for the carbonyl component of 0.2–0.59 [95, 96].[8] A kinetic isotope effect of 1.053 has also been observed in the Wittig reaction of *iso*-propylidenetriphenylphosphorane with benzophenone with a ^{14}C-labelled carbonyl group [95]. Thus in the rate-determining step the bonding is changing at the carbonyl carbon. The decomposition of each of the *cis* and *trans* OPAs (independently generated form β-HPSs) derived from (ethylidene) ethyldiphenylphosphorane and cyclohexyl methyl ketone has been shown to be first order, and to have a positive entropy of activation [88]. The results from investigations on the kinetics of reactions of non-stabilised ylides thus support the operation of a rate-determining cycloreversion as the final step of the reaction.
- *Z*-selectivity increases in line with the increasing steric bulk of the aldehyde in reactions with a common ylide; so tertiary aldehydes show higher *Z*-selectivity than secondary aldehydes, which show higher selectivity than primary aldehydes. Likewise, increasing the steric bulk on the *P*-phenyl groups of triphenylphosphine-derived ylides (e.g. using tri(*ortho*-tolyl)phosphine derived ylides) results in increased *Z*-selectivity. Thus, the factors that should exacerbate the need to relieve 1–3 steric interactions in a puckered cycloaddition TS show increased *Z*-selectivity, as this mechanism would predict.
- The rate of OPA pseudorotation is much greater than the rate of OPA decomposition to alkene and phosphine oxide both for dibenzophosphole-derived OPAs [72, 91] and for unconstrained OPAs [88, 91]. The rate of OPA decomposition is thus independent of the rate of OPA pseudorotation.
- The mechanism is in agreement with the computational results of Aggarwal, Harvey and co-workers [48–50].

1.4.4.3 Semi-stabilised Ylides

The Wittig reactions of semi-stabilised ylides are proposed to proceed by an irreversible [2 + 2] cycloaddition to give OPA, with C–C bonding being more advanced at the TS than P–O bonding. The initially formed OPA with oxygen in

[8] See Ref. [3] pp. 290, 291.

an axial position undergoes pseudorotation to place the ylidic carbon in an apical position in the phosphorus-centred trigonal bipyramid. This OPA then undergoes stereospecific and irreversible cycloreversion to alkene and phosphine oxide [84]. The TS is thought to occur later along the reaction coordinate than in reactions of non-stabilised ylides. A consequence of this is that the phosphorus substituents (spectator ligands, ylidic carbon and aldehyde oxygen) are in a tighter, more nearly trigonal bipyramidal arrangement about phosphorus than is the case in TSs of reactions of non-stabilised ylides. Also the shape of the sp^2-hybridised phenyl or vinyl substituent at carbon-2 of the forming ring is postulated to result in less severe steric interactions between all of the substituents in the TS compared to the alkyl group in the same position in the TSs derived from non-stabilised ylides [85]. It can be envisaged that the unsaturated group could orient itself to avoid steric interactions in a way that an alkyl group could not in such TSs. This may in turn have an effect on how the substituents on phosphorus are disposed. The upshot is that there is less of an energetic advantage to TS puckering, which in reactions of non-stabilised ylides neatly minimises both 1–3 and 1–2 steric interactions. The somewhat decreased importance of 1–3 (and presumably also of 2–3) interactions means that parallel (or nearly parallel) approach of the ylide and aldehyde can be competitive energetically with a TS involving the puckered approach of the reagents. In such a scenario the minimisation of 1–2 interactions dictates that a planar TS should be selective for *trans*-OPA. The *cis*-selective TS is likely to be puckered, and thus to be similar in appearance to the *cis*-selective TS in reactions of non-stabilised ylides (see Fig. 1.8a), albeit with the phosphorus spectator ligands being in a more compact arrangement.

The computational findings of Aggarwal, Harvey and coworkers [49] indicate the above mechanism to be in operation in the reaction of benzylidenetriphenylphosphorane with benzaldehyde at the B3LYP/6-31G*(THF) level of theory. The *cis*-OPA selective cycloaddition pathway is marginally favoured, and the barrier to OPA decomposition is lower than both the corresponding barrier in the reaction of ethylidenetriphenylphosphorane with benzaldehyde, and the barrier to OPA formation, so OPA formation is rate determining. This is in keeping with the experimental observation for semi-stabilised ylides that OPAs cannot be observed spectroscopically except in exceptional circumstances [60]. The barrier to OPA reversal to Wittig starting materials is also higher than in the case of the non-stabilised ylide, and coupled with the lower activation energy for OPA decomposition to alkene and phosphine oxide, this indicates that OPA formation is irreversible. The *cis*-selective TS is found to be puckered (although somewhat less so than the *cis*-selective TS for the non-stabilised ylide) and the *trans*-selective TS to show slight puckering (with a *negative* angle). As with non-stabilised ylides, hydrogen bonding between the carbonyl oxygen and a *P*-phenyl C–H is found to play a role in the structure of the TSs. The calculations indicate that the postulated pseudorotation of the initially formed OPA occurs to give a higher energy pseudorotamer with the P–C bond apical, which then undergoes [2 + 2] cycloreversion to give alkene and phosphine oxide.

The cycloaddition mechanism proposed by Vedejs, and verified computationally by Aggarwal, Harvey and co-workers for Wittig reactions of triphenylphosphine-derived semi-stabilised ylides accounts well for the observed alkene diastereoselectivity. Decreased steric interactions in the planar *trans*-selective cycloaddition TS mean that it becomes competitive with the puckered *cis*-selective TS and thus poor selectivity is observed.

In reactions of semi-stabilised ylides for which one or more of the *P*-phenyl groups are replaced by alkyl group(s), 1–3 interactions become less important, as the *P*-alkyl groups effectively free up space around phosphorus in the cycloaddition TS in a way that is not possible with three *P*-phenyl groups. This is thought to be more to do with the shape of the *P*-alkyl group and its effect on how the other substituents on phosphorus are oriented (so its effect on the shape of the "phosphonium" moiety as a whole) than its relative steric bulk *per se*. As a consequence of the above, the main steric interaction that destabilises the nearly planar *trans*-selective TS for triphenylphosphine-derived ylides is dramatically reduced for reactions of these ylides, and so it is energetically favoured over the *cis*-selective TS due to its much smaller 1–2 interactions. This results in much higher kinetic selectivity for the *trans*-OPA in the cycloaddition step, which is why high *E*-selectivity is observed for such semi-stabilised ylides. Computational results on the reaction of benzylidenetrimethylphosphorane with benzaldehyde indicate OPA formation is rate-determining and irreversible, and that the cycloaddition TS is indeed later (i.e. bond formation and rearrangement of substituent geometries about the reactive centres are more advanced) than for the corresponding reaction of a non-stabilised ylide [49]. OPA formation was found to be exothermic, albeit not to the same extent as the in reaction of ethylidenetriphenylphosphorane. The *cis*-selective TS shows only slight puckering, with a positive angle, while the energetically favoured *trans*-selective TS is also slightly puckered but in a negative sense. There is a possibility that the decrease in the puckering in the *cis*-selective TS is as a result of a reduction of the unfavourable electrostatic interaction in the "flatter" conformation between the C–O and ylide C–Ph bond dipoles (see later).

The importance of 1–3 interactions is emphasised by the different selectivities observed in reactions of a given semi-stabilised ylide with primary and tertiary aldehydes respectively. The latter generally show much greater *Z*-selectivity than the former as a result of more pronounced 1–3 interactions.

The mechanism described above for the Li-salt free Wittig reaction of semi-stabilised ylides is consistent with the following facts:

- Where OPAs can be observed spectroscopically (i.e. dibenzophosphole-derived OPAs, where the rate of decomposition is retarded sufficiently to make OPA decomposition rate determining), their formation has been shown to be under kinetic control, and stereospecific conversion to alkene has been proven [60].
- Stereospecific conversion of β-HPS (by deprotonation) [60, 83] or epoxide (by nucleophilic cleavage with phosphide, and methylation of the resulting

β-oxidophosphine) [90] to alkene via (presumed) betaine and OPA intermediates has been demonstrated.

- Attempted crossover experiments on Wittig reactions of semi-stabilised ylides gave no crossover product [83, 97], although as previously mentioned, this may not be meaningful if the Wittig reaction is already complete by the time the crossover reactant is added.
- No intermediate is observed in Wittig reactions of semi-stabilised ylides (except for DBP ylides). Kinetic studies show the reaction to be overall second order, and to have a positive ρ value for the carbonyl reagent [98–100]. Negative activation volumes [99] and entropies [98] were also reported, which would be in line with a cycloaddition mechanism. Reactions with substituted benzaldehydes showed a kinetic isotope effect at the carbonyl carbon (^{14}C labelled in the study concerned), indicating that the bonding is changing at the carbonyl centre in the rate determining step [100].
- No betaines derived from semi-stabilised ylides have been detected in the course of spectroscopic monitoring of Wittig reactions [70, 83], β-HPS deprotonation reactions or reactions involving epoxide ring opening and phosphorus quaternisation [60, 84].
- The reaction of *P*-chiral semi-stabilised ylide benzylidene-ethylmethylphenylphospohorane with benzaldehyde gave phosphine oxide in which there is retention of configuration at phosphorus. This indicates the operation of a *syn*-elimination of the phosphine oxide from a cyclic intermediate or TS with trigonal bipyramidal phosphorus [76].
- The above points are consistent with an irreversible, rate-determining cycloaddition of ylide and aldehyde to give OPA where the aldehyde carbonyl is acting as an electrophilic centre, followed by an irreversible, stereospecific *syn*-cycloreversion of OPA to alkene and phosphine oxide.
- The experimentally observed selectivities are well accounted for by the proposed mechanism. It is consistent with the facts that the most *Z*-selective reactions of semi-stabilised ylides involve tertiary aldehydes or bulky triarylphosphine-derived ylides, and that the most *E*-selective reactions of these ylides involve non-bulky aldehydes or dialkylphenylphosphine, trialkylphosphine or methyldibenzophosphole-derived ylides.

1.4.4.4 Stabilised Ylides

Stabilised ylides react through a relatively late TS, in which rehybridisation about phosphorus to trigonal bipyramidal and about the ylidic carbon and aldehyde carbonyl centres to tetrahedral is near completion. Vedejs's original proposal contended that this meant the cycloaddition TS was constrained to being planar, and hence 1–2 interactions would generally be the dominating steric interactions, which would be consistent with the generally observed high *E*-selectivity in Wittig reactions of stabilised ylides [60, 85]. Some small inconsistencies did, however, exist with this proposed mechanism, for example it could not adequately explain the high *E*-selectivity observed in reactions of α-alkyl-α-carbonyl disubstituted stabilised ylides. The cycloaddition mechanism for stabilised ylides proposed by Vedejs and co-workers has thus been modified slightly to take into account dipole–dipole interactions in the cycloaddition TS, in line with findings in the computational papers of Aggarwal, Harvey and co-workers [48, 49].

Their calculations on the reaction of (methoxycarbonylmethylidene)-triphenylphosphorane with benzaldehyde at the B3LYP/6-31G*(THF) level of theory indicate rate-determining *endothermic* cycloaddition of ylide and aldehyde to give OPA, which undergoes facile pseudorotation to place the ylidic carbon in an axial position in the phosphorus-centred trigonal bipyramid, and then cycloreverts to alkene and phosphine oxide, as predicted by the Vedejs mechanism. The cycloaddition TS is indeed found in their calculations to be relatively late (i.e. bond formation is quite advanced), and the barrier to OPA decomposition is very low in comparison with the barrier to reversal to ylide and aldehyde, and with the OPA cycloreversion barrier in reactions of non-stabilised and semi-stabilised ylides. Thus OPA formation is irreversible, and the OPA intermediate does not accumulate, which is consistent with experimental observations. Interestingly, the *trans*-OPA with the carbon in the apical position is determined in these calculations to be more stable than the pseudorotamer with oxygen in the apical position.

Where the calculated mechanism differs from the mechanism proposed by Vedejs is in the shape of the cycloaddition TSs. The *trans*-selective TS (shown in Fig. 1.10a) is found to be puckered, but importantly this puckering is in the opposite sense (−40.1°) to that proposed for the *cis*-OPA selective TS in reactions of non-stabilised ylides (Fig. 1.10a) [48, 49]. This results in a TS that has an electrostatically favourable antiparallel orientation of the carbonyl C–O and ylide C–C(O) bond dipoles. Minimisation of both 1–2 and in particular 1–3 steric interactions then dictates that the large aldehyde substituent (R^1) should be pseudo-equatorial, and so this TS is selective for *trans*-OPA. Puckered *cis*-TSs were found to be disfavoured for this reaction—a TS with a negative puckering angle and thus favourable relative orientation of reactant dipoles (Fig. 1.10c) suffers from strong 1–3 (as well as significant 1–2) interactions, while one with the opposite sense of puckering (Fig. 1.10d) has an electrostatically disfavoured disposition of reactant dipoles. The lowest energy *cis*-selective TS was found to be planar (Fig. 1.10b)—it is not particularly disfavoured electrostatically, but is much higher in energy than the *trans*-selective TS as it suffers from large 1–2 interactions and lacks the electrostatically favoured antiparallel orientation of reactant dipoles that is present in the latter TS. Thus high *E*-selectivity is observed in general in Wittig reactions of stabilised ylides in non-polar or polar-aprotic media, with selectivity being extremely high for ylides in which phosphorus bears bulky substituents (e.g. triphenylphosphine-derived ylides), as the possibility of large 1–3 interactions dictates that these discriminate particularly well against puckering of the *cis*-selective TS.

The computational results on the reaction of (methoxycarbonylmethylidene)-trimethylphosphorane with benzaldehyde (B3LYP/6-31G*(THF) level of theory) indicates a similar shape for the cycloaddition TSs to the above reactions, and

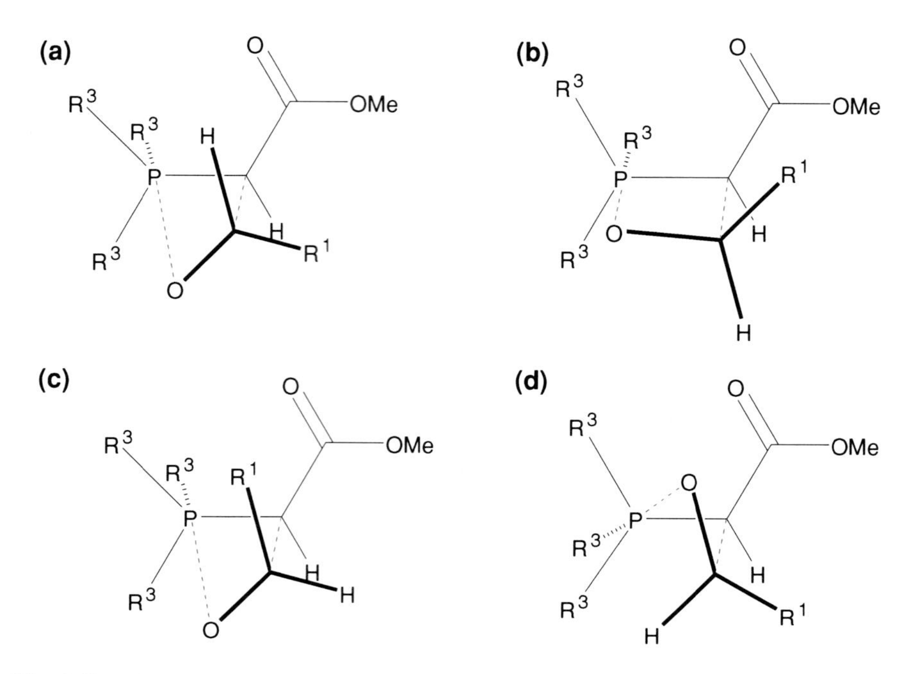

Fig. 1.10 **a** Favoured *trans*-selective TS showing negative puckering angle, **b** Lowest energy *cis*-selective TS, **c** and **d** Higher energy *cis*-selective TSs

indeed a similar energetic advantage for the *trans*-selective TS over the *cis*, which is in keeping with the observed high *E*-selectivity in reactions of trialkylphosphine-derived ylides. However, it may be that in certain circumstances, if the phosphorus substituents are not as bulky as triphenylphosphine-derived ylides (e.g. for methyldiphenylphosphine-derived ylides), then a different *cis*-selective TS with an antiparallel orientation of reactant dipoles is possible (Fig. 1.10c) since placement of the large substituent in the pseudo-axial position is not discriminated against to the same extent by 1–3 interactions, and so somewhat lower *E*-selectivity observed [4].

Diminished *E*-selectivity, and even predominant *Z*-selectivity, has been observed in reactions of stabilised ylides in methanol. A reasonable explanation is that this results from solvent-induced decrease in the importance of the interaction of reactant dipoles in the cycloaddition TSs. In this scenario, the factors governing TS geometry may be quite similar to those in reactions of semi-stabilised ylides.

This mechanism, in taking account of reactant dipole–dipole interactions in the cycloaddition TS, allows a rationalisation of the consistently high *E*-selectivity observed in Wittig reactions of α-alkyl-α-carbonyl disubstituted stabilised ylides. In reactions of such ylides, 1–2 steric interactions alone cannot account for the observed bias in selectivity towards *trans*-OPA and *E*-alkene. The presence of the additional ylide α-substituent, however, does not alter the operation of the dipole–dipole interaction, so the *trans*-selective TS adopts a similar conformation to that

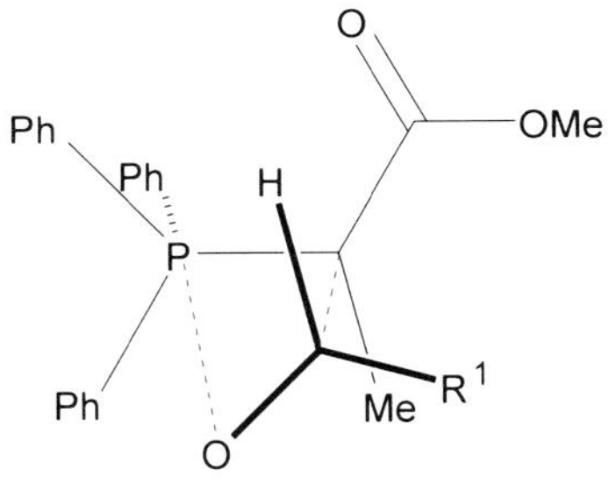

Fig. 1.11 Proposed *trans*-selective TS in Wittig reactions of α,α-disubstituted stabilised ylides, with electrostatically favourable relative orientation of carbonyl C–O and ylide C–C(O) bond dipoles

in reactions of the mono-substituted ylide (see Fig. 1.11), and is similarly favoured energetically over the *cis*-selective TS for the same reasons as are present in reactions of the mono-substituted ylide.

Further experimental evidence for the operation of the cycloaddition mechanism in Wittig reactions of stabilised ylides has been presented recently in a publication quantifying the relative nucleophilicity of phosphonate carbanions (Wadsworth-Emmons reagents), phosphinoxy carbanions and stabilised phosphonium ylides towards substituted quinone methides (Michael acceptors), benzhydrilium ions (carbocations) and benzaldehydes [101]. The nucleophiles are obliged to react by a straightforward nucleophilic substitution with the former two electrophiles. The relative reactivity of all three types of phosphorus ylide towards the benzaldehydes was found to be systematically significantly lower than towards the carbocations and Michael acceptors. Based on this, the authors concluded that the ylides all react with carbonyl compounds through an asynchronous concerted [2 + 2] cycloaddition. The ρ values determined in this study for the benzaldehydes in their reactions with each phosphorus nucleophile (2.9 for the phosphonium ylide, 3.4–3.6 for the phosphonate carbanions, and 2.7–2.8 for the phosphinoxy carbanions) are consistent with the operation of a cycloaddition mechanism with C–C bonding being more advanced than P–O bonding in the reactions of all three nucleophiles with carbonyl species. The value found for the reactions of the phosphonium ylide is consistent with literature precedent [69].

The mechanism described above for the Li-salt free Wittig reaction of stabilised ylides is consistent with the following facts:

- Stereospecific conversion of β-HPS (by deprotonation) to alkene via (presumed) betaine and OPA intermediates has been demonstrated. Thus OPAs are formed under kinetic control, and the stereoselectivity is decided in the C–C bond forming step [60].
- Reactions are second order (first order in each of ylide and carbonyl species), and alkene appears at the same rate as ylide is consumed [67, 102].
- ρ values for the carbonyl reactants have been found to be positive (so it is acting as an electrophile in the rate determining step) [67, 101] and for ylides have been shown to be negative (so it is acting as a nucleophile) [67], and their magnitudes are consistent with the operation of an asynchronous cycloaddition mechanism.

- The nucleophilicity of the stabilised ylide (methoxycarbonylmethylidene)-trimethylphosphorane towards benzaldehydes has been found to be quantifiably different from its nucleophilicity towards carbocations and Michael acceptors, with which it must react by nucleophilic addition [101].
- The entropy of activation for reactions of stabilised ylides is large and negative [67, 102]. There are no strong solvent effects in the reaction,[9] and indeed it has been observed that the rate of the reaction of (fluorenylidene)ethyldiphenyl-phosphorane with *p*-nitrobenzaldehyde is *slower* in acetone or DMF than in benzene [102]. Thus, the TS of the rate determining step is not polar and is highly ordered, characteristic of a cycloaddition TS.
- No OPA or betaine derived from a stabilised ylide has been detected during spectroscopic monitoring of Wittig reactions [70] or β-HPS deprotonations [60].

1.5 Aldehydes Bearing Remote Heteroatoms: Influence of These Substituents on Selectivity in the Wittig Reaction

The experimentally observed selectivity trends in Wittig reactions of various different ylide types have been described in Sect. 1.2. There exist, however, many Wittig reactions in which, in the context of these trends, the stereoselectivity is anomalous. A very comprehensive account of Wittig reactions that show anomalous stereoselectivity is included in the review of Maryanoff and Reitz [2]. The work conducted in the present project has been largely concerned with the unusual selectivity observed in Wittig reactions of aldehydes bearing heteroatom (lone-pair bearing) substituents remote from the carbonyl group. In particular, investigations have been carried out by our group [103] and others [100, 104, 105] into the unusually high *Z*-selectivity in reactions of semi-stabilised benzylides with benzaldehydes bearing one or more *ortho* halogen substituents. Similar high *Z*-selectivity has also been observed in reactions involving *ortho*-methoxy substituted benzaldehyde. There are also many reports in the literature of enhanced *Z*-selectivity in Wittig reactions of aldehydes bearing a heteroatom substituent (typically an oxygen) on the β carbon relative to the carbonyl group (i.e. similarly disposed relative to the carbonyl as the *ortho*-heteroatoms in benzaldehydes). We believe the high *Z*-selectivity observed in all of these reactions to be as a consequence of the same effect, which will be discussed in detail in Chap. 2. The results from the previous publication of Gilheany and co-workers and those of Yamataka and co-workers and Harrowven and co-workers will be discussed more fully below, followed by a discussion of the high *Z*-selectivity observed in reactions of aliphatic aldehydes bearing suitably placed heteroatom substituents,

---

[9] See Ref. [3] pp. 294–296.

1.5.1 *Wittig Reactions of Benzylidenetriphenylphosphoranes with ortho-Heteroatom Substituted Benzaldehydes*

A number of years ago, an investigation by Gilheany and co-workers was carried out on the systematically high *Z*-selectivity observed in the Wittig reactions of benzylides with *ortho*-substituted benzaldehydes in mixed chloroform/water solvent [103]. At the time of that publication, strong *ortho*-effects were already well known in organophosphorus chemistry [106, 107] including in the Wittig reaction [108, 109] and it was also known that *Z*-selectivity in stilbene synthesis could be induced by *ortho*-substituents with heteroatom lone pairs on the aldehyde [100, 110]. The results from the original paper published by the Gilheany group are shown in Table 1.3, along with selected results from the papers of Yamataka and co-workers [100], and Harrowven and co-workers [104, 105].

Similar trends can be observed in reactions with the same (or similarly substituted) reactants in both mixed $CHCl_3/H_2O$ solvent and THF solvent. There are some observations of note on the reactions conducted in $CHCl_3/H_2O$:

1. Benzylidenetriphenylphosphorane (generated in situ from the precursor phosphonium salt) reacted with benzaldehyde in biphasic chloroform-water solvent to give poor selectivity for either the *E* or the *Z* alkene (see Table 1.3 entries 11 and 12), depending on the counter-ion of the starting phosphonium salt used.
2. The same ylide reacted with 2-halobenzaldehydes with high *Z*-selectivity (entry 1).
3. The reactions of *ortho*-halogen substituted benzylides with benzaldehyde showed no selectivity or moderate *E*-selectivity (entries 15, 17–20), depending on the identity of the halogen substituents.
4. It was thus particularly surprising that reactions of *ortho*-halobenzylides with *ortho*-halobenzaldehydes showed even higher *Z*-selectivity (*Z/E* up to 95:5, entries 23, 27, 30, 34, 35) than the corresponding reactions of the unsubstituted benzylide.
5. The effect was shown to be specific to lone pair-bearing *ortho*-substituents by the poor selectivity observed in reactions of 2-methylbenzaldehyde (entries 6 and 30).

This *ortho* effect was subsequently employed by our group [111] and others others [104, 105, 112] to synthesise *Z*-stilbenes with high selectivity. Some reactions with more sterically bulky 2,6-disubstituted benzylides were also carried out. The reactions of these ylides with benzaldehyde and also with 2,6-dihalobenzaldehyde were found to be highly *E*-selective (for example entries 17 and 25), while the reaction of 2,6-dichlorobenzylidenetriphenylphosphorane with 2-chlorobenzaldehyde showed moderate *E*-selectivity (entry 26)—which still indicates a significant increase in *Z*-selectivity over the reaction with the unsubstituted aldehyde (entry 17).

The results of Yamataka and co-workers and co-wokers [100] and Harrowven and co-workers [104, 105] show that *Z*-selectivity is also observed for reactions of *ortho*-heteroatom substituted benzaldehydes in THF using a non-nucleophilic base

Table 1.3 *Z/E* selectivity in Wittig reactions of benzylides with benzaldehydes[a,b]

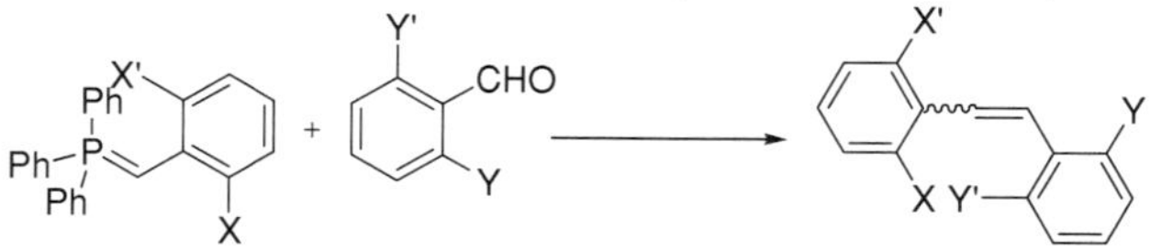

Entry	Ylide X =	Ald Y =	*Z/E* ratio
1	H	Cl	77:23
2	H	Cl	86:14[c]
3	H	Cl	92:8[d]
4	H	Br	85:15
5	H	F	70:30
6	H	Me	57:43
7	H	Me	37:63[c]
8	H	Me	57:43[d]
9	H	OMe	84:16[c]
10	H	OMe	94:6[d]
11	H	H	40:60
12	H	H	61:39
13	H	H	43:57[c]
14	H	H	64:36[d]
15	Cl	H	23:77
16	Cl	H	44:56[c]
17	2,6-Cl	H	1:99
18	Br	H	50:50
19	F	H	26:74
20	Me	H	67:33
21	Me	H	31:69[c]
22	OMe	H	34:66[c]
23	Cl	Cl	87:13
24	Cl	Cl	90:10[f]
25	2,6-Cl	2,6-Cl	1:99
26	2,6-Cl	Cl	40:60
27	Br	Br	95:5
28	Br	Br	88:12[f]
29	F	F	90:10
30	Me	Cl	90:10
31	Cl	Me	70:30
32	Br	OMe	90:10[f]
33	OMe	OMe	80:20[f]
34	Br	F	95:5

(continued)

Table 1.3 (continued)

Entry	Ylide X =	Ald Y =	*Z/E* ratio
35	F	Br	95:5
36	I	I	90:10[f]
37	Cl	I	92:8[e]
38	Br	I	91:9[e]

[a] Unless otherwise specified, the reactions were carried out in $CHCl_3/H_2O$ at 20 °C, and the ylide was generated from the corresponding phosphonium chloride or bromide in situ by the action of NaOH
[b] X' = Y' = H unless otherwise specified
[c] From Ref. [100]. Reaction carried out in THF at 0 °C in THF. Ylide generated in situ from phosphonium bromide salt using NaHMDS
[d] From Ref. [100]. Reaction carried out in THF at −72 °C in THF. Ylide generated in situ from phosphonium bromide salt using NaHMDS
[e] From Ref. [105]. Reaction carried out in THF at 0 °C in THF. Ylide generated in situ from phosphonium bromide salt using ^{t} BuOK
[f] From Ref. [105]. Reaction carried out in THF at 0 °C. Ylide generated in situ from phosphonium bromide or chloride using ^{t} BuOK

to generate the ylide. That higher *Z*-selectivity is observed in the reactions of Yamataka and co-workers at −72 °C than at 0 °C qualitatively indicates that the *ortho* effect is of kinetic origin.

The experiments in $CHCl_3/H_2O$ solvent described above were carried out under conditions for which the operation of kinetic control in Wittig reactions of semi-stabilised ylides has not been proven. Consequently, it was not possible to comment definitively on the source and mechanistic significance of the unexpectedly high *Z*-selectivity based on these results. However, the factors that could lead to the observed selectivity, of which one or more could simultaneously be in operation, were delineated:

1. The presence of the *ortho*-halogen gives rise to a lowering of the energy of the *cis*-selective TS leading to increased kinetic selectivity for the *cis*-OPA and hence the *Z*-alkene;
2. The presence of the *ortho*-halogen (an electron withdrawing group) on the benzaldehyde causes it to be more reactive than unsubstituted benzaldehyde and thus to react through an earlier TS. If this were the case, the geometry of the phosphorus substituents about the phosphorus atom might retain their *pseudo*-tetrahedral geometry so that a puckered *cis*-selective TS would be energetically favoured for the same reasons as in reactions of alkylidenetriphenylphosphoranes. This would lead to kinetic selectivity for the *Z*-alkene;
3. The presence of the *ortho*-halogen allows for the existence of a bonding interaction between the halogen and phosphorus in the OPA, which is favourable in the case of the *cis*-OPA, but not so in the case of the *trans*-OPA due to the resulting very significant 2–3 steric interactions engendered in the latter structure. By itself, increased thermodynamic stability in the product of the cycloaddition step should have no bearing on the selectivity in a kinetically

controlled process, which is controlled rather by the relative rates of formation of the isomers. However, in this case the Hammond postulate implies that the cycloaddition TS would be made more reactant-like (earlier) by the relative stability of the *cis*-OPA, which combined with point (2) would mean increased kinetic selectivity for the *cis*-OPA. Increased stability in the *cis*-OPA could also result in predominant *Z*-selectivity if OPA formation was reversible and the decomposition of *cis*-OPA was faster than that of *trans*-OPA.

Of the possible means by which the *ortho*-halogen could lower the energy of the *cis*-selective TS [see point (1)], it was thought (although not explicitly stated in the paper) that there might be a phosphorus-heteroatom bond in the TS itself (not just in the OPA as proposed in point (3)), which would stabilise the *cis* but not the *trans*-selective TS due to much greater 2–3 interactions in the latter. Yamataka and co-workers had previously proposed the existence of a phosphorus-heteroatom interaction to explain selectivity in stilbene syntheses from *ortho*-heteroatom substituted benzaldehydes [100]. His proposal involved the initial formation of a "σ complex" in which the heteroatom and carbonyl oxygen (presumably via its lone pair(s)) would chelate phosphorus *prior* to the formation of the cycloaddition TS—so in essence the σ complex would guide the reactants into a *cis*-selective TS, but the interaction would not remain in the TS itself. This proposal ignores the fact that the path by which a TS is reached in a kinetically controlled process is immaterial to whether or not the reaction proceeds through that TS. The existence of the σ complex would have no effect on the energy of the subsequent TS, and so could have no effect on kinetic selectivity in the reaction. The interaction referred to above is similar to, but fundamentally different in a very important way from the proposal of Yamataka and co-workers, in that it is present in the cycloaddition TS, and can thus contribute to lowering the energy of that TS. The carbonyl group interacts with the ylide through its π-symmetry orbitals, and the halogen interacts with the ylidic phosphorus through a lone pair. The acceptor orbital on the ylide would be a P–C σ^* orbital.

The results of Yamataka and co-workers included a highly *Z*-selective synthesis of 2-methoxystilbene from 2-methoxybenzaldehyde and benzylidenetriphenylphosphorane (Table 1.3 entries 9 and 10) [100], and indeed Harrowven and co-workers showed several examples of reactions of *ortho*-methoxy substituted benzaldehydes which were highly *Z*-selective (of which two examples are shown in Table 1.3 entries 32 and 33) [104, 105]. This result strongly suggested that the rationale behind point (2) above did not hold. 2-methoxybenzaldehyde is *less* electrophilic at the carbonyl group than benzaldehyde, and thus if the two were reacted with the same ylide, 2-methoxybenzaldehyde could not be reacting by an identical mechanism through an earlier TS than benzaldehyde.

In order to be able to make a definitive comment on the mechanism by which the *ortho*-effect is exerted, further reactions of semi-stabilised ylides needed to be conducted under conditions of strict kinetic control. Kinetic control has been proven in the reactions of methyldiphenylphosphine-derived semi-stabilised ylides [69] (but not directly in reactions of triphenylphosphine-derived ylides).

Consequently, it was decided at the beginning of the present project to conduct Wittig reactions of benzylidenemethyldiphenylphosphoranes (both the parent ylide and various *ortho*-substituted derivatives) with *ortho*-heteroatom substituted benzaldehydes (as well as with "controls" benzaldehyde and 2-methylbenzaldehyde) in THF at −78 °C using a non-nucleophilic base such as NaHMDS or KHMDS to generate the ylide in situ. In this manner, one could truly test whether the *ortho* effect is of kinetic origin, and the reason for the amplification of the effect on reacting with *ortho*-substituted benzylides could be more fully investigated. Perhaps the most important experiments proposed in undertaking the present work involved 2-methoxybenzaldehyde, as these would show whether this apparently electronic effect was exerted by mesomeric interaction with the carbonyl or in a through-space manner (i.e. by the formation of a phosphorus-heteroatom bond in the *cis*-selective TS).

1.5.2 Wittig Reactions of Aliphatic Aldehydes Bearing a β-Heteroatom Substituent

Enhanced *Z*-selectivity is a known effect in Wittig reactions of aldehydes bearing a heteroatom substituent (typically an oxygen) on the *β* carbon relative to the carbonyl group (i.e. similarly disposed relative to the carbonyl as the *ortho*-heteroatoms in benzaldehydes). There are many literature examples of this phenomenon (although the origin of the high *Z*-selectivity has not been identified), and details of the examples up to 1989 are present in the section entitled "Wittig reactions with Anomalous Stereochemistry" in the review of Maryanoff and Reitz [2]. In many of these examples, the carbonyl and the *β*-substituent of the aldehyde are substituents on a ring, and high *Z*-selectivity is observed only if the carbonyl and *β*-heteroatom are oriented *cis* with respect to each other, and it is highest in alcohol solvents [2, 4, 113–116]. High *E*-selectivity is observed if the carbonyl and *β*-heteroatom have *trans* relative orientation, or if there is no *β*-heteroatom [113, 114]. Selected literature examples are shown in Table 1.4 and discussed below.

From the results in Table 1.4 it can be seen that reactions of aldehydes bearing a *β*-heteroatom oriented *cis* relative to the carbonyl show significantly higher *Z*-selectivity (in some cases there is close to total selectivity for the *Z*-alkene) than reactions in the same solvent of aldehydes lacking such a substituent (compare, for example, entries 1, 2 and 9 with entry 7, and entry 11 with entry 12). It is clear that the high *Z*-selectivity is dependent on the presence of the *β*-heteroatom, and that the effect exerted by the *β*-heteroatom is electronic, not steric, as the reaction in entry 9 of Table 1.4 is completely *E*-selective.

The operation of the effect seems to be quite solvent-dependent, with *Z*-selectivity being highest in alcohol solvents. The reaction of 1,2-O-isopropylidene-3-O-methyl-α-D-xylopentodialdofuranose-(1,4) (the aldehyde of entries 5–8 in Table 1.4) with (ethoxycarbonylmethylidene)triphenylphosphorane at room

Table 1.4 Wittig reactions of stabilised ylides with aliphatic aldehydes under various conditions[a,b]

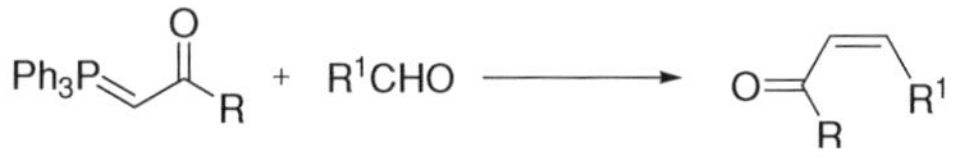

$Ph_3P{=}CHC(O)R$ + R^1CHO ⟶ $O{=}C(R)CH{=}CHR^1$

Entry	1	2	3	4	5[c]	6[d]	7	8
R^1CHO	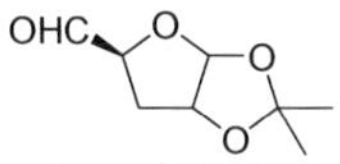	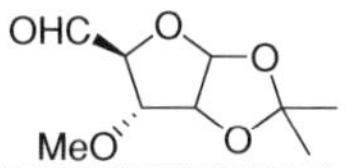	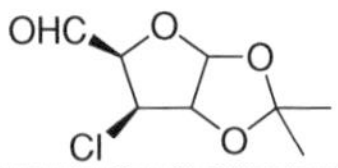			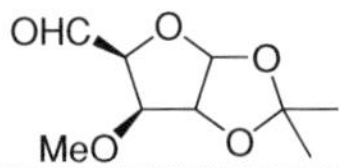		
Ylide R	Me	Me	Me	Me	Me	OMe	OEt	OEt
Z/E ratio	0:100	0:100	25:75	7:93	30:70	58:42	60:40	92:8
Solvent	$CHCl_3$ or C_6H_6	$CHCl_3$ or C_6H_6	$CHCl_3$	C_6H_6	$CHCl_3$	THF	$CHCl_3$	MeOH

Entry	9	10[e]	11[d]	12	13	14	15[f]
R^1CHO	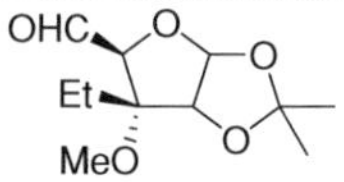	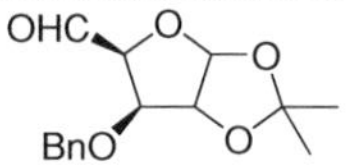		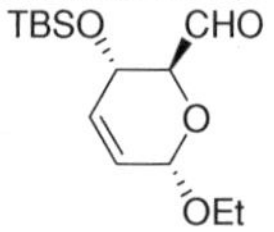	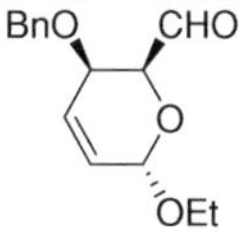	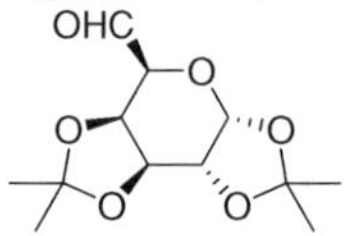	
Ylide R	Me	OMe	OMe	OMe	OMe	OMe	OMe
Z/E ratio	0:100	High *Z*[g]	60:40	1:3.5	100:1	20:1	High *Z*
Solvent	$CHCl_3$	MeOH	THF	MeOH	MeOH	MeOH	MeOH

[a] Entries 1–5 and 7–9 are from Ref. [113], and entries 12–14 are from Ref. [114]
[b] Unless otherwise noted, the reactions were carried out in the designated solvent at 20 °C
[c] The same reaction in benzene solvent gave alkene with a *Z/E* ratio of 5:95. See Ref. [99]
[d] Result from Ref. [117] Aldehyde generated in situ by periodic acid treatment of precursor *iso*-propylidene-protected diol 3-substituted 1,2,5,6-di-*O*-*iso*-propylidene-α-D-glucose. Ylide generated in situ from phosphonium salt and *n*-BuLi. Reaction conducted at 0 °C
[e] Result from Ref. [116]. Reaction carried out at 0 °C
[f] Result from Ref. [115]. Reaction carried out at 4 °C
[g] The reaction of the aldehyde epimeric at C-3 in methanol is quoted as giving "roughly equal proportions" of *Z* and *E* alkene. See Ref. [116]

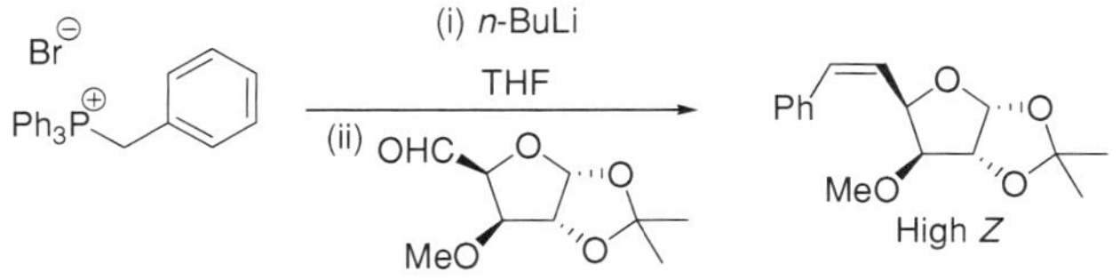

Scheme 1.21 Wittig reaction of bezylidenetriphenylphosphorane (generated in situ from benzyltriphenylphosphonium bromide) with 1,2-O-isopropylidene-3-O-methyl-α-D-xylopentodialdofuranose-(1,4)

temperature was investigated in several solvents, giving a *Z/E* ratio of 14:86 in DMF, 20:80 in benzene, 46:54 in hexane, 47:53 in acetone, 53:47 in carbon tetrachloride, 60:40 in chloroform, 92:8 in methanol [113], and 58:42 (1.4:1) in THF [117]. *Z*-selectivity was also observed to be higher at lower temperatures [114].

All of the Wittig reactions of *β*-heteroatom substituted aldehydes discussed so far have involved stabilised ylides, but there is also at least one literature example involving a semi-stabilised ylide [118]. This report involves the reaction of 1,2-O-isopropylidene-3-O-methyl-α-D-xylopentodialdofuranose-(1,4) with benzylidene-triphenylphosphorane (generated from the parent phosphonium salt using *n*-BuLi) in THF (see Scheme 1.21). The *Z/E* ratio for this reaction at 25 °C was found to be 84:16, and at 0 °C was 95:5. The observed high *Z*-selectivity is in keeping with the results for reactions of stabilised ylides, and, significantly, it occurs in aprotic solvent THF. The similarity of this result to those obtained in the reactions of semi-stabilised ylides with *ortho*-heteroatom substituted benzaldehydes suggests that the observed selectivity is as a result of the same effect. If this is the case, then the selectivity must be a consequence of a through-space electronic interaction of the heteroatom, perhaps with the phosphorus, in the cycloaddition TS. The high *Z*-selectivity would strongly suggest this to be a kinetically controlled process.

An important observation is that in all known examples of reactions of the type described in this section, the stereochemistry of the α-carbon in the aldehyde is uniformly retained in the alkene product. This is slightly surprising because epimerisation at this carbon centre might be expected in the conditions under which the reactions are conducted.

References

1. Wittig G, Geissler G (1953) Justus Liebigs Ann Chem 580:44
2. Maryanoff BE, Reitz AB (1989) Chem Rev 89:863
3. Johnson AW (1993) Ylides and Imines of Phosphorus. Wiley; New York, Chapters 8 and 9; pp 221–305
4. Vedejs E, Peterson MJ (1994) In: Eliel EL, Wilen SH (eds) Topics in Stereochemistry, vol 21. Wiley: New York

5. Vedejs E Peterson MJ (1996) In: Snieckus V (ed) Advances in Carbanion Chemistry, vol 2. JAI, Greenwich
6. Dobado JA, Martínez-García H, Molina J, Sundberg RM (2000) J Am Chem Soc 122:1144
7. Yufit DS, Howard JAK, Davidson MG (2000) J Chem Soc Perkin Trans 2:249
8. Gilheany DG (1994) Chem Rev 94:1339
9. Leyssens T, Peeters D (2008) J Org Chem 73:2725
10. Dobado JA, Martínez-García H, Molina J, Sundberg RM (1998) J Am Chem Soc 120:8461
11. Chesnut DB (2003) J Phys Chem A 107:4307
12. Kocher N, Leusser Murso A, Stalke D (2004) Chem Eur J 10:3622
13. Johnson AW (1993) Ylides and imines of phosphorus. Wiley, New York, Chapter 10, pp 307–358
14. Aggarwal VK, Fulton JR, Sheldon CG, de Vicente J (2003) J Am Chem Soc 125:6034
15. Liu D-N, Tian SK (2009) Chem Eur J 15:4538
16. Zhou R, Wang C, Song H, He Z (2010) Org Lett 12:976
17. Fan R-H, Hou X-L, Dai L-X (2004) J Org Chem 69:689
18. Lebel H, Pacquet V, Proulx C (2001) Angew Chem Int Ed 40:2887
19. Ramazani A, Kazemizadeh AR, Ahmadi E, Noshiranzadeh N, Souldozi A (2008) Curr Org Chem 12:59
20. Schwartz BD, Williams CM, Anders E, Bernhardt PV (2008) Tetrahedron 64:6482
21. McNulty J, Keskar K (2008) Tetrahedron Lett 49:7054
22. Oh JS, Kim BH, Kim YG (2004) Tetrahedron Lett 45:3925
23. Appel M, Blaurock S, Berger S (2002) Eur J Org Chem 1143
24. Wang ZG, Zhang GT, Guzei I, Verkade JG (2001) J Org Chem 61:3521
25. Nishimura Y, Shiraishi T, Yamaguchi M (2008) Tetrahedron Lett 49:3492
26. Bera R, Dhananjaya G, Singh SN, Kumar R, Mukkanti K, Pal M (2009) Tetrahedron 65:1300–1305: describes the use of microwave conditions and is a good leading reference to other variations
27. "On" or "in" water: Tiwari S, Kumar A (2008) Chem Commun 4445–4447; Wu JL, Li D, Zhang D (2005) Synth Comm 2543–2551; see also [28]; phase-transfer conditions: Pascariu A, Ilia G, Bora A, Iliescu S, Popa A, Dehelean G, Pacureanu L (2003) Central Eur J Chem 1:491–534; Moussaoui Y, Said K, Ben Salem R (2006) ARKIVOC, Part (xii):1–22; in a ball mill: Balema VP, Wiench JW, Pruski M, Pecharsky VK (2002) J Am Chem Soc 124:6244–6245; ultrasound assisted: Wu LQ, Yang CG, Yang LM, Yang LJ (2009) J Chem Res (S) 183–185; green Wittig reactions: Nguyen KC, Weizman H (2007) J Chem Ed 84:119–121; Martin E, Kellen-Yuen C (2007) J Chem Ed 84:2004–2006
28. El-Batta A, Jiang CC, Zhao W, Anness R, Cooksy AL, Bergdahl M (2007) J Org Chem 72:5244
29. McNulty J, Das P (2009) Eur J Org Chem 4031
30. Orsini F, Sello G, Fumagalli T (2006) Synlett 1717
31. Li CY, Zhu BH, Ye LW, Jing Q, Sun XL, Tang Y, Shen Q (2007) Tetrahedron 63:8046. Enanatioselective Wittig reactions of chiral ylides with ketenes to give allenes, including one pot reactions involving in situ generation of ylide from phosphine
32. Xu S, Zou W, Wu G, Song H, He Z (2010) Org Lett 12:3556
33. Wu J, Yue C (2006) Synth Commun 36:2939
34. Choudary BM, Mahendar K, Kantam ML, Ranganath KVS, Athar T (2006) Adv Synth Catal 348:1977
35. Lee EY, Kim Y, Lee JS, Park J (2009) Eur J Org Chem 2943
36. Okada H, Mori T, Saikawa Y, Nakata M (2009) Tetrahedron Lett 50:1276
37. Wiktelius D, Luthman K (2007) Org Biomol Chem 5:603
38. Molander GA, Oliveira RA (2008) Tetrahedron Lett 49:1266
39. Rothman JH (2007) J Org Chem 72:3945
40. Hisler K, Tripoli R, Murphy JA (2006) Tetrahedron Lett 47:6293
41. Ye L-W, Han X, Sun X-L, Tang Y (2008) Tetrahedron 64:8149
42. Feist H, Langer P (2008) Synthesis 24:3877

43. O'Brien CJ, Tellez JL, Nixon ZS, Kang LJ, Carter AL, Kunkel SR, Przeworski KC, Chass GC (2009) Angew Chem Int Ed 48: 6836. See also Fairlamb IJS (2009) ChemSusChem 2:1021 for a review on the topic of the phosphine-catalysed Wittig reaction
44. Appel R, Loos R, Mayr H (2009) J Am Chem Soc 131:714
45. Ghosh A, Chakratborty I, Adarsh NN, Lahiri S (2010) Tetrahedron 66:164
46. Wang ZG, Zhang GT, Guzei I, Verkade JG (2001) J Org Chem 61:3521
47. Nishimura Y, Shiraishi T, Yamaguchi M (2008) Tetrahedron Lett 49:3492
48. Robiette R, Richardson J, Aggarwal VK, Harvey JN (2005) J Am Chem Soc 127:13468
49. Robiette R, Richardson J, Aggarwal VK, Harvey JN (2006) J Am Chem Soc 128:2394
50. Harvey JN (2010) Faraday Discuss 145:487
51. Fu Y, Wang H-J, Chong S-S, Guo Q-X, Liu L (2009) J Org Chem 74:810
52. Tosic O, Mattay J (2011) Eur J Org Chem 371
53. Hong B-C, Nimje RY, Lin C-W, Liao J-H (2011) Org Lett 13:1278
54. Hodgson DM, Arif T (2011) Chem Commun 47:2685
55. Lynch JE, Zanatta SD, White JM, Rizzacasa MA (2011) Chem Eur J 17:297
56. Dong D-J, Li Y, Wang J-Q, Tian S-K (2011) Chem Commun 47:2158
57. Dong D-J, Li HH, Tian S-K (2010) J Am Chem Soc 132:5018
58. Fang F, Li Y, Tian S-K (2011) Eur J Org Chem 1084
59. Vedejs E, Marth CF (1987) Tetrahedron Lett 28:3445
60. Vedejs E, Fleck T (1989) J Am Chem Soc 111:5861
61. Wittig G, Schöllkopf U (1954) Chem Ber 87:1318
62. Seth M, Senn HM, Ziegler HM (2005) J Phys Chem A 109:5136
63. Wittig G, Haag A (1963) Chem Ber 96:1535
64. Schlosser M, Christmann KF (1967) Liebigs Ann Chem 708:1
65. Vedejs E, Meier GP, Snoble KAJ (1981) J Am Chem Soc 103:2823
66. Wittig G, Haag W (1954) Chem Ber 88:1654
67. Speziale AJ, Bissing DE (1963) J Am Chem Soc 85:3878
68. Jones ME, Trippett S (1966) J Chem Soc (C) 1090
69. Vedejs E, Marth CF, Ruggeri R (1988) J Am Chem Soc 110:3940
70. Maryanoff BE, Reitz AB, Mutter MS, Inners RR, Almond HR Jr, Whittle RR, Olofson RA (1986) J Am Chem Soc 108:7664
71. Reitz AB, Mutter MS, Maryanoff BE (1873) J Am Chem Soc 1984:106
72. Vedejs E, Marth CF (1990) J Am Chem Soc 112:3905
73. Bergelson LD, Shemyakin MM (1963) Tetrahedron 19:149
74. Schneider WP (1969) J Chem Soc Chem Commun 785
75. Schweizer EE, Crouse DM, Minami T, Wehman A (1971) J Chem Soc Chem Commun 1000
76. Blade-Font A, Vanderwerf CA, McEwen WE (1960) J Am Chem Soc 82:2396
77. Smith DH, Trippett S (1972) J Chem Soc Chem Commun 191
78. Olah GA, Krishnamurthy VV (1982) J Am Chem Soc 104:3987
79. Yamataka H, Nagareda K, Hanafusa T, Nagase S (1989) Tetrahedron Lett 30:7187
80. He G-X, Bruice TC (1991) J Am Chem Soc 113:2747
81. Bestmann HJ (1980) Pure Appl Chem 52:771
82. McEwen WE, Beaver BD, Cooney JV (1985) Phosphorus Sulfur 25:255
83. Ward WJ, McEwen WE (1990) J Org Chem 55:493
84. Schlosser M, Schaub B (1982) J Am Chem Soc 104:5821
85. Vedejs E, Marth CF (1988) J Am Chem Soc 110:3948
86. Vedejs E, Snoble KAJ (1973) J Am Chem Soc 95:5778
87. Maryanoff BE, Reitz AB, Mutter MS, Inners RR, Almond HR Jr (1985) J Am Chem Soc 1068:107
88. Bangerter F, Karpf M, Meier LA, Rys P, Skrabal P (1998) J Am Chem Soc 120:10653
89. Vedejs E, Snoble KAJ, Fuchs PL (1973) J Org Chem 38:1178
90. Vedejs E, Fuchs PL (1973) J Am Chem Soc 95:822
91. Vedejs E, Marth CF (1989) J Am Chem Soc 111:1519

92. Westheimer FH (1968) Acc Chem Res 1:70
93. Albright TA, Burdett JK, Whangbo M-H (1985) Orbital Interactions in Chemistry. Wiley, New York
94. Kay PB, Trippett S (1986) J Chem Res (S) 62
95. Yamataka H, Nagareda K, Takai Y, Sawada M, Hanafusa T (1988) J Org Chem 53:3877
96. Yamataka H, Nagareda K, Tsutomu T, Ando K, Hanafusa T, Nagase S (1993) J Am Chem Soc 115:8570
97. Carins SM, McEwen WE (1986) Tetrahedron Lett 27:1541
98. Donxia L, Dexian W, Yaoshong L, Huaming Z (1986) Tetrahedron 42:4161
99. Isaacs NS, Abed OH (1986) Tetrahedron Lett 27:995
100. Yamataka H, Nagareda K, Ando K, Hanafusa T (1992) J Org Chem 57:2865
101. Appel R, Loos R, Mayr H (2009) J Am Chem Soc 131:704
102. Frøyen P (1972) Acta Chem Scand 26:2163
103. Dunne EC, Coyne EJ, Crowley PB, Gilheany DG (2002) Tetrahedron Lett 43:2449
104. Harrowven DC, Guy IL, Howell M, Packham G (2006) Synlett 2977
105. Harrowven DC, Guy IL, Nanson L (2006) Angew Chem Int Ed 45:2242
106. McEwen WE, James AB, Knapczyk JW, Killingstad VL, Shiau W-I, Shore S, Smith JH (1978) J Am Chem Soc 100:7304
107. Keldsen GL, McEwen WE (1978) J Am Chem Soc 100:7312
108. McEwen WE, Cooney JV (1983) J Org Chem 48:983
109. Zhang X, Schlosser M (1993) Tetrahedron Lett 1925:34
110. Cushman M, Nagarathnam D, Gopal D, Chakraborti AK, Lin CM, Hamel EJ (1991) J Med Chem 34:2579–2588
111. Robinson SB, Dunne EC, O'Mahony CP, Garcia-Ruiz V, Gilheany DG in preparation
112. Takenaka N, Sarangthem RS, Captain B (2008) Angew Chem Int Ed 47:9708–9710
113. Tronchet JMJ, Gentile B (1979) Helv Chim Acta 62:2091
114. Valverde S, Martin-Lomas M, Herradon B, Garcia Ochoa S (1987) Tetrahedron 43:1987
115. Brimacombe JS, Hanna R, Kabir AKMS, Bennett F, Taylor ID (1986) J Chem Soc Perkin Trans 1 815
116. Brimacombe JS, Kabir AKMS (1986) Carbohydr Res 150:35
117. Sartillo-Piscil F, Vargas M, Anaya de Parrodi C, Quintero L (2003) Tetrahedron Lett 44:3919
118. Gurjar MK, Khaladkar TP, Borhade RG, Murugan A (2003) Tetrahedron Lett 44:5183

Chapter 2
Wittig Reactions of Aldehydes Bearing a β-Heteroatom Substituent

2.1 Fragility of Alkene Stereochemistry: The True *Z/E* Ratio

In this work, the kinetic selectivity of the OPA forming step in Wittig reactions of semi-stabilised and stabilised ylides is inferred from the observed *Z/E* ratio of the alkene product. It is thus very important to be sure that the alkene *Z/E* ratio is truly reflective of the kinetic OPA *cis/trans* ratio, and to be aware of possible means by which there may arise a non-correspondence between the two ratios. Changes may occur to the *Z/E* ratio both during and after the Wittig reaction. The latter problem is prosaic but pernicious. It is not sufficiently recognised that *Z*-1,2-disubstituted alkenes are quite easily converted, under a variety of conditions, to a *Z/E* mixture and sometimes completely to the *E*-isomer. Therefore the *Z/E* ratio resulting from the reaction is fragile and can be affected by the presence of acids (especially benzoic acid) [1], strong bases [2], the chromatographic stationary phase used in purification, the solvent, heat and sunlight [3]. Previous literature reports have been identified in a publication by the Gilheany group [4], and also in the review of Vedejs and Peterson where there was undoubtedly isomerisation in favour of the *E*-alkene subsequent to the actual Wittig reaction. This can be synthetically convenient, for example, it may contribute to the results of recently reported microwave accelerated reactions, e.g. [5]. Indeed, in the course of this project, isomerisation of certain *Z*-alkenes has been observed simply by allowing the sample to stand for a period—this isomerisation is most likely to have been caused by light. *Z*-4-arylbut-3-en-2-ones were found to be particularly susceptible, to the point where they could not be purified from a reaction mixture due to almost complete conversion to the *E*-isomer. Others—notably 2-iodostilbene,

P. A. Byrne, *Investigation of Reactions Involving Pentacoordinate Intermediates*, Springer Theses, DOI: 10.1007/978-3-642-32045-3_2,

2,2′-diiodostilbene, and 2,2′-difluorostilbene[1]—were also found to undergo partial, if not complete isomerisation when allowed to stand in the light. Extensive precautions have been taken in this study to ensure that the observed *Z/E* ratio is truly reflective of that rendered by the Wittig reaction in question. Thus all aldehydes used were checked by NMR prior to use for the presence of the carboxylic acid, the alkene *Z/E* ratio in each reaction was measured immediately after a minimal work-up before chromatography and great care was taken to ensure that the reactions were carried out under conditions for which the operation of kinetic control has been demonstrated. Dry solvents were used for reactions and NMR studies, dry phosphonium salts and non-nucleophilic bases NaHMDS and KHMDS were employed to form the ylides, and both were stored under argon in a glove box. Reaction flasks were flame dried and cooled under vacuum to ensure the exclusion of moisture. The rubber tubing used to conduct nitrogen to the reaction flask from the Schlenk manifold was kept dry by fitting a syringe barrel and needle to the open end of the tubing when it was not in use and inserting the needle through a septum into a flask of solid dry KOH pellets. The hygroscopic KOH gradually dries the rubber tubing of water if the procedure described is continued for a period of time. 2,2′-difluorostilbene is excluded from Table 2.1 since it has been shown to be prone to isomerisation after completion of the Wittig reaction [4].

Several of the above factors are also relevant to the issue of *stereochemical drift*, which is another, more mechanistically significant, source of erosion of the stereochemistry. It describes the well-established phenomenon that, under certain circumstances, the first-formed OPAs from cycloaddition may equilibrate [3, 6, 7], leading to a different (thermodynamic) ratio of *cis/trans* OPAs and therefore a different *Z/E* alkene ratio. Reactions conducted in the presence of additives such as salts that are soluble in the reaction solvent—in particular lithium cation [3, 8] and iodide anion [9, 10] salts—have been shown to give *Z/E* ratios that are altered with respect to reactions conducted in the absence of such additives. Lithium halide with small amounts of alcohol [3], and benzoic acid [1] have also exhibited this effect when present in Wittig reaction mixtures. The addition of methanol to reactions of non-stabilised ylides at low temperature has also been shown to result in increased formation of *E*-alkene compared to reactions conducted by the same procedure without alcohol addition; no such increase is observed if the methanol is added after the reaction mixture has been warmed to room temperature [11]. The role of lithium ion is solvent dependent, with a profound effect being observed for reactions in non-polar solvents, and essentially no effect in solvents that readily complex Li^+ [8, 12]. Hydroxylic solvents and high temperature have also been implicated as possible initiators of OPA equilibration in reactions of aromatic aldehydes [3]. In some cases, it is even possible for stereochemical drift to occur *in the absence of dissolved salts* (Li-containing or otherwise). This has been described in detail in Sect. 1.4.1.1, but briefly, it has been observed for OPAs derived from trialkylphosphonium alkylides and tertiary or aromatic aldehydes [6, 7, 13], and also for

[1] 2,2′-difluorostilbene is extremely prone to isomerisation—see Ref. [4].

OPAs derived from ethylidenetriphenylphosphoranes and aromatic aldehydes (although in the latter of these OPA equilibration only occurs at or above the temperature at which OPA can decompose to alkene) [14]. In all cases, equilibration results in transformation of the *cis*-OPA into the *trans*-OPA, since the observed proportion of *E*-alkene is greater than that of the *trans*-OPA [6, 7].

This is why, in order to discuss the relevance of the results obtained in this project to the Wittig mechanism it must be ensured that the reactions occur under conditions of kinetic control—i.e. conditions for which irreversibility of OPA and alkene formation has been explicitly demonstrated. The "true *Z/E* ratio" is defined as that obtained from a kinetically controlled reaction and preserved subsequently during isolation and measurement.

For reactions of non-stabilised ylides with benzaldehyde, it is not sufficient to rely on *Z/E* ratios to indicate the kinetic selectivity of the OPA forming step, since stereochemical drift has been shown to be in operation by our low temperature NMR and acid quenching experiments with these ylides. Instead, the OPA *cis/trans* ratios have in some cases been obtained directly by low temperature ^{1}H and ^{31}P NMR observation of Wittig reaction mixtures. In others, low temperature acid quenching of the Wittig reaction has been carried out and the *erythro/threo* ratio of the β-hydroxyphosphonium salt (β-HPS) determined to establish the kinetic *cis/trans* ratio of OPA. These two methods of obtaining the kinetic OPA *cis/trans* ratio have been found to be in excellent agreement in the experiments conducted in this project, and in reports from other groups [15].

In general the OPA *cis/trans* ratio for a Wittig reaction must be at least as high as the observed alkene *Z/E* ratio since OPA decomposition is stereospecific and irreversible, reflecting the fact that *cis*-OPAs are normally higher in energy than *trans*-OPAs. As long as the *cis*-OPA is indeed higher in energy than the *trans* isomer, it can be assumed that the reactions that are highly selective for the *Z*-alkene are under dominant or total kinetic control [6]. The consequence is that it is not ordinarily possible to obtain *Z*-selectivity by accident or by intervention of equilibration. Therefore it is apposite that our conclusions (vide infra) on the results of reactions of semi-stabilised and stabilised ylides are dependent on high *Z/E* ratios, which have a strong probability of being the "true" values, given the conditions under which they were carried out, and the fact that they generally show very high *Z*-selectivity. For reactions of non-stabilised ylides the kinetic OPA *cis/trans* ratio has been explicitly determined. The operation of kinetic control (or lack thereof) in these reactions, although not required for the determination of the kinetic selectivity of OPA formation, has been explicitly demonstrated by comparison of kinetic OPA *cis/trans* ratios and alkene *Z/E* ratios.

2.2 *Z/E* Ratios of Alkenes Produced in the Reactions of Benzylides with Benzaldehydes

Kinetic control has been shown to operate in the reactions of methyldiphenylphosphine-derived semi-stabilised ylides by demonstration of stereospecific decomposition of the betaines and OPAs transiently produced in β-HPS deprotonation experiments [16, 17], as described fully in Sect. 1.4.1.2. Thus to prove that the high *Z*-selectivity observed in reactions of benzylides with *ortho*-heteroatom substituted benzaldehydes [4] was due to the selectivity inherent in a kinetically controlled process, reactions of several benzylidenemethyldiphenylphosphoranes with various benzaldehydes were carried out under the experimental conditions for which this operation of kinetic control had been established, i.e. using a non-nucleophilic base to generate the ylide (NaHMDS or KHMDS), in THF (polar aprotic) solvent, in the absence of dissolved salts (especially Li-salts), and at −78 °C. Reactions of similar benzylidenetriphenylphosphoranes were also carried out to allow comparison of the *Z/E* ratios obtained. A minimal work-up was carried out, involving removal of dissolved salts (and *some* phosphine oxide) by a water wash of the crude reaction product dissolved in pentane or cyclohexane. *Z/E* ratios were usually assigned based on integration of signals characteristic of the *Z* and *E* isomers in the ^{1}H NMR spectrum of the material produced after this work-up. In instances where the signals of the alkene isomers could not be unambiguously assigned or identified, the material was purified by elution through a neutral alumina plug using pentane or cyclohexane as solvent. This resulted in the removal of aldehyde and phosphine oxide, which allowed many or all of the signals belonging to each alkene isomer to be assigned and integrated. The stilbenes for which alumina plug purification was necessary were shown not to isomerise in contact with the stationary phase by subjecting samples of known *Z/E* ratio to the purification procedure; the *Z/E* ratio was shown to be the same before and after this procedure. Subsequent to much of this work, a new technique for the removal of phosphine oxide and aldehyde from the Wittig reaction mixture was discovered. Treatment of the crude product with oxalyl chloride caused the phosphine oxide to form chloro phosphonium salt. Addition of cyclohexane to this mixture and cannula filtration resulted in the complete isolation of the alkene after solvent removal. The *Z/E* ratio of the alkene was unaffected by this process. See Chap. 3, and Sect. 4.9 of Chap. 4 for full details.

2.2.1 Reactions of Benzylidenemethyldiphenylphosphoranes with Benzaldehydes

Shown in Table 2.1 are the stilbene *Z/E* ratios (determined by comparison of the integrations of signals assigned to the *Z* and *E* isomers in the ^{1}H NMR of the crude product) found for reactions of benzylidenemethyldiphenylphosphoranes with

Table 2.1 *Z/E* ratio[a] for stilbenes produced in the reactions of benzylidenemethyldiphenylphosphoranes, derived from the parent phosphonium salts[b] (with *ortho*-substituent X) with benzaldehydes (with *ortho*-substituent Y)

(i) NaHMDS or KHMDS THF; (ii) 2-Y-benzaldehyde (CHO), -78 °C

Entry	Ylide X	Ald Y	*Z*:*E* ratio
1	H	Cl	93:7
2	H	Br	94:6
3	H	I	97:3
4	H	F	84:16
5	H	OMe	88:12
6	H	H	15:85
7	Cl	H	34:66
8	Br	H	33:67
9	I	H	28:72
10	F	H	41:59
11[c]	Me	H	7:93
12[c]	OMe	H	12:88
13	Cl	Cl	97:3
14	Br	Br	99:1
15	I	I	>99:1
16	F	F	94:6
17	F	Br	94:6
18	Br	F	84:16
19[c]	Me	Cl	94:6
20[c]	OMe	OMe	48:52
21	Br	OMe	95:5
22[c]	Me	Me	31:69
23	Br	Me	75:25
24	Cl	Me	73:27
25	F	Me	51:49
26	H	Me	33:67
27[c]	OMe	Me	66:34

[a] *Z/E* ratio determined by ^{1}H NMR analysis of the crude product obtained after aqueous work-up. See Sects. 4.3 and 4.4 for full details of the reaction, the base used to generate the ylide, work-up and analyses

[b] Counterion Z = Br^- in all cases except where otherwise noted

[c] Counterion Z = Cl^-

benzaldehydes under the conditions described above. The ylide in each reaction was generated in situ from the parent phosphonium salt using NaHMDS or KHMDS as base.

From the results in Table 2.1 the following observations can be made:

1. The reaction of benzylidenemethyldiphenylphosphorane and benzaldehyde (the unsubstituted reaction partners) gives stilbene with a *Z/E* ratio of 15:85 (see entry 6), in good agreement with literature precedent for this reaction under the same reaction conditions.[2]
2. Reactions of benzylidenemethyldiphenylphosphorane with *ortho*-heteroatom substituted benzaldehydes show very high *Z*-selectivity (entries 1–5). This *Z*-selectivity is observed for benzaldehydes more reactive at the carbonyl than benzaldehyde (entries 1, 2, 4), for 2-methoxybenzaldehyde, which is less reactive than benzaldehyde (entry 5), and for 2-iodobenzaldehyde (entry 3), which is similar in reactivity to benzaldehyde.
3. Reactions of ylides bearing an *ortho*-substituent on the benzylidene group with benzaldehyde show moderate to very high *E*-selectivity (see entries 7–12). *E*–selectivity increases approximately in parallel with the increasing activating ability of the *ortho*-substituent i.e. in the order F, Cl, Br, I, H, OMe, Me—so that more reactive ylides give greater *E*-selectivity.
4. Reactions of *ortho*-substituted benzylides with *ortho*-heteroatom substituted benzaldehydes show equivalent or *even greater* *Z*-selectivity than the reactions mentioned in point 2 (see entries 13–19, 21). This is the case for aldehydes that bear an electron withdrawing substituent and are thus more reactive than benzaldehyde (entries 13, 14, 16–19), for those bearing an electron donating substituent that are therefore less reactive than benzaldehyde (entry 21), and for 2-iodobenzaldehyde (entry 15), which is of similar reactivity to benzaldehyde. In the context of these results, 2-methoxybenzylidenemethyldiphenylphosphorane shows unusually low *Z*-selectivity in its reaction with 2-methoxybenzaldehyde (entry 20), although the same aldehyde was shown to react with 2-bromobenzylidenemethyldiphenylphosphorane with very high *Z*-selectivity (entry 21). That the reaction of 2-methylbenzylidenemethyldiphenylphosphorane with 2-chlorobenzaldehyde shows comparably high *Z*-selectivity to the other reactions shows that the contribution of the ylide benzylidene substituent to the selectivity is of steric origin i.e. it is *not* an electronic effect.
5. The reaction of 2-methylbenzaldehyde with benzylidenemethyldiphenylphosphorane shows moderate *E*-selectivity (entry 26), as does the reaction of this aldehyde with 2-methylbenzylidenemethyldiphenylphosphorane (entry 22). Its reaction with *ortho*-heteroatom substituted benzylides shows poor to moderate *Z*-selectivity (entries 23–25 and 27). That high *Z*-selectivity is not observed in these reactions shows that the effect observed in the reactions of *ortho*-heteroatom substituted benzaldehydes is dependent on the *ortho* substituent being lone-pair bearing—i.e. *the effect is not of steric origin.*

---

[2] See Ref. [3] pp. 61–70.

Table 2.2 *Z/E* ratio[a] for stilbenes produced in the reactions of benzylidenetriphenylphosphoranes derived from the parent phosphonium salts[b] (with *ortho*-substituent X) with benzaldehydes (with *ortho*-substituent Y)

(i) NaHMDS THF (ii) CHO -78 °C

Entry	Ylide X	Ald Y	*Z*:*E* ratio
1	H	Br	87:13
2	H	I	88:12
3[d]	H	H	59:41
4	H	OMe	90:10
5	Br	H	42:58
6[e]	OMe	H	42:58
7[c]	Br	Br	94:6
8	I	I	94:6
9	Me	Cl	95:5
10[e]	OMe	OMe	90:10

[a] *Z/E* ratio determined by ^{1}H NMR analysis of the crude product obtained after aqueous work-up. See Sects. 4.3 and 4.4 for full details
[b] Counterion Z = Br^- in all cases except where otherwise noted
[c] Result cross-checked by normal phase HPLC
[d] From Ref. [3]
[e] Counterion Z = Cl^-

2.2.2 *Reactions of Benzylidenetriphenylphosphoranes with Benzaldehydes*

A similar (but smaller) set of reactions involving various benzaldehydes were also carried out using benzylidenetriphenylphosphoranes under the same experimental conditions. The *Z/E* ratios of the stilbenes produced in these reactions are shown in Table 2.2. The ylide in each reaction was generated in situ from the parent phosphonium salt using NaHMDS as base.

Examination of the results in Table 2.2 reveals the following:

1. The reaction of benzylidenetriphenylphosphorane and benzaldehyde has previously been found to give stilbene with a *Z/E* ratio of 59:41 (see entry 3) [3].
2. Reactions of benzylidenetriphenylphosphorane with *ortho*-heteroatom substituted benzaldehydes show high *Z*-selectivity (entries 1, 2, 4). This is observed for 2-bromobenzaldehyde (entry 1), which is more reactive than benzaldehyde, for 2-iodobenzaldehyde, which is of similar reactivity to benzaldehyde, and for 2-methoxybenzaldehyde (entry 4), which is less reactive than benzaldehyde.

3. Reactions of *ortho*-heteroatom substituted ylides with benzaldehyde show low *E*-selectivity (entries 5 and 6).
4. Reactions of ylides bearing an *ortho*-substituent on the benzylidene group with *ortho*-heteroatom substituted benzaldehydes show even higher *Z*-selectivity than the reactions mentioned in point 2. This is observed for benzaldehydes that are more reactive than benzaldehyde (entries 7 and 9), for 2-methoxybenzaldehyde (entry 10), which is less reactive than benzaldehyde, and for 2-iodobenzaldehyde, whose reactivity is similar to that of benzaldehyde. The result shown in entry 9 (reaction of 2-methylbenzylidenetriphenylphosphorane with 2-chlorobenzaldehyde) shows that it is the increased steric bulk of the *ortho*-substituted ylides of entries 7–10 that causes increased *Z*-selectivity compared to reactions of point 2 (entries 1, 2, 4).

2.2.3 Discussion

The result in Tables 2.1 and 2.2 are entirely consistent with each other, with the exception of Table 2.1 entry 20. Reactions of unsubstituted benzylides with *ortho*-heteroatom substituted benzaldehydes show very high *Z*-selectivity—far higher than that observed in the reaction of the same ylide with benzaldehyde itself. Benzylides with an *ortho*-substituent react with *ortho*-heteroatom substituted benzaldehydes with even higher *Z*-selectivity than the corresponding reactions of the unsubstituted ylides. The average proportion of *Z*-alkene produced in reactions of benzaldehydes lacking a heteroatom substituent, calculated from entries 6–12 and 22–27 of Table 2.1, and entries 3, 5 and 6 of Table 2.2, is 49 %. The same quantity calculated for reactions of *ortho*-heteroatom substituted benzaldehydes using Table 2.1 entries 1–5 and 13–21, and Table 2.2 entries 1, 2, 4, and 7–10 is 90 %, so there is an average jump of 41 % in *Z*-selectivity induced in reactions of semi-stabilised ylides by there being an *ortho*-heteroatom on the benzaldehyde.

The increase in *Z*-selectivity from reactions of unsubstituted benzylides to *ortho*-substituted benzylides appears to be as a result of a steric effect, since it operates whether the ylide *ortho*-substituent is lone-pair bearing or otherwise. The same *ortho*-substituted benzylides react with benzaldehyde with moderate to high *E*-selectivity; this indicates that the geometry of the TSs in this latter set of reactions is completely different to that of the TSs in reactions involving *ortho*-heteroatom substituted benzaldehydes. It is also obvious by examination of the results that methyldiphenylphosphine-derived benzylides consistently show higher *Z*-selectivity than their triphenylphosphine-derived analogues; we believe this to be of mechanistic significance, and it will be discussed more fully later in this section and following sections of this chapter.

Two very important observations on the set of results presented in Tables 2.1 and 2.2 indicate firstly that the high *Z*-selectivity induced by the presence of an *ortho*-heteroatom on the benzaldehyde arises from an electronic effect and not a steric one, and secondly that the electronic effect operates through space and not by electron withdrawal or donation through the aromatic system of the benzaldehyde. The reactions detailed in Table 2.1 entries 22–27 (involving 2-methylbenzaldehyde) show no great *Z*-selectivity, and thus it can be concluded that the presence of an *ortho*-substituent *per se* on the benzaldehyde partner, although necessary, is not sufficient to induce high *Z*-selectivity in Wittig reactions—the *ortho* substituent should bear a lone pair of electrons. 2-Methoxybenzaldehyde, which is *less* reactive than benzaldehyde itself, shows comparably high *Z*-selectivity to reactions of 2-halobenzaldehydes. In the previous work in the laboratory, it had been thought that the explanation for the high *Z*-selectivity laid in the earlier cycloaddition TS, induced by the higher inherent reactivity of the electron-deficient 2-halobenzaldehydes. The cycloaddition TS in a Wittig reaction of 2-methoxybenzaldehyde cannot be earlier than the TS in the reaction of benzaldehyde with the same ylide due to the fundamentally lower reactivity of the former. The same comments apply to the reactions of 2-iodobenzaldehyde, whose reactivity towards nucleophiles is similar to that of benzaldehyde. Thus, *Z*-selectivity cannot be simply due to the *cis*-selective TS being early and hence puckered, given that none of the control reactions with benzaldehyde are *Z*-selective. The observation of *Z*-selectivity in reactions of *ortho*-heteroatom substituted benzaldehydes is therefore the consequence of the existence of a fundamentally different cycloaddition TS in these reactions. Since the *Z*-selective effect is not of steric origin and not an electronic effect conducted through the aromatic system, it must operate through space—that is, there must be a bond formed (transiently or otherwise) in the TS leading to OPA, which causes the *cis*-selective TS to be favoured over the *trans*-selective TS.

It is proposed, in light of the results of Tables 2.1 and 2.2, that in these reactions there exists a bonding interaction between the *ortho*-heteroatom and phosphorus in the TS leading to *cis*-OPA, as shown in Fig. 2.1. The phosphorus atom has low-lying acceptor orbitals ($P{-}R^3$ σ^* orbitals) in the TS structure that could facilitate the existence of such a bond.

The very existence of the putative phosphorus-heteroatom bond requires the heteroatom-bearing aryl group of the benzaldehyde to be close to phosphorus. As a result, this TS has an entirely different set of steric interactions than would be expected for a TS of an analogous reaction involving an aldehyde that lacks an *ortho*-heteroatom substituent. This is so not least because the other substituents joined to phosphorus presumably rearrange (perhaps to pseudo-octahedral geometry) to allow room for the bond to form. Thus 1–3 steric interactions are not an issue in this TS to the same extent that they would be in "normal" Wittig cycloaddition TSs—both because of the rearrangement of the phosphorus-substituents, and because of the existence of the phosphorus-heteroatom bond. In proposing such a TS it is initially impossible to say with certainty whether the TS should be puckered or planar. However, ^{31}P NMR studies of the OPAs produced in reactions of non-stabilised ylides with aldehydes bearing heteroatom in the β-position relative to the carbonyl (which will be described in detail later, in Sect. 2.4) shed some light on the issue. OPAs are constrained to being essentially planar (puckering angle $< 30°$). The chemical shifts of the OPAs in the ^{31}P NMR (δ_P in the range of

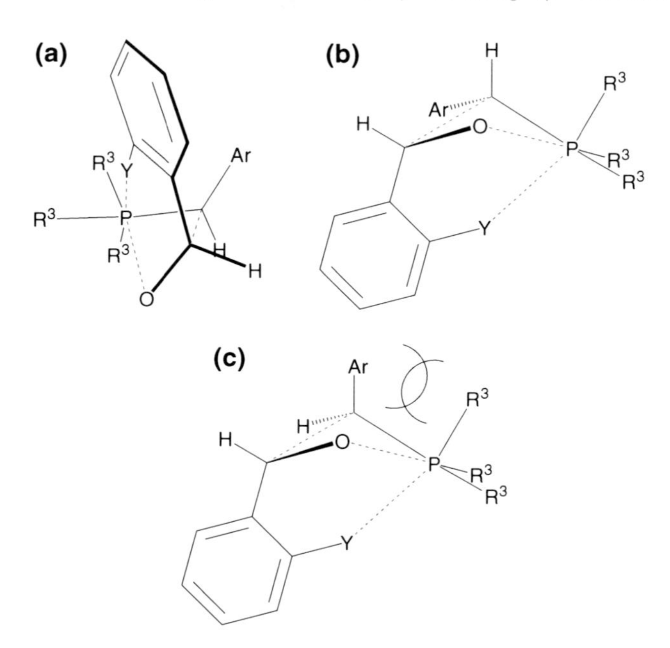

Fig. 2.1 **a** and **b** show different perspectives of the puckered *cis*-selective TS with phosphorus-heteroatom bonding. The ylide α-substituent (Ar) is oriented as shown to minimise 2–3 steric interactions by avoiding the phosphorus R^3 substituents. **c** A *trans*-selective TS with phosphorus-heteroatom bonding suffers from large 2–3 steric interactions

-60 to -75 relative to H_3PO_4) are entirely characteristic of pentacoordinate phosphorus, and so it is clear that no phosphorus-heteroatom bond exists in the planar OPA. We conclude that the geometry of a planar ring is such that the formation of the phosphorus-heteroatom bond is not possible. Based on this observation, it is postulated that the *cis*-selective TSs in reactions of heteroatom substituted aldehydes that show high *Z*-selectivity are puckered in order to facilitate the phosphorus-heteroatom binding interaction.

In a puckered TS involving phosphorus-heteroatom bonding, it seems likely that there should be large steric interactions between the rearranged phosphorus ligands (R^3) and the ylide α-substituent (2–3 interactions), since there are six substituents around phosphorus in this species. Minimisation of 2–3 interactions dictates that the ylide α-substituent points to the same side of the forming ring as the aldehyde substituent. The relatively small 1–2 interactions (*gauche* interaction between ylide and aldehyde substituent) inherent in this conformation are much less significant than the 2–3 interactions present in a TS in which the ylide α-substituent and the aldehyde aryl group point to opposite sides of the forming ring (*trans*-selective TS, see Fig. 2.1c). Thus, the most energetically favourable cycloaddition TS (with 2–3 interactions minimised) is selective for *cis*-OPA (see two schematic representations of this TS from different perspectives in Fig. 2.1a and b), which gives rise ultimately to *Z*-alkene.

Further evidence for the existence of a bonding interaction between phosphorus and the heteroatom can be seen by examination of results from Tables 2.1 and 2.2:

1. The effect increases as the heteroatom polarizability increases and electronegativity decreases (Table 2.1, entries 1–5: i.e. in the order F, O, Cl, Br, I), which correlates with the ability of the heteroatom to bond to phosphorus.
2. The effect increases in line with the increasing length of the carbon-heteroatom bond. The relatively long C–I and C–Br bonds may place these heteroatoms more easily in bonding range of phosphorus.
3. The reaction of 2-fluorobenzylidenemethyldiphenylphosphorane with 2-bromobenzaldehyde is more *Z*-selective than the reaction of 2-bromobenzylidenemethyldiphenylphosphorane with 2-fluorobenzaldehyde (Table 2.1 entries 17 and 18), which again is in line with a greater TS bonding interaction in the former case.
4. There must be some significant effect resulting in the dramatic reversal of selectivity from *E* to *Z* when going from the reaction of 2-methylbenzylidene methyldiphenylphosphorane with benzaldehyde (Table 2.1 entry 11, *Z/E* = 7:93) to its reaction with 2-chlorobenzaldehyde (Table 2.1 entry 19, *Z/E* = 94:6), and indeed to result in such extraordinarily high *Z*-selectivity in the reaction giving 2,2′-diiodostilbene, shown in Table 2.1 entry 15.
5. The latter reaction is overwhelmingly selective for the thermodynamically disfavoured isomer, strongly indicating the operation of kinetic control. Since phosphorus-heteroatom bonding has been shown by NMR experiments to be absent in OPAs produced in the reactions of non-stabilised ylides with 2-bromobenzaldehyde (see Sect. 2.4), and since the operation of stereochemical drift observed in these reactions furnishes an increased proportion of *E*-alkene via *trans*-OPA, the *cis*-OPA must be thermodynamically disfavoured relative to the *trans*-isomer in these reactions of non-stabilised ylides. It can reasonably be assumed that the same is true for OPAs in reactions of semi-stabilised and stabilised ylides.

As alluded to earlier, *Z*-selectivity in reactions of benzylides with *ortho*-heteroatom substituted benzaldehydes is consistently higher when the benzylide bears an *ortho*-substituent compared to the corresponding reaction of the unsubstituted benzylide (see Tables 2.1 and 2.2). A possible explanation for this phenomenon is that the *cis*-selective TS is better able to accommodate the increased steric demands (and especially 2–3 interactions) of the bulkier ylidic substituent than is the *trans*-selective TS, resulting in greater discrimination between the two. The high *Z*-selectivity obtained in the reactions of the *ortho*-heteroatom substituted benzaldehydes with the *ortho*-methyl substituted benzylides (Table 2.1 entry 19, Table 2.2 entry 9) shows that this is indeed a steric effect.

If this subtle augmentation of the remote heteroatom effect in reactions of benzylides with benzaldehydes is genuine, as is suggested by the consistency of the results in Tables 2.1 and 2.2, then it is not consistently reproduced in the corresponding reactions of other types of ylides, as described below. Acetonylides with increased steric bulk on the ylide γ-carbon (relative to phosphorus) do show significantly increased *Z*-selectivity compared to the unsubstituted acetonylide.[3] However, there is no significant effect of ylide steric bulk on selectivity in reactions of non-stabilised[4] and ester-stabilised[5] ylides respectively with *ortho*-heteroatom substituted benzaldehydes. The operation of this effect may be very dependent on the shape of the cycloaddition TSs in these particular reactions.

2.3 *Z/E* Ratios of Alkenes Produced in the Reactions of Stabilised Ylides with Benzaldehydes

Following the above work on the selectivity in reactions of semi-stabilised ylides, the possibility of the operation of similar *ortho*-heteroatom induced *Z*-selectivity in Wittig reactions of stabilised ylides was investigated. The reactions were carried out under identical conditions to those employed in the investigation of reactions of semi-stabilised ylides, with the exception that in many cases the reaction was quenched at −78 °C in order to prove that the Wittig reaction had actually occurred at that temperature (and not in the course of warming up to room temperature). Kinetic control had been shown to operate in the reactions of methyldiphenylphosphine-derived ester-stabilised ylides at 20 °C in both THF and methanol solvents by stereospecific decomposition to *Z*-alkenes of the betaines and OPAs necessarily produced in the deprotonation of *erythro*-β-HPSs derived from stabilised ylides [16]. Kinetic control for reactions in THF at −78 °C could thus be assumed. The ylides employed in the investigation were methyldiphenylphosphine and triphenylphosphine-derived ester and keto-stabilised ylides.

2.3.1 *Reactions of (Alkoxycarbonylmethylidene) methyldiphenylphosphoranes (Ester-Stabilised Ylides)*

The *Z/E* ratios of the alkyl cinnamates produced in the reactions of selected (alkoxycarbonylmethylidene)methyldiphenylphosphoranes with benzaldehydes in THF at −78 °C are shown in Table 2.3. The ylide in each reaction was generated in situ from the parent phosphonium salt using NaHMDS or KHMDS as base. The reactions were quenched at low temperature by the addition of saturated aqueous ammonium chloride solution.

Reactions of the three ester-stabilised ylides investigated with benzaldehyde showed moderate *E*-selectivity (see Table 2.3 entries 1–3). The reactions of the

---

[3] See Sect. 2.3.2.

[4] See Sect. 2.4.

[5] See Sect. 2.3.1.

Table 2.3 *Z/E* ratio for reactions of selected (alkoxycarbonylmethylidene)methyldiphenyl-phosphoranes (generated in situ from the parent phosphonium salts) with selected benzaldehydes[a] Counter-ion Z = Br unless otherwise indicated

Base, THF; (i) 2-Y-benzaldehyde, -78 °C; (ii) NH_4Cl, H_2O, -78 °C

Entry	Base	Ylide X	Aldehyde Y	Enoate *Z/E* ratio
1[b]	NaHMDS	OMe	H	36:64
2	KHMDS	OEt	H	36:64
3	NaHMDS	O(*t*-Bu)	H	40:60
4	KHMDS	OEt	OMe	66:34
5	KHMDS	O(*t*-Bu)	OMe	77:23
6	KHMDS	OMe	Cl	79:21
7	KHMDS	OEt	Cl	77:23
8[b]	NaHMDS	OMe	Br	83:17
9	KHMDS	OEt	Br	83:17
10	NaHMDS	O(*t*-Bu)	Br	85:15

[a] All reactions were carried out at −78 °C, and subsequently quenched with aqueous ammonium chloride at this temperature. *Z/E* ratio determined by ^{1}H NMR analysis of the crude product obtained after aqueous work-up of the reaction mixture. See Sects. 4.3 and 4.5 for full details

[b] Phosphonium salt counter-ion Z = Cl^-

same ylides with *ortho*-heteroatom substituted benzaldehydes exhibit very high *Z*-selectivity, both in reactions in which the aldehyde is more reactive than benzaldehyde itself (Table 2.3 entries 6–10) and in reactions of 2-methoxybenzaldehyde, which is less reactive than benzaldehyde (Table 2.3 entries 4 and 5). Such high *Z*-selectivity for reactions of stabilised ylides in aprotic solvents is unprecedented; its like has only ever previously been reported for reactions of stabilised ylides in methanol[6] [18–22]. *Z*-selectivity is highest for reactions of 2-bromobenzalehyde (Table 2.3 entries 8–10), while reactions of 2-chlorobenzaldehyde (Table 2.3 entries 6 and 7) give similar or slightly enhanced *Z*-selectivity compared to 2-methoxybenzaldehyde (Table 2.3 entries 4 and 5). Thus there is an apparent trend, similar to that observed in the reactions of semi-stabilised ylides, that *Z*-selectivity increases in line with the polarizability of the aldehyde *ortho*-heteroatom.

The steric bulk of the ester alkyl group appears to have no significant effect on the stereo selectivity of the reactions since all three ester stabilised ylides gave very similar results. The very high *Z*-selectivity observed in the reactions of the *ortho*-heteroatom substituted benzaldehydes lends credence to the previously reported evidence that Wittig reactions of ester-stabilised ylides occur under kinetic control [16].

[6] See Ref. [3], pp. 61–70.

Table 2.4 *Z/E* ratio[a] for reactions of selected 2-oxoalkylidenemethyldiphenylphosphoranes (generated in situ from the corresponding phosphonium salts)[b] with selected benzaldehydes

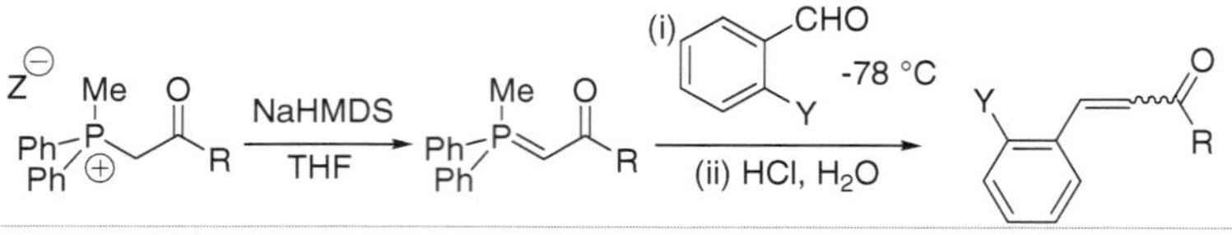

Entry	Ylide R	Aldehyde Y	Temp °C	Enone *Z/E* ratio
1[c]	CH_3	H	−45	19:81
2[d]	CH_3	H	20	20:80
3[c]	CH_3	Cl	−45	33:67
4[d]	CH_3	Br	20	40:60
5[e]	CH_2Cl	H	−78	12:88
6[c,f]	*t*-Bu	H	−78	17:83
7[c]	CH_2Cl	Cl	−78	50:50
8[e]	CH_2Cl	Br	−78	50:50
9[c,f]	*t*-Bu	Br	−78	48:52

[a] *Z/E* ratio determined by ^{1}H NMR analysis of the crude product obtained after aqueous work-up of the reaction mixture. See Sect. 4.6.2 for full details
[b] Phosphonium salt counter-ion Z = Cl^- in all cases except where otherwise noted
[c] Quenched with 5 % aqueous HCl after 20 min at the reaction temperature indicated
[d] Reactions at 20 °C stirred for 4 h, then quenched with 5 % aqueous HCl
[e] Removed from cooling bath after 20 min, and stirred for 12 h at room temperature before being quenched with 5 % aqueous HCl
[f] Counter-ion Z = Br^-

2.3.2 *Reactions of (2-oxoalkylidene)methyldiphenylphosphoranes (Keto-Stabilised Ylides)*

Reactions of a number of keto-stabilised ylides with benzaldehydes were investigated to see if enhanced *Z*-selectivity could be observed in reactions of *ortho*-heteroatom substituted benzaldehydes compared to the reaction of the same ylide with benzaldehyde. The results for ylides derived from methyldiphenylphosphine are shown in Table 2.4. The ylides were generated in situ from the parent phosphonium salt using NaHMDS as base. The reactions were quenched by the addition of acid to ensure that the reaction had occurred at the desired temperature. Some reactions were quenched at low temperature, while others were allowed to warm to room temperature before being quenched.

Acetonylidenemethyldiphenylphosphorane (or (2-oxopropylidene)-methyldiphenylphosphorane, R = CH_3 in the diagram accompanying Table 2.4) was found to be insoluble in THF below −50 °C. Any attempts to carry out Wittig reactions with this ylide below −50 °C (with low temperature quenching) yielded no alkene product. Its reactions were thus carried out at −45 or 20 °C. The reactions of this ylide with benzaldehyde showed high *E*-selectivity (Table 2.4 entries 1 and 2), while its reactions with *ortho*-heteroatom substituted benzaldehydes showed a moderate increase in *Z*-selectivity (up to 30 or 40 % *Z*-isomer produced, see Table 2.4 entries 3 and 4).

Table 2.5 *Z/E* ratio[a] for enones produced in the reactions of selected acetonylidenetriphenylphosphoranes with selected benzaldehydes (all reactions at −78 °C)

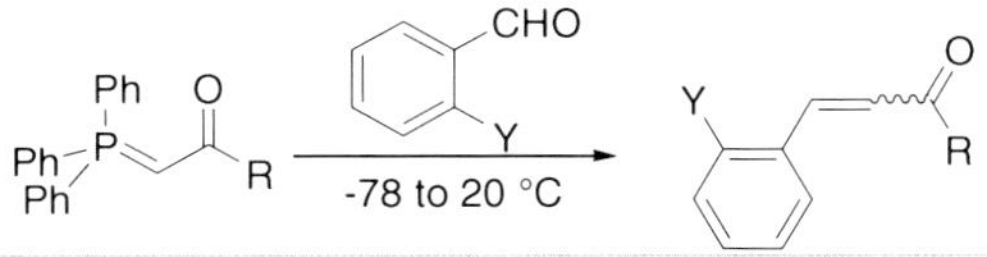

Entry	Ylide R	Aldehyde Y	Enone *Z/E* ratio
1	CH_3	H	3:97
2	CH_3	Br	11:89
3	CH_3	OMe	10:90
4	CH_2OMe	Br	17:83

[a] *Z/E* ratio determined by ^{1}H NMR analysis of the crude product obtained after aqueous work-up of the reaction mixture. See Sect. 4.6.1 for full details

Ylides with one or more substituents on carbon-3 of the 2-oxoalkylidene group gave rise to increased *E*-selectivity in their reactions with benzaldehyde (Table 2.4 entries 5 and 6) relative to the reaction of acetonylidenemethyldiphenylphosphorane with this aldehyde. Reactions of these ylides with *ortho*-heteroatom substituted benzaldehydes (Table 2.4 entries 7–9) resulted in considerably increased *Z*-selectivity by comparison. The trends observed in the series of reactions shown in Table 2.4 thus replicate those observed in reactions of semi-stabilised ylides (see Tables 2.1 and 2.2, Sect. 2.2). That is, the unsubstituted acetonylide reacts with *ortho*-heteroatom substituted benzaldehydes with moderate *Z*-selectivity, 2-oxoalkylidenemethyldiphenylphosphoranes with greater steric bulk at carbon-3 of the oxoalkylidene moiety exhibit significantly increased *Z*-selectivity in their reactions with the same aldehydes, despite the fact that these ylides are more *E*-selective than the unsubstituted acetonylide in their reactions with benzaldehyde itself.

2.3.3 *Reactions of (2-oxoalkylidene)triphenylphosphoranes (Keto-Stabilised Ylides)*

The *Z/E* ratios for alkenes produced in reactions of ylides derived from triphenylphosphine are shown in Table 2.5. The ylides used were pre-formed, and all reactions were carried out at −78 °C but not quenched until they had been allowed to warm to room temperature. Thus the actual reaction temperature is not certain for these reactions.

The reaction of acetonylidenetriphenylphosphorane (or (2-oxopropylidene)-triphenylphosphorane) with benzaldehyde yields the expected very high *E*-selectivity in its reaction with benzaldehyde (Table 2.5 entry 1), while its reactions with *ortho*-heteroatom substituted benzaldehydes are somewhat more *Z*-selective (Table 2.5 entries 2 and 3). The reaction of 3-methoxy-2-oxopropylidenetriphenylphosphorane with 2-bromobenzaldehyde shows a further increase in

Z-selectivity (Table 2.5 entry 4). Thus the trends observed for the reactions of 2-oxoalkylidenemethyldiphenylphosphoranes detailed in Table 2.4 are reproduced here, albeit with far less dramatic shifts in *Z*-selectivity.

2.3.4 Discussion

Kinetic control has been proven in reactions of ester-stabilised ylides [16], and the high *Z*-selectivity observed for the reactions in Table 2.3 strongly implies its operation here. Kinetic control in the reactions of keto-stabilised ylides has not been explicitly proven, but is assumed by analogy. The average proportion of *Z*-alkene produced in the reactions of stabilised ylides with benzaldehydes, calculated from Table 2.3 entries 1–3, Table 2.4 entries 1, 2, 5 and 6 and Table 2.5 entry 1, is 23 %. The average proportion of *Z*-alkene produced in the reaction of these aldehydes with *ortho*-heteroatom substituted benzaldehydes, calculated from Table 2.3 entries 4–10, Table 2.4 entries 3, 4, 7–9 and Table 2.5 entries 2–4, is 54 %, so the *Z*-selectivity is caused to jump 31 % on average by the presence of the *ortho*-heteroatom.

That the electronic effect of the *ortho*-heteroatom on the reactivity of the carbonyl group of the benzaldehyde does not affect the occurrence of high *Z*-selectivity in reactions of stabilised ylides indicates that this selectivity is not a consequence of the cycloaddition TS occurring early on the reaction coordinate. Indeed, the comparatively low reactivity of stabilised ylides means that the TS for the initial reaction between ylide and aldehyde is more or less certain to be relatively late. The highly atypical selectivity observed in these reactions implies the existence of *cis*-selective TS in the step where selectivity is decided that is fundamentally different to the corresponding TS in reactions of aldehydes lacking a suitably placed heteroatom. It is proposed that the reaction proceeds by irreversible cycloaddition, and that in the *cis*-selective cycloaddition TS (shown in Fig. 2.2) there exists a phosphorus-heteroatom bond, similar to that proposed for semi-stabilised ylides (see Fig. 2.1). This TS is favoured over the *trans*-selective TS in reactions of ester-stabilised ylides, and is competitive with the *trans*-selective TS in reactions of keto-stabilised ylides due to the existence of the phosphorus-heteroatom bond and also due to minimisation of 2–3 steric interactions in the *cis*-selective TS. *Z*-selectivity thus arises from a kinetic preference for the formation of *cis*-OPA. This rationale is consistent with *Z*-selectivity being higher with increased polarizability of the *ortho*-heteroatom, as one would expect that the more polarizable a heteroatom is, the greater its capacity to form a bond to phosphorus. The aldehyde carbon-heteroatom bond length may also be important—there is an apparent parallel between the increase in bond length across the series of aldehydes employed and the increase in observed *Z*-selectivity. It may be that the relatively long C–Br bond of 2-bromobenzaldehyde puts the bromine atom in a disposition relative to phosphorus that is more conducive to the establishment of a phosphorus-heteroatom bond than is the case for other aldehydes with shorter

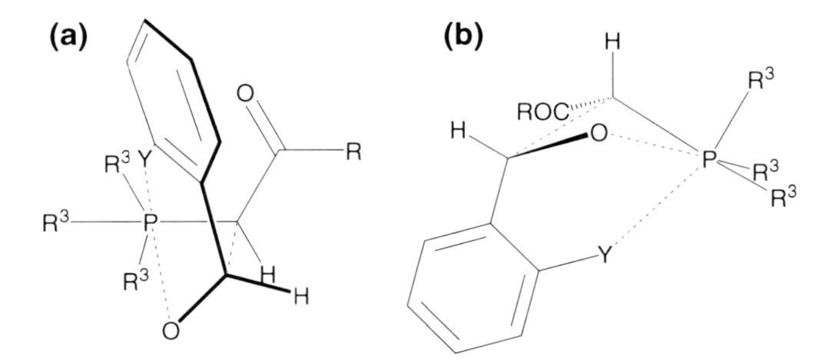

Fig. 2.2 Two views of the proposed *cis*-selective transition state for reactions of stabilised ylides, with phosphorus-heteroatom bond. The *trans*-selective variant of this TS is much higher in energy due to strong 2–3 steric interactions

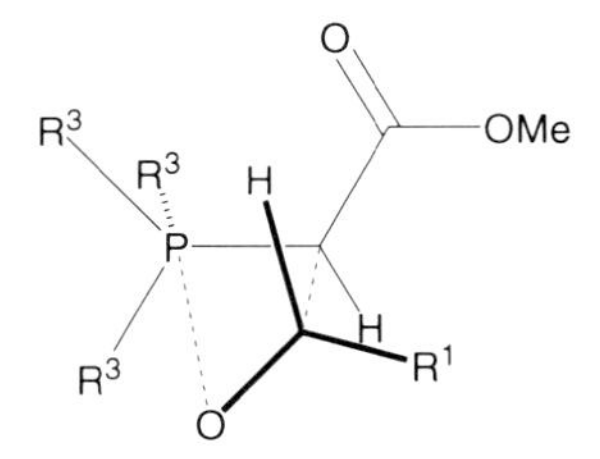

Fig. 2.3 The *trans*-selective transition state found to occur in Wittig reactions of stabilised ylides in the computational findings of Aggarwal, Harvey and co-workers [23–25]. This TS shows a *negative* puckering angle, which is proposed to be as a consequence of the aldehyde C–O and ylide C–C(O) bond dipoles aligning in an electrostatically favourable antiparallel orientation

carbon-heteroatom bonds. The existence of the proposed TS is consistent with selectivity being essentially independent of aldehyde reactivity. The observed selectivity is nigh-on impossible to rationalise outside of the context of the cycloaddition mechanism.

The currently accepted mechanism for Wittig reactions of stabilised ylides is described in Sect. 1.4.3. Vedejs and co-workers were the first to suggest that reactions of stabilised ylides could proceed by irreversible cycloaddition, which implies that the high *E*-selectivity observed in most reactions of stabilised ylides is as a result of a kinetic preference for the formation of *trans*-OPA in the initial cycloaddition step [16]. They were able to prove that the OPAs that are thought to be produced in Wittig reactions of ester-stabilised ylides do not equilibrate. In their computational investigations into the Wittig reaction, Aggarwal, Harvey and co-workers [23–25] found that the kinetically favoured *trans*-selective cycloaddition TS in reactions of stabilised ylides is puckered, with the puckering angle being *negative,* as shown in Fig. 2.3. They proposed that this TS geometry arises as a result of there being an electrostatically favourable antiparallel orientation of the dipoles along the aldehyde C–O bond and the ylide C–C(O) bond in this TS. With the four atoms of the incipient OPA ring disposed as dictated by this electrostatic interaction, the most sterically favourable arrangement of the ylide and aldehyde substituents is to place them, respectively, in pseudo-axial and pseudo-equatorial sites in the forming ring. Hence, this TS gives rise to the formation of *trans*-OPA. See Sect. 1.4.3 for a full discussion of this topic, and in particular of why the *trans*-selective TS is energetically favoured over the *cis*-selective TS.

It is clear, then, that interactions between dipoles of the ylide and aldehyde reactants are very important in dictating transition state geometry, especially in reactions of stabilised ylides. The model shown in Fig. 2.2 for the *cis*-selective TS that is proposed to explain the high *Z*-selectivity observed in reactions of stabilised ylides and *ortho*-heteroatom substituted benzaldehydes also has a negative puckering angle (as indeed does the corresponding model for semi-stabilised ylide shown in Fig. 2.1). Such a TS can thus benefit from the favourable antiparallel orientation of reactant dipoles that normally is available only to *trans*-selective TSs, and *not* to *cis*-selective TSs (see Sect. 1.4.3). This may help to explain why the *Z*-selective route in reactions of stabilised ylides with *ortho*-heteroatom substituted benzaldehydes is competitive with the *E*-selective route, and why it can in some cases be favoured over the *E*-selective route (i.e. in reactions of ester-stabilised ylides). The difference between the *cis*-selective TS in reactions of stabilised ylides with *ortho*-heteroatom substituted benzaldehydes and that in reactions of aldehydes lacking a suitably placed heteroatom is that in the former case it is energetically favourable for the *ortho*-heteroatom substituted aryl group of the aldehyde to be close to phosphorus. There is a bonding interaction between the heteroatom and phosphorus, and also presumably some significant reorganisation of the substituents about phosphorus occurs to facilitate such bonding in the TS of the former case. A *cis*-selective TS in a reaction of an aldehyde lacking a suitably placed heteroatom which had similar geometry, and in particular the same proximity between phosphorus and the large carbonyl substituent of the aldehyde, would suffer greatly from 1–3 steric interactions due to the lack of a bonding interaction and little or no phosphorus-substituent reorganisation (see Fig. 1.10c and the associated discussion in Sect. 1.4.3).

The results for the highly *Z*-selective reactions presented in Tables 2.3 and to a lesser extent those in Table 2.4 strongly imply the operation of kinetic control in these reactions. The energies and structures of the OPA diastereomers formed in these reactions are likely to be very similar to those of the OPA diastereomers from Wittig reactions of benzaldehydes lacking *ortho*-heteroatom substituents (i.e. no phosphorus-heteroatom bonding in the OPAs; see the details on NMR studies of OPAs produced in reactions of non-stabilised ylides in Sect. 2.4). The energy of the *cis*-selective TS is clearly very much lower in reactions of *ortho*-heteroatom substituted benzaldehydes than in reactions of other aldehydes given the high *Z*-selectivity observed in these reactions. Thus, if reversal of OPA to ylide and aldehyde were an issue in Wittig reactions of stabilised ylides in general, it would be *more likely to be in operation in these reactions* than in any others due to the lower activation energy for OPA reversal. Since any equilibrating process would naturally lead to the thermodynamically favoured *trans*-OPA, OPA equilibration is ruled out for these *Z*-selective reactions, and therefore for all other Li-salt free Wittig reactions of stabilised ylides, which are likely to have a higher barrier to OPA reversal than is the case for the *Z*-selective reactions. This, of course, is predicated on the assumption that the *trans*-OPA in reactions of stabilised ylides is thermodynamically favoured.

The results also provide convincing evidence against the involvement of betaines in the Wittig reactions done in the course of the present study, and also against their involvement in Li-salt free Wittig reactions in general. This is significant in particular for reactions of stabilised ylides where, despite the substantial evidence available to support the cycloaddition mechanism, the assumption of the involvement of betaines persists. Many recent reviews concerning Wittig chemistry sit on the fence on the issue, cautiously stating that betaines were initially thought to be the first-formed intermediates but that much evidence has been presented in favour of direct OPA formation.

E-selectivity in reactions of stabilised ylides is explained in the context of the betaine mechanism by the postulation of the existence of an equilibrium between the betaine or OPA intermediates, with one of the *trans*-selective intermediates decomposing faster to produce *E*-alkene as the major product [3]. The non-equilibration of OPAs has been proven beyond doubt for all ylide types, and is confirmed by the results obtained in the course of this project [6, 16]. Thus, the only possibility that remains through which a reaction proceeding by the betaine mechanism could selectively furnish *E*-alkene is if there were an equilibrium between the betaines (by reversal to ylide and aldehyde) and if the *threo*-betaine ring-closes (irreversibly) to OPA faster than does the *erythro*-betaine.

So, in the context of the betaine mechanism, the set of results presented in Table 2.3 would be explained as follows: The reaction of an ester-stabilised ylide with either benzaldehyde or an *ortho*-heteroatom substituted benzaldehyde initially gives *anti-erythro*-betaine for the usual reason of minimisation of steric repulsion in the formation of such a betaine. In order for the reaction of benzaldehyde to be *E*-selective, the *erythro*-betaine would necessarily equilibrate with *threo*-betaine, resulting in proportional enrichment of the latter, which ring-closes faster than *erythro*-betaine to give *trans*-OPA and hence *E*-alkene predominantly. The reactions of the *ortho*-heteroatom substituted benzaldehydes would necessarily involve less betaine equilibration or have kinetically favoured ring-closure to *cis*-OPA in order to explain the high *Z*-selectivity. There is no reason why betaines formed from *ortho*-heteroatom substituted benzaldehydes should be less susceptible to equilibration than benzaldehyde itself, especially as the set of aldehydes involved contains some that are more reactive than benzaldehyde and another that is less reactive. There is also no reason why ring closure should be faster for *erythro*-betaine specifically in the case of *ortho*-heteroatom substituted benzaldehydes. Any contribution that might be available from phosphorus-heteroatom bonding would be equally available to the *threo*-betaine. Also, it seems highly likely that the TS in the ring-closing process of the *threo*-betaine would benefit from the same advantages that the *trans*-OPA does with regard to minimisation of steric interactions, regardless of the nature of the species involved. The two phenomena cannot simultaneously be explained by the betaine mechanism, and thus it is concluded that betaines are not involved at all in Li-salt free Wittig reactions of stabilised ylides, whether they are *Z*-selective or otherwise.

2.4 Oxaphosphetane *cis/trans* Ratios in Reactions of Non-stabilised Ylides with Benzaldehydes

2.4.1 The Determination of OPA cis/trans *Ratios in Reactions of Non-stabilised Ylides*

An investigation was also carried out to determine whether there was a similar *ortho*-heteroatom-induced effect in operation in reactions of non-stabilised ylides with benzaldehydes. Since it is generally possible to observe spectroscopically the mixture of OPAs produced in Wittig reactions of non-stabilised ylides, an opportunity was available in studying these reactions to measure the kinetic OPA *cis/trans* ratio directly rather than inferring it from alkene *Z/E* ratios, as was (necessarily) done for reactions of semi-stabilised and stabilised ylides. Since the heteroatom effect is proposed to operate in the putative irreversible cycloaddition step of the Wittig reaction, it was therefore appropriate to ascertain the kinetic selectivity of the OPA-forming process directly. Furthermore, Wittig reactions of ethylidenetriphenylphosphorane have previously been shown *not* to be under kinetic control i.e. the OPAs formed in the Wittig reaction equilibrate by reversal to ylide and aldehyde [14]. The investigations undertaken in this project have shown that OPAs produced in Wittig reactions of ethylides in general (i.e. not just those derived from triphenylphosphine) are prone to equilibration at or above the temperature at which alkene formation begins to occur. Consequently, the activation energy for Wittig reversal is comparable to that for cycloreversion to alkene and phosphine oxide, at least for the *cis*-OPA. At the time that this project started, it was generally believed (based mainly on evidence from reactions of ethylidenetriphenylphosphorane and alkylidenetrialkylphosphoranes) [6, 7, 14] that reactions of all alkylides with aromatic aldehydes might in principle be susceptible to equilibration in a similar way. The experiments carried out in the present project, coupled with some previously reported results for Wittig reactions of *n*-butylidenetriphenylphosphorane with benzaldehyde [7, 13, 15], show that Wittig reversal may not generally be an issue in Wittig reactions of longer chain alkylides. The details of these findings will be discussed more fully later in this section, but for now it will suffice to say that at the outset of this project we thought that OPA *cis/trans* ratios could not be reliably inferred from alkene *Z/E* ratios for reactions of *any* alkylide with an aromatic aldehyde, and so it was deemed particularly appropriate to directly observe OPA *cis/trans* ratios for the reactions of interest.

A further complication that is present in reactions of non-stabilised ylides is that the most widely used of these ylides, alkylidenetriphenylphosphoranes, generally react with very high *Z*-selectivity. Hence in certain cases it would be quite difficult even to demonstrate improved *Z*-selectivity since the reaction with benzaldehyde would be likely to be highly *Z*-selective itself. For this reason, reactions of non-stabilised ylides derived from other phosphines were also investigated.

In contrast to reactions of semi-stabilised and stabilised ylides, in reactions of non-stabilised ylides OPA cycloreversion to alkene and phosphine oxide is the rate determining step. Consequently, the OPAs produced in Wittig reactions of non-stabilised ylides have a finite lifetime. The ratio of OPA diastereomers produced at temperatures well below the temperature at which decomposition occurs is generally invariant for OPAs derived from triphenylphosphine, alkyldiphenylphosphines and *P*-phenyl-5*H*-dibenzophosphole. Furthermore, all known instances of OPA equilibration result in an increase in the proportion of *trans*-OPA at the expense of *cis*-OPA (which may sometimes be judged by there being a greater proportion of *E*-isomer in the alkene product than there was of the *trans*-isomer in the OPA intermediate), so for any reaction that gives a high proportion of *cis*-OPA or *Z*-alkene, the low temperature OPA *cis/trans* ratio corresponds to the kinetic OPA *cis/trans* ratio.

Experimentally, the kinetic OPA *cis/trans* ratio may conveniently be determined by either of two means, which have been found by us and others [15, 18] to give identical results. One is low temperature NMR observation of the OPA mixture, which involves either carrying the reaction out in an NMR tube or transferring the reaction mixture at low temperature to an NMR tube under an inert atmosphere and then obtaining NMR spectra of the intermediate. ^{31}P NMR is particularly useful for this method, since the OPAs appear in a region of the spectrum that is highly diagnostic of pentacoordinate phosphorus species (-60 to -80 ppm). The second is low temperature acid quenching of the reaction mixture to give β-hydroxyphosphonium salt (β-HPS), whose *erythro/threo* ratio exactly matches the OPA *cis/trans* ratio. This method requires transfer of the reaction mixture at low temperature into an acid solution also at low temperature, and is experimentally quite complex, but the analysis of the product is often more straightforward since β-hydroxyphosphonium salts are not air sensitive.

2.4.2 Determination of the Kinetic OPA cis/trans Ratio in Wittig Reactions of Non-stabilised Ylides by Low Temperature Acid Quenching

Low temperature acid quenching was carried out for Wittig reactions of several non-stabilised ylides with benzaldehydes. In most of these reactions, the ylide was generated at room temperature from the precursor phosphonium bromide salt. However, *P*-(ethylidene)-*P*-phenyldibenzophospholane proved to be considerably more demanding to handle than other ylides. The optimised procedure for reactions with this ylide involved adding dry phosphonium salt and solid KHMDS together in the right proportions (an "instant ylide mix") in a Schlenk flask inside a glove box under argon, and then transferring the Schlenk flask to a Schlenk manifold, where it was charged with nitrogen by the pump and fill technique (see Chap. 4 for full details) [26]. The mixture of white solids was cooled to −25 °C and dry THF was added slowly to give a maroon solution of ylide, which was not stirred until the septum on the Schlenk flask was replaced with a stopper, and the tap was closed. The solution was then cooled to −45 °C, at which temperature it was stirred for 15–20 min. It was then further cooled to −78 °C, the reaction temperature. Attempting to generate this ylide at too low a temperature resulted in very slow deprotonation, while at too high a temperature it was observed to decompose (perhaps by hydrolysis) quite readily.

The Wittig reactions were all carried out at −78 °C, followed 15 min later by cannula transfer of the reaction mixture into a solution of HCl in THF/methanol cooled to −78 °C to quench the OPA. It has previously been reported that the β-HPS counter-ion is the counter-ion from the starting phosphonium salt (and not that from the acid) [6], and it has been confirmed from a crystal structure of the *erythro*-β-HPS of entry 2 in Table 2.6 (obtained by crystallisation of the crude product from acetonitrile) that bromide is the counter-ion, not chloride. The crude β-HPS product was analysed by a series of NMR techniques (^{1}H, ^{31}P, ^{13}C, COSY, TOCSY, ^{1}H-^{31}P HMBC, HSQC, HMBC) in order to assist assignment of signals in the ^{1}H and ^{31}P NMR spectra. Of particular aid in the assignment of diastereomeric ratios for β-HPSs or OPAs generated from β-HPSs was a two-dimensional ^{1}H-^{31}P HMBC NMR technique. This allows all of the hydrogen nuclei in a molecule that undergo long-range coupling to phosphorus to be identified as being in the same molecule as that phosphorus and each other. Another NMR technique that gives the same result is selectively decoupled ^{1}H{^{31}P} NMR, in which a ^{1}H spectrum is obtained while selectively decoupling from a single signal in the ^{31}P spectrum. The signals in the ^{1}H{^{31}P} spectrum whose coupling to phosphorus had been removed could then be identified by their decreased multiplicity, and so all of the phosphorus-coupled proton signals belonging to each diastereomer could be determined. This is essentially the one-dimensional NMR equivalent of the ^{1}H-^{31}P HMBC technique. In this way, multiple signals can be assigned to belong to one diastereomer or another, and a diastereomeric ratio can be determined by integration of all of the signals in the ^{1}H spectrum and the single peak in the ^{31}P spectrum that can be unambiguously assigned to each diastereomer. The diastereomeric ratios determined in this way are shown in Table 2.6.

Establishing which set of signals belongs to which diastereomer was done in a number of ways. Most of the Wittig reactions under consideration are predominantly *Z*-selective, and therefore the major β-HPS produced in that reaction can be taken to be the *erythro*-isomer. The OPA generated by deprotonation of the (1-(2-bromophenyl)-1-hydroxy-3-methylbut-2-yl)phenyldibenzophospholium bromide produced in the low temperature acid quenching of the reaction shown in Table 2.6 entry 8 was proven to be *cis* by 1D-NOESY NMR (see Sect. 2.4.5 for further details), and therefore must have been derived from *erythro*-β-HPS. The major diastereomer of (1-(2-bromophenyl)-1-hydroxyprop-2-yl)triphenylphosphonium bromide produced in the reaction of ethylidenetriphenylphosphorane and

Table 2.6 Wittig reactions of non-stabilised ylides (generated from phosphonium bromide salt using NaHMDS) with benzaldehydes to give OPA (initially), and subsequently β-HPS after low temperature acid quenching of OPA

$R^aR^bPhP{=}CHR^2 + R^1CHO \xrightarrow[-78\ °C]{THF}$ OPA ($R^bR^aPhP{-}O$ ring, R^2, R^1) $\xrightarrow[-78\ °C]{HCl}$ $R^bR^aPhP^{\oplus}$ … OH, $Br^{\ominus}$ (R^2, R^1)

Entry	Ylide R^a	Ylide R^b	Ylide R^2	Aldehyde	β-HPS *erythro/threo* ratio[a]	Alkene *Z/E* ratio[b]
1	Ph	Ph	Me	PhCHO	90:10	85:15
2	Ph	Ph	Me	2-BrC_6H_4CHO	95:5	75:25
3	Ph	Et	Me	PhCHO	54:46	32:68
4	Ph	Et	Me	2-BrC_6H_4CHO	64:36	55:45
5	DBP system		Me	PhCHO	72:28	53:47[d]
6	DBP system		Me	2-BrC_6H_4CHO	94:6[c]	81:19[d]
7	DBP system		*i*-Pr	PhCHO	89:11	90:10[e]
8	DBP system		*i*-Pr	2-BrC_6H_4CHO	94:6	91:9[e]
9	Ph	Ph	*i*-Pr	PhCHO	100:0	–
10	Ph	Ph	*i*-Pr	2-BrC_6H_4CHO	100:0	82:18

[a] All reactions were carried out at −78 °C, and subsequently quenched by cannulation of the reaction mixture into HCl solution in THF/methanol. The *erythro/threo* ratio was determined by ^{1}H and ^{31}P NMR of the crude product after minimal aqueous work-up
[b] Unless otherwise specified, this is the *Z/E* ratio of the corresponding unquenched Wittig reaction allowed to warm to room temperature after 15 min at −78 °C, as determined by integration of characteristic signals in the ^{1}H NMR of the crude product after aqueous work-up
[c] The *threo*-β-HPS could not be unambiguously assigned, as it is present to only a small degree. A large amount of ylide hydrolysis appears to have occurred based on the quantity of phosphine oxide (non-Wittig product) and unreacted 2-bromobenzaldehyde present in the crude spectrum. A peak of matching chemical shift to the *threo*-β-HPS of entry 5 ($\delta_P = 33.8$) is present in the crude spectrum ($\delta_P = 34.0$) that integrates to give the *erythro/trheo* ratio shown. Also, direct monitoring at −20 °C of the OPA produced in the Wittig reaction (carried out at −78 °C) reveals a kinetic OPA *cis/trans* ratio of 94:6
[d] *Z/E* ratio of the corresponding unquenched Wittig reaction heated to 80 °C for 2 h after stirring for 15 min at −78 °C, as determined by integration of characteristic signals in the ^{1}H NMR of the crude product
[e] *Z/E* ratio of alkene produced in β-HPS deprotonation experiment in toluene-*d8*

2-bromobenzaldehyde (see Table 2.6 entry 2) was also recrystallised, and was shown to be of the *erythro*-configuration by obtaining a crystal structure of the solid.[7] NMR characterisation of the isomers of (1-hydroxy-1-phenylprop-2-yl) triphenylphosphonium bromide has been reported [7]. The major diastereomer of β-HPS (assigned to be *erythro* in each case) produced in each reaction of

[7] CCDC-883627 contain the X-ray crystallographic data for this compound. This data can be obtained free of charge from The Cambridge Crystallographic Data Centre via http://www.ccdc.cam.ac.uk/data_request/cif. An image of the crystal structure can be found in Sect. 4.7.2.

(ethylidene)ethyldiphenylphosphorane (see Table 2.6 entries 3 and 4) was isolated by careful recrystallisation of the crude product. The spectral characteristics of the β-HPSs derived from the reactions of (ethylidene)ethyldiphenylphosphorane with benzaldehydes are sufficiently similar to those of the definitively characterised β-HPSs that it is possible to assign each set of diastereomeric signals to one isomer or another of the former β-HPSs by analogy with the spectra of the known β-HPSs. In particular, the coupling constant $^3J_{PH}$ between phosphorus and the OCH signal is typically larger for the *threo* isomer (9 Hz up to 14.5 Hz) than for the *erythro* (6–8 Hz), the *erythro* OCH signal invariably has a higher chemical shift than the corresponding signal of the *threo*-isomer, and the double doublet arising from resonance of the PCHCH_3 protons typically has a higher chemical shift for the *erythro*-isomer than for the *threo*-isomer. Assignment of the configuration of the *erythro*-β-HPSs produced in low temperature acid quenching reactions of (*iso*-butylidene)triphenylphosphorane (see Table 2.6 entries 9 and 10) were done by analogy with (1-hydroxy-1-phenylpent-2-yl)triphenylphosphonium bromide [7]. Assignments made in this manner invariably agreed with assignments made on the basis of the selectivity of the corresponding unquenched Wittig reaction.

The *erythro/threo* ratios obtained from the ^{1}H and ^{31}P spectra of the crude β-HPS product invariably agreed very closely with each other. Other products present included some phosphine oxide (from ylide hydrolysis) and phosphonium salt (from unreacted ylide).

Ethylidenetriphenylphosphorane reacted with benzaldehyde to give the expected high *erythro*-selectivity in the β-HPS quench product (Table 2.6 entry 1). The same ylide reacted with 2-bromobenzaldehyde showing increased selectivity for *erythro*-β-HPS (Table 2.6 entry 2). (Ethylidene)ethyldiphenylphosphorane reacted with benzaldehyde with marginal selectivity for the *erythro* isomer (Table 2.6 entry 3), and its reaction with 2-bromobenzaldehyde showed moderate *erythro*-selectivity (Table 2.6 entry 4).

P-(ethylidene)phenyldibenzophospholane reacted with benzaldehyde to give β-HPS with an *erythro/threo* ratio of 72:28 (Table 2.6 entry 5), while the diastereomeric ratio of β-HPS produced in the reaction of this ylide with 2-bromobenzaldehyde showed very high *erythro*-selectivity (Table 2.6 entry 6). The reactions of *P*-(*iso*-butylidene)-*P*-phenyldibenzophospholane also showed increased *erythro*- selectivity with 2-bromobenzaldehyde (Table 2.6 entry 8) compared with benzaldehyde (Table 2.6 entry 7). These *erythro/threo* ratios shown in Table 2.6 entries 4–8 agree exactly with the kinetic OPA *cis/trans* ratios determined for the same reactions by ^{31}P NMR analysis of the Wittig reaction mixtures (see Table 2.7 below). These results confirm the occurrence of increased kinetic selectivity for *cis*-OPA in Wittig reactions of *ortho*-heteroatom substituted benzaldehydes.

No *threo*-β-HPS could be detected in the crude products obtained after low temperature acid quenching of the Wittig reactions of (*iso*-butylidene)triphenylphosphorane with either benzaldehyde or 2-bromobenzaldehdye, and so it can be inferred that OPA formation in these reactions is completely selective for *cis*-OPA.

Discussion of the *Z/E* ratios of the alkenes produced in these reactions is deferred until Sect. 2.4.4.

The Wittig reaction of (ethylidene)ethyldiphenylphosphorane with benzaldehyde (unquenched) is predominantly *E*-selective (see Table 2.6 entry 3), and so it is not possible to infer the stereochemistry of the (1-hydroxy-1-phenylprop-2-yl)ethyldiphenylphosphonium bromide Wittig reaction quench product from the observed selectivity of the unquenched Wittig reaction. Crystals of the major diastereomer of this β-HPS (isolated from the crude product by crystallisation) were deprotonated at −78 °C under an inert atmosphere to form OPA, which was then allowed to warm to 20 °C to decompose to alkene and phosphine oxide. ^{1}H NMR analysis of the crude product showed it to be *Z*-1-phenylprop-1-ene, thus confirming the configuration of the precursor β-HPS to be *erythro*, in agreement with the assignment made based on comparison of its spectral characteristics with those of characterised β-HPSs.[8]

2.4.3 Determination of the Kinetic OPA cis/trans Ratio in Wittig Reactions of Non-stabilised Ylides by ^{31}P NMR

OPA adducts from Wittig reactions of *P*-(alkylidene)-*P*-phenyldibenzophospholanes are generally stable at room temperature and above; indeed they must be heated to effect alkene formation. This confers significant practical advantages to the NMR observation of these OPAs compared to unconstrained analogues. The relative stability of dibenzophosphole-derived OPAs is kinetic. OPA decomposition must necessarily go through TS in which the DBP ring C–P–C bond angle is in the process of being stretched from 94° in the trigonal bipyramidal OPA (the five-membered ring spans axial and equatorial sites) to the 104–107° angle present in the phosphine oxide product [27], which induces considerable angle strain in the five-membered ring. This results in a dramatically increased activation energy for the cycloreversion process compared to unconstrained OPAs. The convenience of being able in principle to determine kinetic OPA selectivity in Wittig reactions at room temperature by using DBP-derived ylides was, however, somewhat offset by the difficulty of handing the ylides themselves. They are exceptionally sensitive to the presence of any water, and many of our initial experiments were marred by the presence of large amounts of phosphine oxide derived from ylide hydrolysis. No alkene product was observed, and the same phosphine oxides as had been observed in the OPA generation experiments were produced even without the addition of aldehyde, indicating that the phosphine oxide was not derived from a Wittig reaction or hydrolysis of a Wittig intermediate.

---

[8] This assignment has been confirmed by X-ray crystallographic analysis, which is yet to be published. See CCDC-883627 for the full crystallographic data.

Table 2.7 Wittig reactions of non-stabilised ylides (generated from phosphonium bromide salt using NaHMDS or KHMDS) with benzaldehydes to give OPA (initially), and subsequently alkene[a]

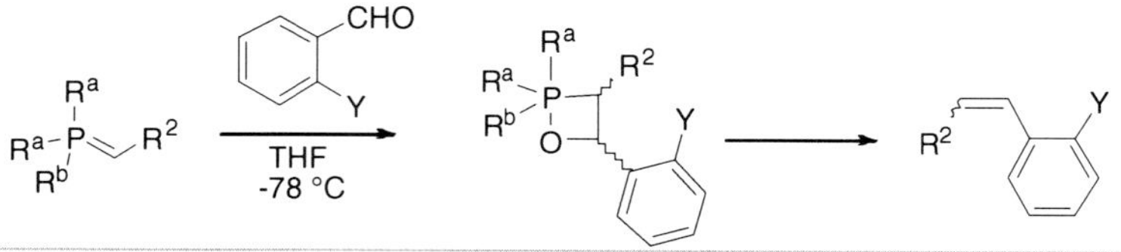

Entry	Base	Ylide			Aldehyde Y	OPA *cis/trans* ratio	Alkene *Z/E* ratio[b]
		R^a	R^b	R^2			
1[c]	KHMDS	DBP	Ph	Me	H	71:29	53:47
2[c]	KHMDS	DBP	Ph	Me	Br	94:6	82:18
3	NaHMDS	DBP	Ph	*i*-Pr	H	89:11	89:11[d]
4	NaHMDS	DBP	Ph	*i*-Pr	Br	94:6	91:9[d]
5	KHMDS	Ph	Et	Me	Br	64:36[e]	56:44
6	KHMDS	Ph	Et	Me	Br	–[f]	32:68

[a] All reactions were carried out at −78 °C. OPA *cis/trans* ratios were determined by ^{31}P NMR (obtained at 30 °C unless otherwise indicated) after cannula filtration of the reaction mixture into an NMR under an inert atmosphere and addition of toluene-*d8*. The ^{31}P NMR spectra all indicated the presence of relatively small amounts of phosphine oxide by-product, which were shown to be derived directly from the ylide by control reactions in which no aldehyde was added, and by the fact that no alkene product could be observed by NMR prior to heating

[b] Alkene *Z/E* ratios were determined by integration of characteristic signals in the ^{1}H NMR of the crude product. DBP-derived OPAs were heated to 80 °C for 2 h to effect alkene formation, while the $EtPh_2P$-derived OPA began to decompose to alkene and phosphine oxide at ca. −10 °C

[c] The ylide was generated at −20 °C, and then stirred for 0.5 h at −45 °C before cooling to −78 °C for the reaction. The OPA generated in this reaction was monitored by ^{31}P NMR at −20 °C

[d] This alkene was obtained from experiments involving deprotonation of β-HPS of *erythro/threo* ratio matching the indicated OPA *cis/trans* ratio, as described in Sect. 2.4.2

[e] OPA *cis/trans* ratio determined at −40 °C

[f] OPA signals have same ^{31}P chemical shifts both at −40 and at −20 °C [7], so no diastereomeric ratio could be determined

In general for reactions of non-stabilised ylides it has been found that conducting the experiments in long, thin Schlenk tubes with minimal nitrogen flow is optimal in order to reduce the flux through the reaction flask. Tubing connecting the flask to the nitrogen source (which is distributed to various flasks using a Schlenk manifold) must be dried. This was done by fitting a syringe barrel with an attached needle to the tubing when not in use, and keeping the needle inserted into a sealed flask of dried solid potassium hydroxide, with the needle tips embedded in the KOH. The ylide *P*-(*iso*-butylidene)-*P*-phenyldibenzophospholane proved to be relatively robust—it could be generated at room temperature, and its OPA adducts could be monitored by NMR at room temperature. (Ethylidene)ethyldiphenylphosphorane was also generated at room temperature. The exceptionally moisture sensitive *P*-(ethylidene)-*P*-phenyldibenzophospholane was generated by the more cautious procedure referred to in Sect. 2.4.2. Full details of the experimental procedures employed are given in Chap. 4.

OPA formation was carried out at −78 °C in all cases. All manipulations were carried out under an inert atmosphere. After OPA formation was complete (as judged by the fading of the colour of the ylide as the aldehyde was added), the solution was usually stirred for approximately 10 min at −78 °C. OPAs derived form *P*-(ethylidene)-*P*-phenyldibenzophospholane were transferred by cannula filtration into an NMR tube pre-cooled to ca. −50 °C. NMR observation was then carried out at −20 °C. OPAs derived from (ethylidene)ethyldiphenylphosphorane were transferred in the same manner to an NMR tube pre-cooled to −78 °C, and were monitored by NMR initially at −40 °C or lower (and thereafter by variable temperature NMR in one case). In this way minimal non-Wittig production of phosphine oxide was observed. Solutions of OPAs derived from *P*-(*iso*-butylidene)-*P*-phenyldibenzophospholane were allowed to warm to room temperature before being passed though a cannula filter into an NMR tube under nitrogen, and NMR observation was performed at 30 °C. Initial OPA generation experiments involved removal of the THF reaction solvent and addition of anhydrous toluene-*d8* to allow analysis of the OPA by ^{1}H and ^{13}C NMR and various 2D NMR techniques. However, significant amounts of phosphine oxide were produced during these experiments. Much of this was likely to have come from ylide hydrolysis, but there still remains the possibility that OPA may form phosphine oxide by a hydrolytic (non-Wittig) process. In order to minimise this, and thereby maximise the yield of OPA, a sample of the reaction mixture was simply transferred (via cannula filter to remove inorganic salts) into an NMR tube. Enough toluene-*d8* was then added to allow the NMR spectrometer to find a deuterium lock, thus obviating the need for solvent removal. Although meaning that the kinetic OPA *cis/trans* ratios were determined based on ^{31}P NMR only, this precaution along with the others employed—drying the Schlenk manifold tubing with KOH, closing the Schlenk flask taps to nitrogen flow whenever possible, and generating *P*-(ethylidene)phenyldibenzophospholane by the rather exacting procedure described above—resulted in OPA being successfully transferred to the NMR in very high yield. The only other major species appearing in the ^{31}P NMR was the ylide at ca. −10 ppm. This method allowed definitive determination of the kinetic OPA *cis/trans* ratio.

The OPA signals appeared around −60 to −70 ppm in the ^{31}P NMR. Assignment of the signals to one diastereomer or the other was done based on the *Z/E* ratio of the alkene product after OPA cycloreversion. Since all of the reactions in which assignments were done in this manner were predominantly *Z*-selective, it can be concluded that the major diastereomer of OPA produced as intermediate in the Wittig reaction is the *cis*-isomer. The kinetic OPA *cis/trans* ratios determined by ^{31}P NMR for the reactions of non-stabilised ylides derived from *P*-phenyldibenzophosphole are shown in Table 2.7.

P-(ethylidene)-*P*-phenyldibenzophospholane reacted with benzaldehyde to give an OPA *cis/trans* ratio of 71:29 (Table 2.7 entry 1). This ylide reacted with 2-bromobenzaldehyde with very high selectivity for the *cis*-OPA (entry 2). *P*-(*iso*-

butylidene)-*P*-phenyldibenzophospholane reacted with benzaldehyde with high selectivity for the *cis*-OPA (entry 3) but the *cis*-selectivity in its reaction with 2-bromobenzaldehyde was even higher (entry 4). These results were also reproduced by low temperature acid quenching experiments (see Sect. 2.4.2 and particularly Table 2.6 above). The reaction of (ethylidene)ethyldiphenylphosphorane and 2-bromobenzaldehyde was observed to form OPA with a *cis/trans* ratio of 64:36 (entry 5). This is identical to the β-HPS *erythro/threo* ratio obtained by low temperature acid quenching of this Wittig reaction (see Table 2.6 entry 4 above). The OPA signals in the reaction of (ethylidene)ethyldiphenylphosphorane with benzaldehyde could not be resolved (a single, relatively broad peak was observed at −61.6 ppm in the ^{31}P NMR), which is consistent with a similar report for the OPA produced in the reaction of ethylidenetriphenylphosphorane with benzaldehyde [7]. The *erythro/threo* ratio of the β-HPS produced by low temperature acid quenching of the reaction of (ethylidene)ethyldiphenylphosphorane with benzaldehyde was 54:46 (see Table 2.6 entry 3 above).

It can be seen from these results that selectivity for *cis*-OPA is increased for the reaction of each ylide with 2-bromobenzaldehyde relative to its reaction with benzaldehyde. This demonstrates the operation of the *ortho*-heteroatom effect in Wittig reactions of non-stabilised ylides. Discussion of the *Z/E* ratios of the alkenes produced in these reactions is deferred until Sect. 2.4.4.

2.4.4 Investigation of the Operation of Stereochemical Drift in Reactions of Non-Stabilised Ylides

The kinetic OPA *cis/trans* ratios of reactions of a number of non-stabilised ylides have been determined by the methods described in Sects. 2.4.2 and 2.4.3. In order to aid assignment of the diastereomeric OPA signals, and in order to test for the possible operation of stereochemical drift in these reactions, the *Z/E* ratios of the alkene products of the relevant Wittig reactions were determined. The *Z/E* ratios for the Wittig reactions of benzaldehyde and 2-bromobenzaldehyde respectively with of each of ethylidenetriphenylphosphorane, (ethylidene)ethyldiphenylphosphorane and *P*-ethylidene-*P*-phenyldibenzophospholane are shown in both Tables 2.6 and 2.7. Comparison of the *Z/E* ratio of the alkene produced in each reaction with the relevant kinetic OPA *cis/trans* ratio (compare the alkene *Z/E* and β-HPS *erythro-threo* ratios of Table 2.6 entries 1–6; also compare the alkene *Z/E* and OPA *cis/trans* ratios of Table 2.7 entries 1, 2, 5 and 6) clearly shows that stereochemical drift is in operation in these reactions, since there is a non-correspondence between the alkene *Z/E* ratio and the β-HPS *erythro-threo* ratio or OPA *cis/trans* ratio in favour of the *E*-alkene, as would be expected. Despite this, the *Z*-alkene is still the predominant product in all cases but one, and so for each of these the major OPA isomer can be assigned to be the *cis*-isomer.

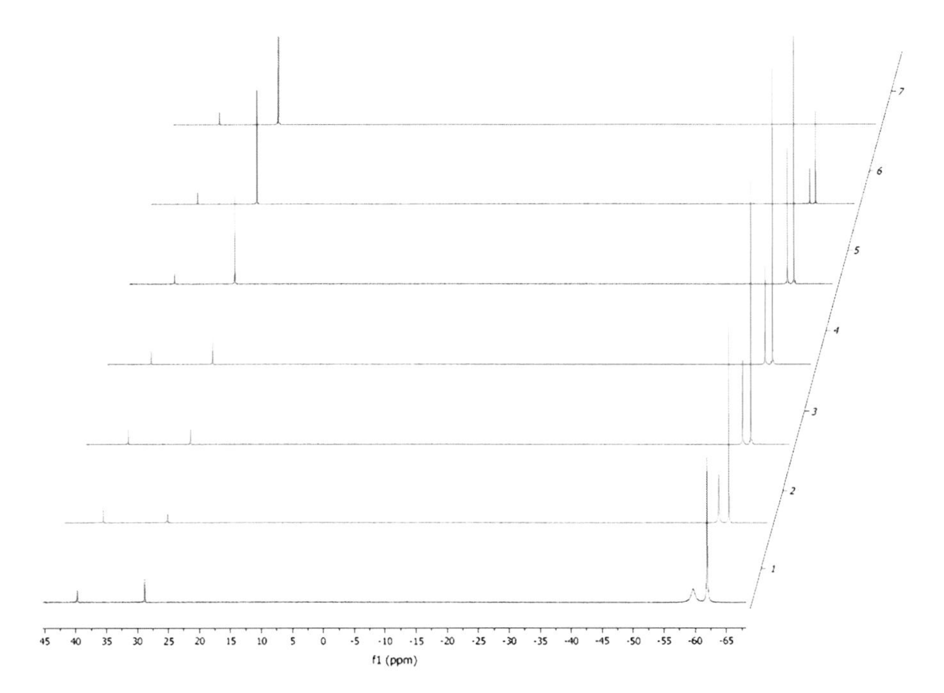

Fig. 2.4 Set of stacked ^{31}P spectra at −70 (*bottom spectrum*), −40, −20, −10, 0, 10 and 20 °C (*top spectrum*) for the reaction of (ethylidene)ethyldiphenylphosphorane with 2-bromobenzaldehyde in THF (NMR sample spiked with a small quantity of toluene-*d8*)

The OPAs produced in the reaction of (ethylidene)ethyldiphenylphosphorane with 2-bromobenzaldehyde (Table 2.7 entry 5) were monitored by variable temperature ^{31}P NMR. The spectra obtained at −70, 40, −20, −10, 0, 10 and 20 °C are shown in the stacked array of Fig. 2.4. A small amount of phosphine oxide is present (ethyltriphenylphosphine oxide at $\delta = 29$, diethylphenylphosphine oxide at $\delta = 40$) even at low temperature (ca. 10 % of total amount of phosphorus present), resulting from a small amount of ylide hydrolysis. It can be seen that the relative proportion of ethyltriphenylphosphine oxide begins to increase between −10 and 0 °C, and then grows further at higher temperatures, until the OPAs have vanished completely at 20 °C. Although the relative amount of OPA decreases as the temperature goes up, the *cis/trans* ratio remains invariant at 64:36 in all recorded spectra. However, the final alkene *Z/E* ratio (determined from the same sample dissolved in $CDCl_3$) was unsurprisingly found to be 56:44, as had previously been observed. This demonstrates that no interconversion of OPA occurs until at least 10 °C in this reaction, in keeping with an earlier observation by Vedejs et al. that OPA interconversion did not take place below the temperature at which alkene formation had also begun to occur in the reaction of ethylidenetriphenylphosphorane and benzaldehyde [14].

The β-HPS crude product from the acid quenching of each of the reactions of *P*-(*iso*-butylidene)-*P*-phenyldibenzophospholane (Table 2.6 entries 7 and 8) was precipitated (along with *P*-(*iso*-butyl)-*P*-phenyldibenzophospholium salt and some phosphine oxide) from chloroform/ethyl acetate to remove residual aldehyde, and was then deprotonated at low temperature in toluene-*d8*. This OPA produced was characterised by a series of NMR techniques. The *cis/trans* ratio of the OPA matched that of the precursor β-HPS. Heating of this OPA to 80 °C for two hours gave alkene (whose *Z/E* ratio is also indicated in Table 2.6 entries 7 and 8). The close correspondence between the OPA *cis/trans* ratio and the alkene *Z/E* ratio indicates minimal, if any, stereochemical drift in these reactions. More details on the NMR studies of the OPAs produced in these reactions are given in Sect. 2.4.5.

The β-HPS (1-(2-bromophenyl)-1-hydroxy-3-methylbut-1-yl)triphenylphosphonium bromide produced by low temperature acid quenching of the reaction of (isobutylidene)triphenylphosphorane and 2-bromobenzaldehyde was found to be made up exclusively of the *erythro*-isomer, but the corresponding unquenched Wittig reaction gave alkene with a *Z/E* ratio of 82:18 (see Table 2.6 entry 10), clearly indicating the operation of stereochemical drift in this reaction.

The operation of stereochemical drift observed here in reactions of ethylides with aromatic aldehydes is in keeping with literature precedents [7, 14]. Stereochemical drift in reaction of longer chain alkylides appears to depend on the exact nature of the ylide; OPAs derived from *P*-(isobutylidene)-*P*-phenyldibenzophospholane with benzaldehydes decompose with negligible stereochemical drift, while those derived from (isobutylidene)triphenylphosphorane appear not to decompose stereospecifically. Kinetic control has also been demonstrated for reactions of longer chain alkylides with these aldehydes (Table 2.1 entries 7–9) [7, 13, 15].

2.4.5 *NMR Observation of Oxaphosphetanes and β-Hydroxyphosphonium Salts Derived from Non-stabilised Ylides: Experimental Techniques and Further Information Acquired*

The primary goal of the NMR observations of OPAs produced in Wittig reactions of *ortho*-heteroatom benzaldehydes (which have already been described in Sect. 2.4.1 above) was to establish the kinetic OPA *cis/trans* ratio and thus the kinetic selectivity of the Wittig reaction in question. A subsidiary goal was to determine whether there was any evidence suggesting that the OPAs produced in these reactions contained a phosphorus-heteroatom bond like the one that is proposed to exist in the transition state leading to OPA. The ^{31}P NMR chemical shifts found for OPAs generated in Wittig reactions of *P*-(alkylidene)-*P*-phenyldibenzophospholanes are in the range −60 to −75 ppm, which is the same range previously reported for OPAs with pentacoordinate phosphorus [6, 16, 28, 29, 30]. An OPA with a phosphorus-heteroatom bond (and thus hexacoordinate phosphorus) would be expected to have a significantly more negative chemical shift in the ^{31}P NMR, analogous to oxaphosphetanides, which have previously been reported to have chemical shifts below −100 ppm [31, 32]. Hence it can be surmised that the phosphorus in the OPAs produced in these reactions is pentacoordinate, i.e. not bearing a phosphorus-heteroatom bond. We must conclude therefore that the proposed bonding interaction, for which there is a very substantial body of evidence presented here, exists only in the *transition state* leading to OPA. Furthermore, we may postulate that due to geometrical constraints, a planar structure (such as an OPA or a planar TS) does not allow a phosphorus-heteroatom bond. The logical extension of this proposal is that the cycloaddition TS in these reactions must be puckered, as has already been suggested above for the TS model for reactions of semi-stabilised and stabilised ylides.

As was briefly mentioned in Sect. 2.4.4, the β-HPSs of Table 2.7 entries 7 and 8, produced by the method described in Sect. 2.4.2, were subsequently precipitated (along with any remaining starting phosphonium salt and some phosphine oxide) from chloroform/ethyl acetate to remove aldehyde. The sample thus obtained was treated with NaHMDS in toluene-*d8* at −78 °C under a nitrogen atmosphere to generate OPA (as a mixture of diastereomers), which was observed by NMR after cannula filtration into an NMR tube under a nitrogen atmosphere. The OPA was typically characterised by ^{1}H, ^{31}P, COSY, TOCSY, 1D and 2D NOESY, ^{13}C, HSQC and HMBC NMR techniques. The *cis/trans* ratio of OPA produced by β-HPS deprotonation does not necessarily correspond to the *erythro/threo* ratio of the β-HPS (which itself *does* correspond to the kinetic ratio of OPA diastereomers produced in the Wittig reaction), but in all cases investigated here the ratios were found to be identical.

The OPA produced from (1-(2-bromophenyl)-1-hydroxy-3-methylbut-2-yl)-*P*-phenyldibenzophospholium bromide (Table 2.7 entry 8) was characterised by ^{1}H-^{31}P HMBC NMR (as well as several other techniques) at −20 °C, which in tandem with the COSY spectrum of the reaction mixture facilitated the assignment of signals in the ^{1}H NMR to *trans*-OPA (the minor diastereomer in this case). Each of the OPA isomers was shown to be in its most stable pseudorotameric form(s) by ^{13}C NMR. Since the dibenzophosphole system in these OPAs must span axial and equatorial sites, there is only one possible axial site that remains. The OPA ring must also span axial and equatorial sites, and therefore the remaining axial site may only be occupied by the ring oxygen or the carbon at ring position 3. The large one bond coupling constant $^{1}J_{PC}$ for ring carbon 3 signal (82.0 Hz for the *trans*-OPA ring C−3 at $\delta = 76.1$, 85.9 Hz for the *cis*-OPA ring C−3 at $\delta = 73.3$) is indicative of an aliphatic carbon in an equatorial position in a phosphorus-centred trigonal bipyramid [29, 33], and so oxygen must be in the axial position, as would be expected for an electronegative element. The major diastereomer was also shown to be the *cis*-OPA by 1D NOESY spectroscopy. Irradiation of the sample at the resonant frequency of the OC*H* proton ($\delta = 5.50$) resulted in an NOE response from the PC*H* proton ($\delta = 5.19$), as shown in Fig. 2.5. There is also NOE contact with some of the hydrogens of the dibenzophosphole group, and at longer mixing times with the *iso*-propyl CH_3 hydrogens.

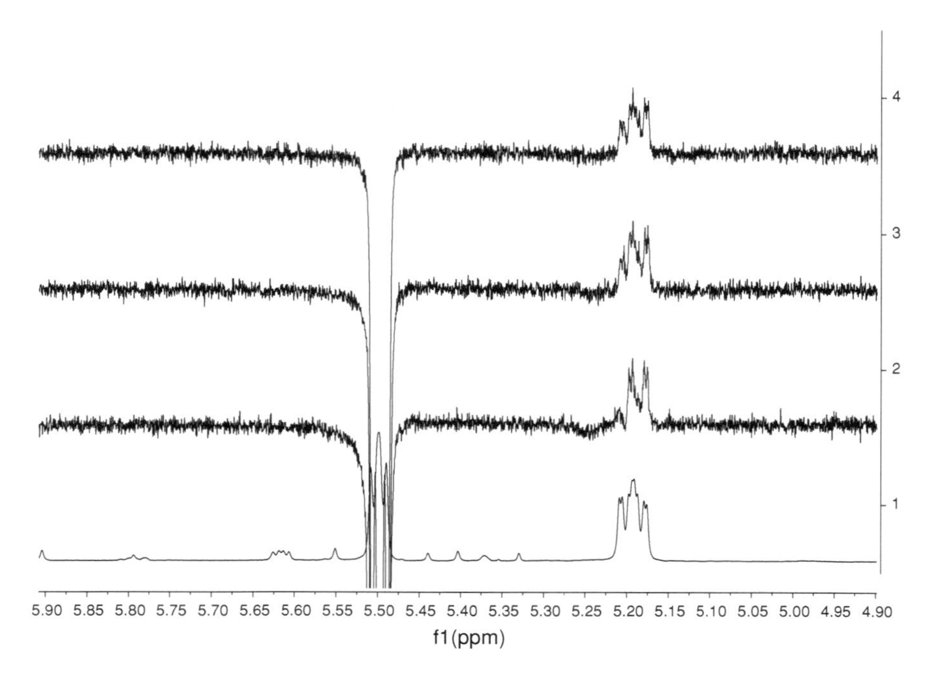

Fig. 2.5 1D NOESY experiment on the OPA generated from (1-(2-bromophenyl)-1-hydroxy-3-methylbut-2-yl)phenyldibenzophospholium bromide, irradiated at the resonant frequency of the OC*H* proton. The original ^{1}H spectrum is shown at the bottom, and also the spectra obtained with three different NOE mixing times (0.5, 1.0 and 1.5 s)

2.4.5.1 Description of the ^{1}H- ^{31}P HMBC NMR Technique

The use of ^{1}H-^{31}P HMBC to assign signals in each of the crude ^{1}H and ^{31}P spectra in the β-HPS produced in the low temperature acid quenching of the Wittig reaction of *iso*-butylidenephenyldibenzophospholane with benzaldehyde will now be described as an example of the technique. In Fig. 2.6 is shown the ^{1}H-^{31}P HMBC spectrum of the crude product optimised for a ^{1}H-^{31}P coupling constant $J_{\mathrm{PH}} = 6$ Hz.

In Fig. 2.7 is shown a close-up from the ^{1}H-^{31}P HMBC spectrum optimised for 6 Hz coupling. The two large signals in the ^{1}H spectrum (which are the highest field of the three signals shown) are shown by COSY and TOCSY spectra to be coupled; the signal at $\delta = 5.39$ is assigned to the OC*H* of the major diastereomer (shows heavily roofed double doublet due to splitting by phosphorus and PC*H*), and the signal at $\delta = 5.15$ is assigned to the PC*H* of the major diastereomer. As can be seen from the spectrum in Fig. 2.7, both couple to the phosphorus signal at $\delta_{\mathrm{P}} = 31.7$, confirming that these three signals belong to one compound. The double doublet at $\delta = 5.49$ is assigned to the C*H*OH of the minor diastereomer (it couples with PC*H* at $\delta = 4.03$). This is shown by the ^{1}H-^{31}P HMBC spectrum in Fig. 2.7 to be coupled to the phosphorus signal at $\delta_{\mathrm{P}} = 30.6$, confirming it to be the phosphorus

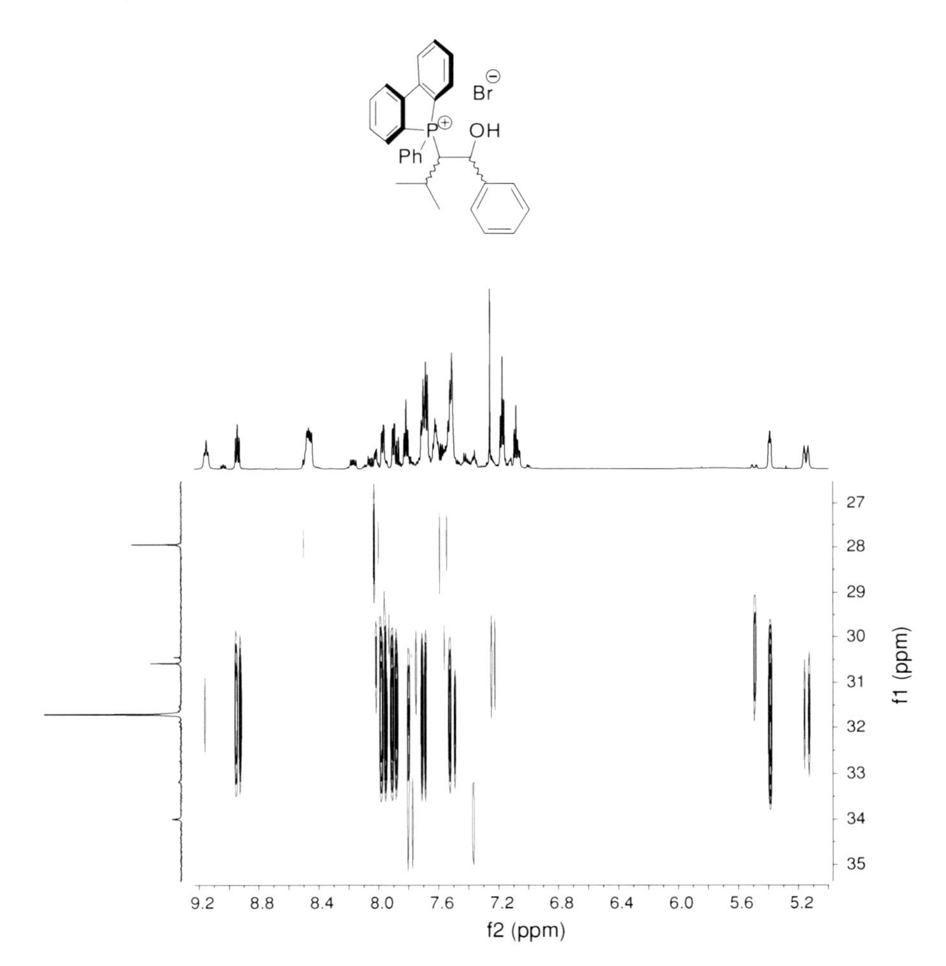

Fig. 2.6 ^{1}H-^{31}P HMBC spectrum of the crude *P*-(1-hydroxy-1-phenyl-3-methylbut-2-yl)-*P*-phenyldibenzophospholium bromide from low temperature acid quenching of the reaction of *P*-(isobutylidene)-*P*-phenyldibenzophospholane and benzaldehyde optimised for a coupling constant $J_{PH} = 6$ Hz. On the abscissa is the ^{1}H spectrum, and on the ordinate is the ^{31}P spectrum

signal of the minor diastereomer. The diastereomeric ratio obtained by comparison of the integrals of all the baseline-separated signals belonging to the major and minor diastereomers respectively in the ^{1}H NMR spectrum is 89:11, which is agreed upon by comparison of the integrals of the major and minor diastereomer signals in the ^{31}P spectrum. Since deprotonation of this β-HPS sample gave alkene with a *Z/E* ratio of 90:10, it can be surmised that the major β-HPS diastereomer is *erythro*, and hence the *erythro/threo* ratio is 89:11.

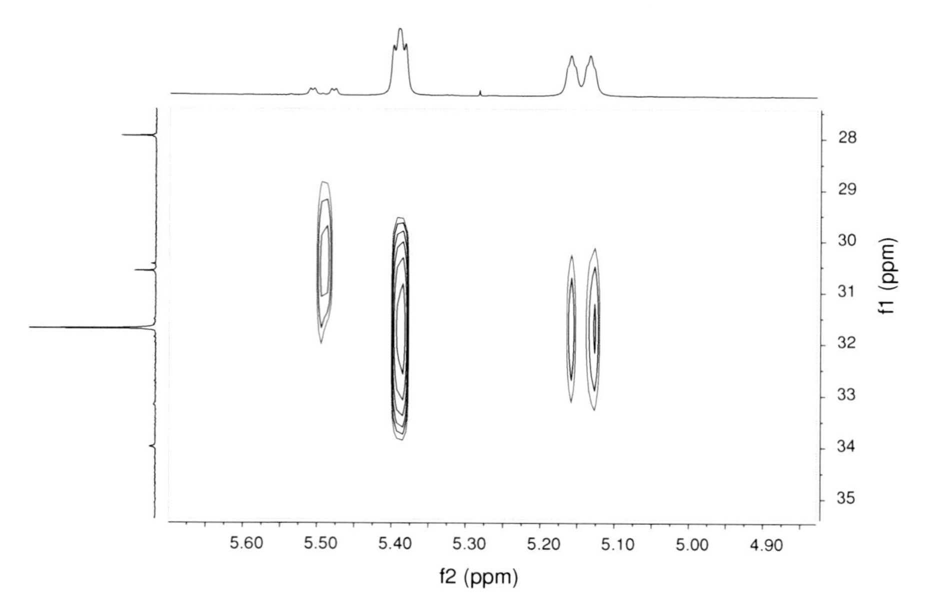

Fig. 2.7 Close-up on a region of the ^{1}H-^{31}P HMBC spectrum of the crude product from low temperature acid quenching of reaction of *P*-(isobutylidene)-*P*-phenyldibenzophospholane and benzaldehyde optimised for a coupling constant $J_{PH} = 6$ Hz. On the abscissa is the ^{1}H spectrum, and on the ordinate is the ^{31}P spectrum

2.4.5.2 Description of the ^{1}H{^{31}P} Selective Decoupling Technique

The second technique that can be used to establish connectivity between the atoms giving rise to signals in the ^{1}H and ^{31}P NMR spectra is selectively decoupled ^{1}H{^{31}P} NMR. This technique involves the selective irradiation the NMR sample with a pulse equal to the resonant frequency of a particular ^{31}P signal while acquiring ^{1}H NMR spectrum. Any signals in the ^{1}H spectrum that are coupled to the specific ^{31}P peak are decoupled and can be identified by their reduced multiplicity compared to the non-decoupled ^{1}H spectrum. As an example of this technique, it will now be described how it was used to determine connectivity between the atoms giving rise to signals in the ^{1}H and ^{31}P spectra of the crude (1-(2-bromophenyl)-1-hydroxyprop-2-yl)ethyldiphenylphosphonium bromide product obtained from the low temperature acid quenching of the Wittig reaction of *P*-(ethylidene)ethyldiphenylphosphorane and 2-bromobenzaldehyde.

Shown in Fig. 2.8 is a section of the ^{31}P spectrum of the crude product of acid quenching of the reaction of *P*-(ethylidene)ethyldiphenylphosphorane and 2-bromobenzaldehyde. There are no other signals in the spectrum. The peaks at $\delta = 32.5$ and $\delta = 34.2$ are indicated by the selective decoupling experiments to be diethyldiphenylphosphonium bromide (starting phosphonium salt) and phosphine oxide (likely to be ethyldiphenylphosphine oxide, whose ^{31}P NMR chemical shift has previously been reported as $\delta = 34$ ppm) [24].

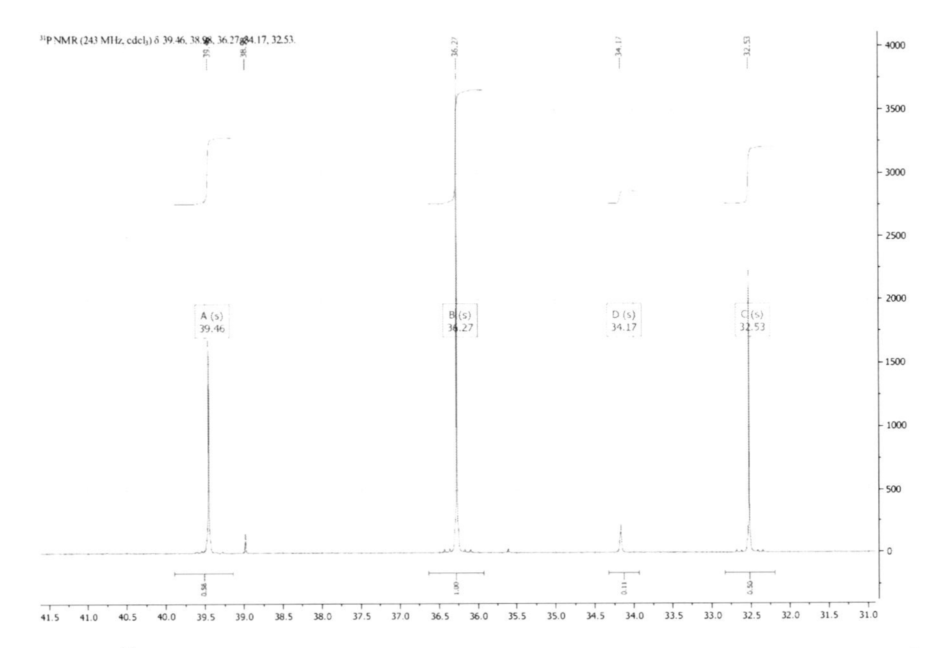

Fig. 2.8 ^{31}P NMR spectrum of the crude product of acid quenching of the reaction of (ethylidene)ethyldiphenylphosphorane and 2-bromobenzaldehyde

Two stacked spectra are shown in Fig. 2.9. The lower of these is the non-decoupled ^{1}H spectrum, which is very useful for comparison with the selectively decoupled spectrum. Running a ^{1}H spectrum while selectively irradiating the sample at the frequency of the peak at $\delta = 36.3$ gives the upper spectrum. These spectra show that the phosphorus coupling to the signals at $\delta = 5.81$ (double doublet collapses to doublet, OC*H* of major diastereomer), $\delta = 1.31$ (double triplet collapses to triplet, PCH_2CH_3 of major diastereomer), and the partially obscured signal at $\delta = 1.05$ (double doublet collapses to doublet, $PCHCH_3$ of major diastereomer) is removed, indicating that these signals are all in the same molecule as the phosphorus that gives the signal at $\delta = 36.3$ in the ^{31}P NMR. The connectivity of the molecule giving rise to this network of signals is confirmed by COSY and TOCSY spectra. The $PCHCH_3$ signal of this diastereomer at $\delta = 3.67$ (not shown in Fig. 2.9) is also observed to collapse from a double quartet of doublets to a quartet of doublets. The diastereotopic PCH_2CH_3 hydrogens of the major diastereomer appear at approximately $\delta = 3.48$ and $\delta = 3.82$, overlapping with the corresponding signals for the minor diastereomer at $\delta = 3.54$ and $\delta = 3.89$. The integrations of the various signals (where baseline separated) are equal to those of the other signals assigned to this compound, while the integrations of the combined signals of each of the overlapping PCH_2CH_3 signals add up to the sum of the integrations of two major protons and two minor protons.

In Fig. 2.10 is shown the ^{1}H spectrum of the same sample selectively decoupled from the peak at $\delta = 39.5$ (upper of the two spectra), again compared with the

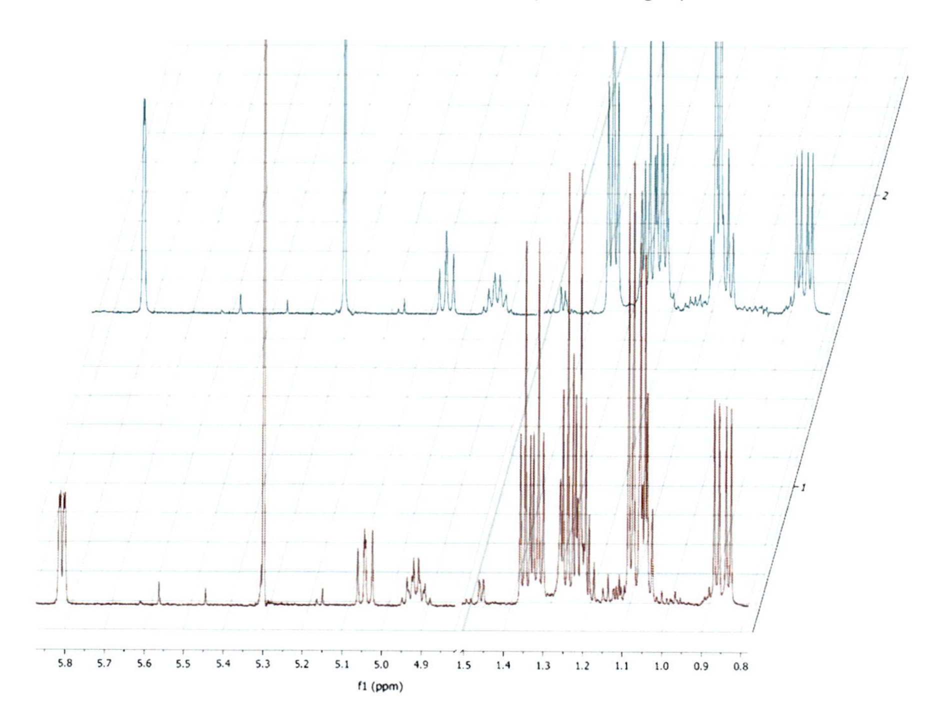

Fig. 2.9 ^{1}H spectra on the crude product of acid quenching of the reaction of (ethylidene)ethyldiphenylphosphorane and 2-bromobenzaldehyde. The bottom spectrum is the non-decoupled ^{1}H spectrum. The top spectrum is selectively decoupled from the peak at $\delta = 36.3$ in the ^{31}P spectrum. The region $\delta = 1.5$–4.9 is not shown

non-decoupled spectrum (lower spectrum). These spectra demonstrate that the phosphorus coupling to the signals at $\delta = 5.02$ (double doublet collapses to doublet, minor diastereomer C*H*OH), $\delta = 4.89$ (multiplet shows simplified coupling, minor diastereomer PC*H*) and $\delta = 0.83$ (double doublet collapses to doublet, minor diastereomer PCHCH_3) is eliminated by decoupling from the peak at $\delta = 39.5$ in the ^{31}P spectrum. This indicates that all of the hydrogens and the phosphorus involved are part of the same molecule. The assignments of these signals are confirmed by the relative magnitudes of their integrations, and by the coupling patterns indicated in the COSY and TOCSY spectra of the crude β-HPS.

In the experiment giving entry 2 in Table 2.7, (1-(2-bromophenyl)-1-hydroxyprop-2-yl)triphenylphosphonium bromide was synthesised by low temperature acid quenching of the Wittig reaction of ethylidenetriphenylphosphorane and 2-bromobenzaldehyde. The diastereomeric ratio was obtained using selective phosphorus decoupling of ^{1}H NMR spectra of the crude product, as described above. The major β-HPS isomer was then isolated by recrystallisation of the crude product from chloroform/ethyl acetate, and confirmed to be the *erythro* isomer by X-ray diffraction. Based on the NMR spectral data of *erythro*-(1-(2-bromophenyl)-1-hydroxyprop-2-yl)triphenylphosphonium bromide, and previous

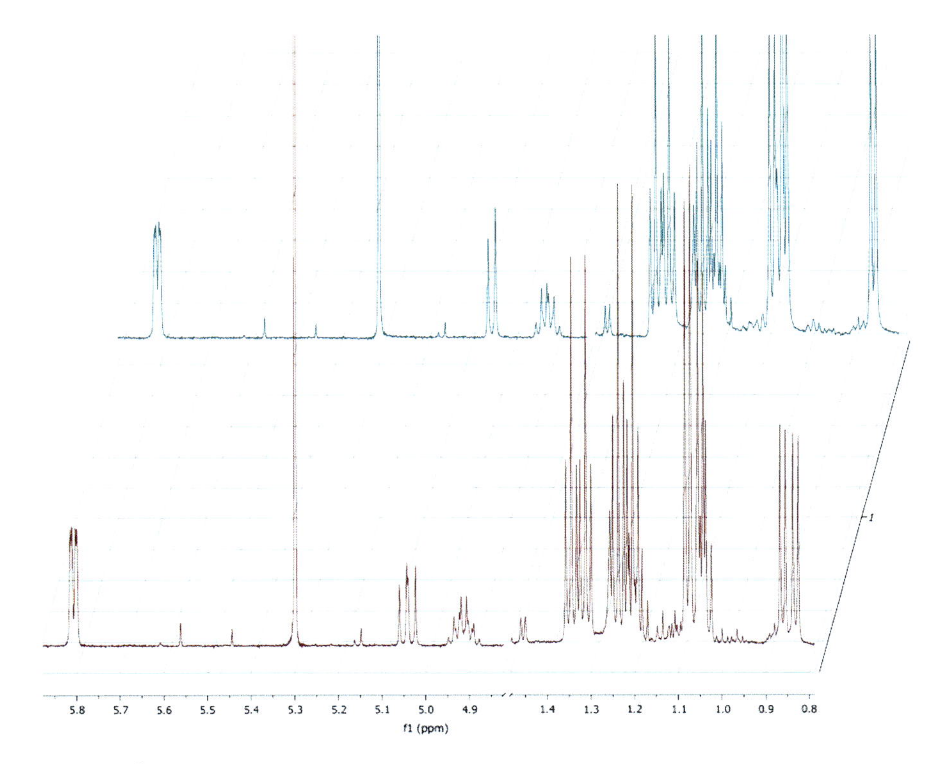

Fig. 2.10 ^{1}H spectra on the crude product of acid quenching of the reaction of (ethylidene)ethyldiphenylphosphorane and 2-bromobenzaldehyde. The bottom spectrum is the non-decoupled ^{1}H spectrum. The top spectrum is selectively decoupled from the peak at $\delta = 39.5$ in the ^{31}P spectrum. The region $\delta = 1.5$–4.9 is not shown

characterisations in the literature of the *erythro* and *threo* isomers of the closely related β-HPS (1-hydroxy-1-phenylpent-2-yl)triphenylphosphonium bromide [7], the major diastereomer in the experiment described above was assigned to be *erythro*-(1-(2-bromophenyl)-1-hydroxyprop-2-yl)ethyldiphenylphosphonium bromide, and the minor diastereomer was assigned to be the *threo* isomer. Integration of the signals belonging to these species in the ^{1}H and ^{31}P spectra of the crude product allows the *erythro/threo* ratio to be determined to be 65:35. This corresponds to the kinetic OPA *cis/trans* selectivity in the Wittig reaction.

Care must be taken in establishing connectivity by the selectively decoupled ^{1}H{^{31}P} NMR technique, as if there is another signal close to the one for which narrowband decoupling is intended then ^{1}H signals coupled to this second ^{31}P signal may also show reduced multiplicity and could thus be erroneously assigned to the wrong compound. This problem can usually be anticipated based on the signal dispersion in the ^{31}P NMR spectrum, and can often be detected by the ^{1}H signals being only *partially* decoupled—so for example a double doublet might collapse to a pseudo-triplet or a heavily roofed AB-type double doublet.

2.4.5.3 Variable Temperature ^{31}P NMR Monitoring of OPAs Produced in Reactions of P-(ethylidene)-P-phenyldibenzophospholane

The OPAs produced in the Wittig reactions of dibenzophosphole-derived ylide *P*-(ethylidene)-*P*-phenyldibenzophospholane with benzaldehyde and 2-bromobenzaldehyde respectively (the same reactions as were quenched at low temperature for entries 5 and 6 of Table 2.7) were monitored by variable temperature ^{31}P NMR. All manipulations were carried out under an inert atmosphere. The ylide was generated in situ from the precursor phosphonium salt (by the procedure described in Sect. 2.4.2) and reacted with aldehyde at −78 °C. The solution of OPA produced was stirred for 10 min. Approximately 0.6 ml of the OPA solution was then cannula filtered into an NMR tube (in a long Schlenk flask) at −40 °C and toluene-*d8* was added. The NMR tube was kept at −40 °C until such time as it could be placed in the NMR spectrometer at −20 °C. ^{31}P spectra were taken at every 10 °C between −20 and 40 °C, allowing sufficient time at each temperature for the solution to equilibrate before acquisition. The *cis*-OPA was determined to be the major OPA diastereomer in each reaction by subsequent decomposition of the mixture of OPAs to predominantly *Z*-alkene at higher temperature, and by comparison with the low temperature acid quenching experiments of Table 2.6 (entries 5 and 6). In each reaction, four major species were observed in the ^{31}P NMR—OPA (*cis* and *trans*, $\delta = -60$ to -70), ylide (integrates for ca. 15 % vs. *cis*-OPA, $\delta = -10.0$), and phosphine oxides (two signals, each integrate for ca. 5 % vs. *cis*-OPA). There were also up to two minor signals in the OPA region, each integrating for ca. 1 % vs. the *cis*-OPA. It is proposed that the small signals in the OPA region observed here indicate that OPA pseudorotamers have been resolved in these spectra. The resolution of OPA pseudorotamers by low temperature NMR has been reported previously for OPAs derived from unconstrained phosphines and also from *P*-methyl-5*H*-dibenzophosphole [29, 30, 35]. Bangerter et al. concluded in their publication that dibenzophosphole-derived OPAs which show only one set of signals by low temperature ^{13}C NMR have one pseudorotamer (with an apical oxygen) that is the predominant solution structure, rather than the single set of signals being the average of two rapidly converting pseudorotamers (see Sect. 1.4.3 for further details) [35]. The major pseudorotamer for each of the *cis* and *trans*-OPAs in each case is very likely to be one of the O-apical pseudorotamers. The presence of more than one pseudorotamer for each of the *cis* and *trans*-OPAs may in principle affect the assigned *cis/trans* ratios, but because the minor pseudorotamers are present to such a small extent this effect is negligible. In any case, the accuracy of the OPA *cis/trans* ratios determined by ^{31}P NMR (Table 2.7) has been confirmed by the determination of the *erythro/threo* ratios of the β-HPSs produced by low temperature acid quenching of the same reactions (Table 2.6).

It can be seen from the set of stacked spectra shown in Fig. 2.11 for the reaction of *P*-(ethylidene)-*P*-phenyldibenzophospholane with 2-bromobenzaldehyde that very significant broadening of the OPA peaks occurs as the temperature is raised, with the *trans*-OPA spreading into the baseline to such an extent that above 10 °C

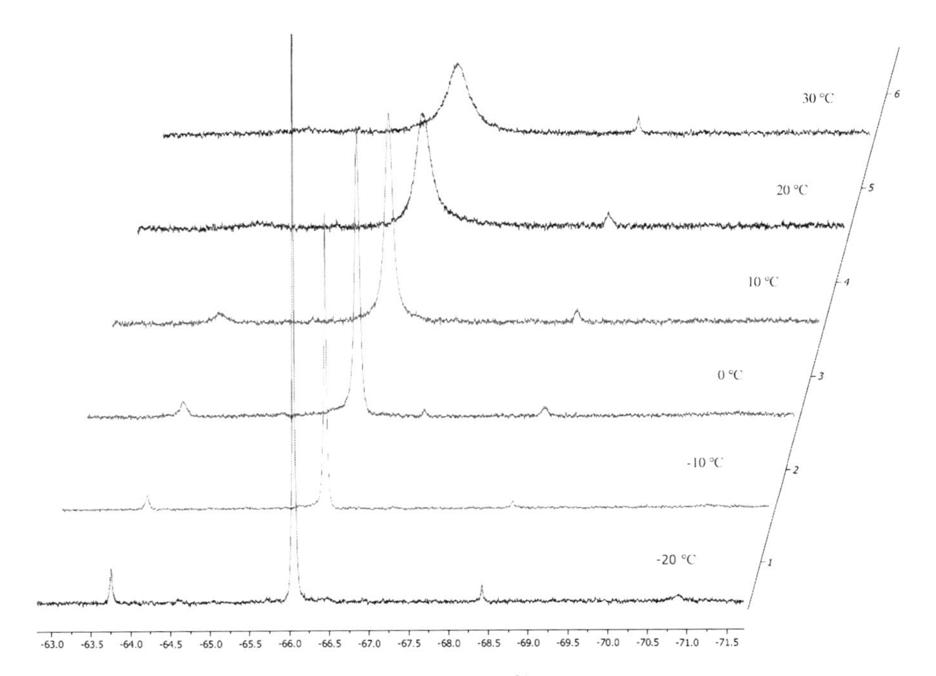

Fig. 2.11 OPA region of Variable Temperature ^{31}P NMR of the reaction of ethylidene phenyldibenzophospholane with 2-bromobenzaldehyde

its presence can barely be detected. The peaks of the phosphine oxide and ylide (not shown) remain the same shape at each temperature for which a ^{31}P spectrum was obtained. The relative proportions of OPA, ylide, and phosphine oxide are also invariant over the whole range of temperatures, indicating at least that no decomposition to alkene and phosphine oxide (which is irreversible) is occurring. The relative proportions of the *cis* and *trans*-OPA can only be ascertained up to 10 °C due to signal broadening, which results in the two signals overlapping to some degree. Up to 10 °C the OPA *cis/trans* ratio is invariant at 94:6.

A very similar set of observations can also be made about the corresponding experiment involving *P*-(ethylidene)-*P*-phenyldibenzophospholane and benzaldehyde, with the exception that significantly more *trans*-OPA is produced in that reaction, and it appears that the *cis*-OPA signal does not broaden to the same extent as the *trans*-OPA, nor indeed to the same extent as the *cis*-OPA in Fig. 2.11.

2.4.6 Discussion

The results presented in Tables 2.6 and 2.7 show that the reactions of non-stabilised ylides with *ortho*-heteroatom substituted benzaldehydes are more selective for *cis*-OPA than are reactions of the same ylides with benzaldehyde. Thus the *ortho*-heteroatom effect is demonstrated for non-stabilised ylides. The absence of phosphorus-heteroatom bonding in the OPAs intermediates of these reactions has been demonstrated by NMR. This implies that a planar structure cannot engage in phosphorus-heteroatom bonding. Consequently, the *cis*-selective TS with this bonding interaction is likely to be puckered. It is thus postulated that the TS for the cycloaddition step in these reactions is entirely analogous to those shown in Figs. 2.1 and 2.2.

The average selectivity of the Wittig cycloaddition step for *cis*-OPA in reactions of non-stabilised ylides with benzaldehyde is 76 % (calculated from Table 2.6 entries 1, 3, 5, and 7), while in reactions of the same ylides with 2-bromobenzaldeyde it is 87 % (calculated from Table 2.6 entries 2, 4, 6, and 8). Thus, even though reactions of non-stabilised ylides with benzaldehdyes show a relatively high inherent selectivity for *cis*-OPA, the reactions of *ortho*-heteroatom substituted benzaldehyde are on average 11 % more *cis*-selective.

2.5 Reactions of an Aliphatic Aldehyde Showing Increased Z-Selectivity Due to a "β-Heteroatom Effect"

The results described above are common to all three ylide classes. They are self-consistent and can all be explained by the same transition state arguments. Therefore they argue strongly for a common mechanism for Wittig reactions of all ylide types. However, it could be argued that the effect is solely confined to *ortho*-heteroatom benzaldehydes and might not extend to other aldehydes. Many Wittig reactions of aliphatic aldehydes bearing heteroatoms on the carbon β to the carbonyl have been shown to give alkene with anomalously high *Z*-selectivity. A number of examples are given in Sect. 1.5.2. Consequently, it was necessary to investigate whether high *Z* or *cis*-selectivity would be observed in Wittig reactions under our conditions of strict kinetic control with an aldehyde that had previously been shown to induce high *Z*-selectivity in its reactions with semi-stabilised and stabilised ylides, 1,2-O-isopropylidene-3-O-methyl-α-D-xylopentodialdofuranose-(1,4) [18, 19]. The carbonyl group of this aldehyde is a substituent on a five-membered ring and there is a β-methoxy substituent oriented *cis* with respect to the carbonyl, so the relative disposition of the carbonyl and the heteroatom is almost identical to that in *ortho*-heteroatom substituted benzaldehydes. The *Z/E* ratios for the alkenes produced in the reactions of this aldehyde with representative non-stabilised, semi-stabilised and stabilised ylides at −78 °C in THF under Li-salt free conditions are shown in Table 2.8. *P*-(isobutylidene)-*P*-phenyldibenzophospholane was also reacted with cyclopentanecarboxaldehyde under the same conditions in order to ascertain its selectivity with a non-heteroatom-bearing aliphatic aldehyde. In addition, the kinetic OPA *cis/trans* ratio for the reactions of the non-stabilised dibenzophosphole-derived ylide with these aldehydes was determined by ^{31}P NMR.

Table 2.8 *Z/E* ratios for Wittig reactions of non-stabilised, semi-stabilised and stabilised ylides (generated in situ from phosphonium salt using NaHMDS) with aliphatic aldehydes. OPA *cis/trans* ratios were also determined for reactions of non-stabilised ylides

Entry	Aldehyde R^1	Ylide R^2	Ylide R_3P group	Alkene Z/E ratio
1		Ph	$MePh_2P$	95:5
2		2-BrC6H4	$MePh_2P$	95:5
3		COOMe	$MePh_2P$	79:21[a]
4		COO(t-Bu)	$MePh_2P$	79:21[a]
5	MeO	COOEt	$MePh_2P$	79:21[a]
6		i-Pr	PhDBP	90:10[b]
7	cyclopentyl	i-Pr	PhDBP	43:57[c]

[a] Reactions of stabilised ylides were quenched at −78 °C by addition of aqueous NH_4Cl in order to ensure the reaction had occurred at this temperature
[b] The kinetic OPA *cis/trans* ratio was observed by ^{31}P NMR at −20 °C and found to be 94:6
[c] The kinetic OPA *cis/trans* ratio was observed by ^{31}P NMR at 30 °C and found to be 45:55

1,2-O-isopropylidene-3-O-methyl-α-D-xylopentodialdofuranose-(1,4) was found to react with semi-stabilised ylides with very high *Z*-selectivity (Table 2.8 entries 1 and 2). This aldehyde also reacted with ester-stabilised ylides (Table 2.8 entries 3–5) with remarkably high *Z*-selectivity, particularly for reactions of stabilised ylides in an aprotic solvent. The kinetic OPA *cis/trans* ratio for the reaction of *P*-(isobutylidene)-*P*-phenyldibenzophospholane and 1,2-O-isopropylidene-3-O-methyl-α-D-xylopentodialdofuranose-(1,4) was determined by ^{31}P NMR at −20 °C and found to strongly favour the *cis*-OPA.

A sample of *β*-HPS heavily enriched in the *erythro*-isomer was obtained by low temperature acid quenching of the same reaction. Deprotonation of this *β*-HPS at low temperature and subsequent heating of the OPA to 80 °C for 2 h gave alkene with a *Z/E* ratio of 90:10 (as indicated in Table 2.8 entry 6). The high stereospecificity of OPA decomposition to alkene indicates that OPA formation is under dominant kinetic control. *P*-(isobutylidene)-*P*-phenyldibenzophospholane was also reacted with cyclopentanecarboxaldehyde, and the kinetic OPA *cis/trans* ratio was determined by ^{31}P NMR at 30 °C. This ratio was found to be almost the same as the *Z/E* ratio of the alkene produced after heating of the OPA to effect its decomposition to alkene and phosphine oxide (Table 2.8 entry 7). Kinetic control is thus in operation in this reaction, and stereospecific OPA decomposition is also proven. The results of Table 2.8 entries 6 and 7 show that there is a very dramatic jump in selectivity for *cis*-OPA, and hence *Z*-alkene, in reactions of *P*-(isobutylidene)-*P*-phenyldibenzophospholane in its reactions with aldehydes bearing a suitably disposed β-heteroatom substituent.

The *Z*-selectivity observed in the reactions of 1,2-O-isopropylidene-3-O-methyl-α-D-xylopentodialdofuranose-(1,4) with representative stabilised, semi-stabilised and non-stabilised ylides is of comparable magnitude to that observed in the corresponding reactions of *ortho*-heteroatom substituted benzaldehydes with the same ylides. This selectivity is far greater than what would be expected for reactions of these ylides involving an aldehyde that lacks a suitably disposed β-heteroatom. The magnitude of the enhancement of *Z*-selectivity in these reactions is emphasised by the moderate *E*-selectivity observed in the kinetically controlled reaction of *P*-(isobutylidene)-*P*-phenyldibenzophospholane with cyclopentane-carboxaldehyde, which lacks a β-heteroatom substituent. There is no apparent influence on the magnitude of the *Z*-selectivity in reactions of 1,2-O-isopropylidene-3-O-methyl-α-D-xylopentodialdofuranose-(1,4) with representatives of a certain class of ylide from the steric bulk of the substituent on the ylide α-carbon. The high *Z*-selectivity observed in these reactions strongly implies that reactions all occur under kinetic control, especially given that they were carried out under conditions for which the operation of kinetic control in Wittig reactions has been verified. Indeed, the operation of dominant kinetic control was directly demonstrated for the reactions of the non-stabilised ylide.

In all of the alkenes in Table 2.8 derived from 1,2-O-isopropylidene-3-O-methyl-α-D-xylopentodialdofuranose-(1,4), no coupling is observed between H-2 and H-3 (as confirmed by multiplicity of these signals by ^{1}H NMR, and the lack of cross-signals in gCOSY and TOCSY spectra), exactly as in the precursor aldehyde. This implies that the conformation of the ring is such that H-2 and H-3 are oriented at approximately 90° with respect to each other. Examination of the NOESY spectrum of the crude product from the reaction of the aldehyde with benzylidenemethyldiphenylphosphorane (shown in Fig. 2.12 for a sample containing a significant amount of the aldehyde starting material, blue = phase down, yellow = phase up) shows that H-2 experiences NOE contact with *both* H-3 and OMe, which are geminal substituents. This is consistent with it having a dihedral angle close to 90° with each of H-3 and OMe. The NOESY spectrum shows NOE contact between H-3 and H-4, indicating that they remain *cis* to each other, as in

Fig. 2.12 NOESY spectra of the alkene derived from the reaction of benzylidenemethyldiphenylphosphorane and 1,2-O-isopropylidene-3-O-methyl-α-D-xylopentodialdofuranose-(1,4). **a** Close-up on 3–5 ppm region of the 2D NOESY spectrum, showing NOE contact for alkene H-3 with H-4, thus showing relative *cis* geometry of these hydrogens, and for H-2 (whose signal overlaps with aldehyde H-2 in ^{1}H spectrum) with both H-3 and OCH_3, which are geminal substituents. Note that an identical set of interactions are shown in this spectrum to be present in the aldehyde starting material. **b** Close-up on the alkene region of the 2D NOESY spectrum, showing NOE contact between the *Z*-alkene hydrogens ▶

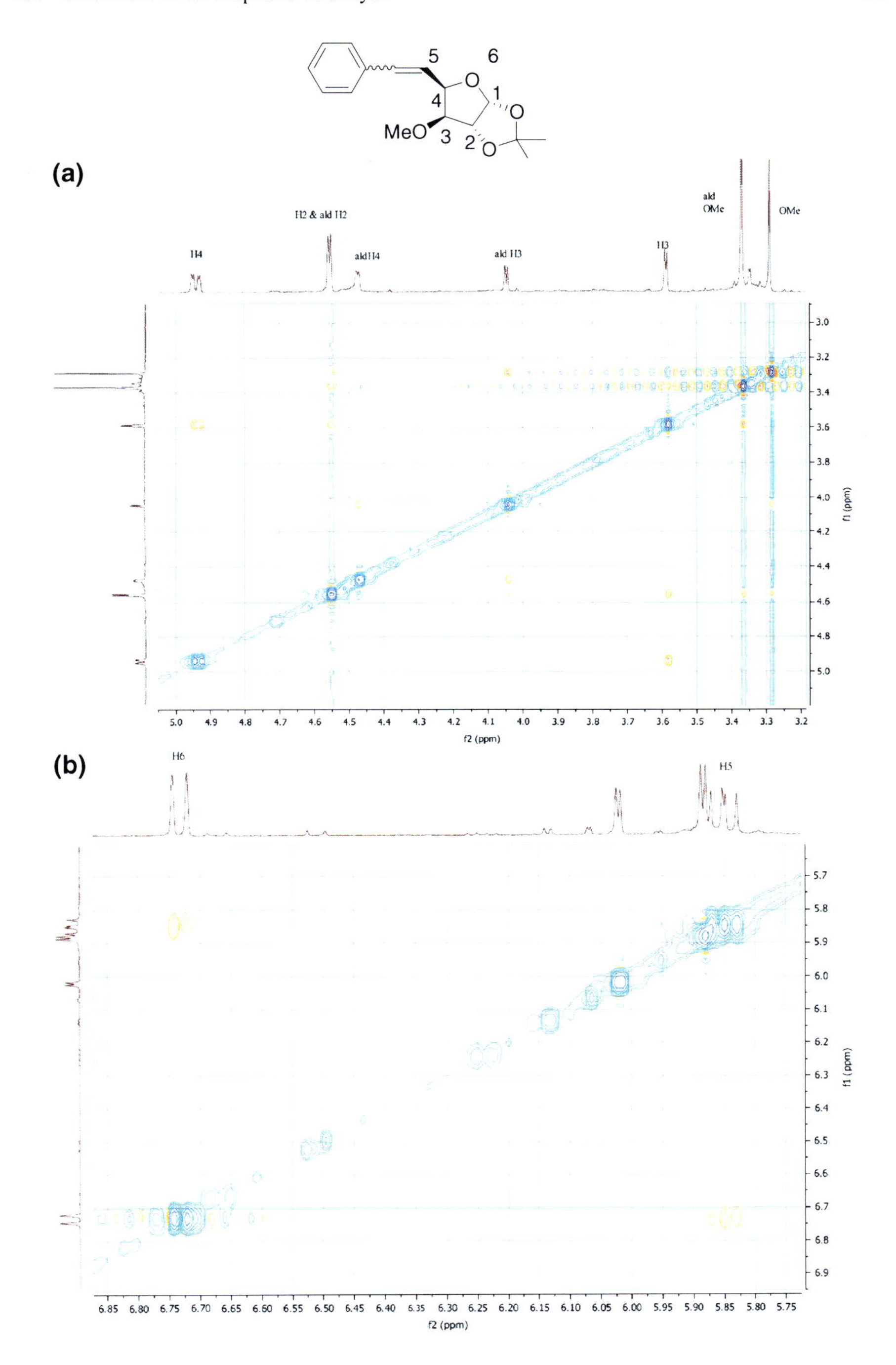
5
6
O
1
4
O
MeO
3
2
O
(a)
H2 & ald H2
ald
OMe
OMe
H4
ald H4
ald H3
H3
f2 (ppm)
f1 (ppm)
(b)
H6
H5
f2 (ppm)
f1 (ppm)

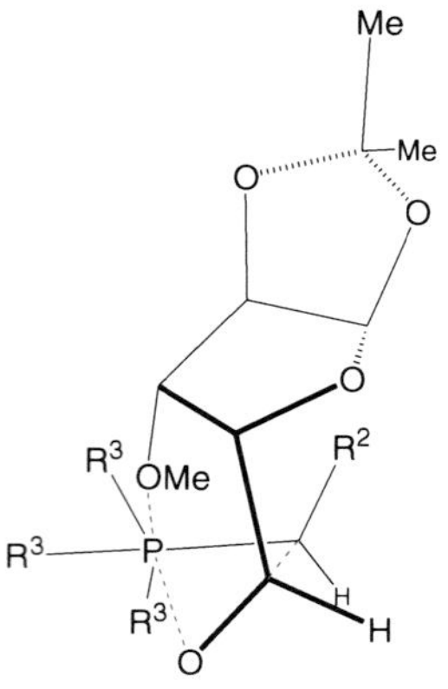

Fig. 2.13 Diagram of the proposed *cis*-selective transition state in reactions of 1,2-O-isopropylidene-3-O-methyl-α-D-xylopentodialdofuranose-(1,4)

the aldehyde, and thus that there was no epimerisation at the aldehyde α-carbon under the reaction conditions. The NOESY spectrum also confirms that the alkene is of *Z*-configuration, since there is NOE contact between H-5 and H-6.

The conservation of the relative stereochemistry of the aldehyde in the alkene product (and therefore the non-epimerisation at the aldehyde α-carbon) and was confirmed in a similar manner using 1D or 2D NOESY for the *Z*-isomer of the alkene of Table 2.8 entry 2, for the *E*-isomer of alkene of entry 4, and for both the *E* and *Z* isomers of the alkene of entry 3 (see Sect. 4.8 for full details).

Given that the same trends are observed in the reactions of this aliphatic aldehyde that bears a β-heteroatom substituent as are observed in reactions of *ortho*-heteroatom substituted benzaldehydes, it is reasonable to conclude that the same effect is in operation in both types of reaction. The kinetic preference for *cis*-OPA formation in these reactions is thus postulated to be as a consequence of the existence of a phosphorus-heteroatom bond in the *cis*-selective cycloaddition TS. As with the TS model proposed for reactions of *ortho*-heteroatom substituted benzaldehydes, this TS is thought to be *cis*-selective due to the bonding interactions and the minimisation of 2–3 steric interactions that comes about if the reactants are arranged in the cycloaddition TS as shown in Fig. 2.13. This TS would also benefit from a favourable antiparallel orientation of the ylide C–C(R^2) and aldehyde C = O bond dipoles, especially for reactions of stabilised ylides (with R^2 = COOR and R = alkyl).

2.6 Conclusions

All of the reactions described in this chapter were carried out under conditions for which the operation of kinetic control in Wittig reactions has been proven. In the reactions involving non-stabilised ylides the irreversibility of OPA formation at the temperature at which the *cis/trans* ratio was measured was directly demonstrated by NMR. Thus it can be concluded that each of the results is truly indicative of the kinetic selectivity for *cis* or *trans* OPA in that reaction. A phenomenon that has its origins in the existence of a bonding interaction in the TS leading to OPA and results in the formation of *cis*-OPA and hence *Z*-alkene has been discovered in Wittig reactions of aldehydes bearing a suitably oriented β-heteroatom. This effect has been shown to be in operation in reactions of representatives of all three major types of ylide, and for both aromatic and aliphatic aldehydes.

That an effect that is undoubtedly common to Li-salt free Wittig reactions of all types of ylide has been uncovered indicates at the very least that a common mechanism is in operation in all such reactions. The reactions described are all under kinetic control, and thus there is no contribution to the selectivity from equilibration of intermediates. In particular, the high *Z*-selectivity observed in the reactions of stabilised ylides indicates that these are under kinetic control. It is unlikely that there is any difference in the OPA intermediates produced in the reactions of β-heteroatom substituted aldehydes and those produced in reactions of aldehydes lacking such a substituent based on the NMR studies of the OPAs produced in reactions of non-stabilised ylides. Furthermore, the *cis*-selective TS is clearly lower in energy in reactions of β-heteroatom substituted aldehydes than in reactions of analogous unsubstituted aldehydes, and thus if OPA equilibration by Wittig reversal was of significant mechanistic significance, it would be *more likely* to exert an effect in these reactions than otherwise since the barrier to OPA reversal is lower than it would otherwise be.

The results presented here also provide convincing evidence that there is no role played by betaines *in any Li-salt free Wittig reaction.* An increased tendency towards betaine equilibration has previously been suggested as a possible reason why reactions of stabilised ylides are so consistently *E*-selective. This possibility would also require ring closure to OPA to be faster for *threo*-betaine than for the *erythro*-isomer, given that OPA equilibration has been ruled out for reactions of all ylide types [6, 16]. Betaine equilibration is highly unlikely to be in operation in the reaction of one aldehyde and not in another that only differs in having a β-heteroatom substituent. The operation of kinetic control in the reactions of stabilised ylides reported here implies that all Li-salt free Wittig reactions of stabilised ylides are irreversible processes.

Since the involvement of betaine intermediates or equilibration of OPAs have been demonstrated not to play a part in the Wittig reactions of β-heteroatom substituted aldehydes,[9] it can be concluded that these reactions occur by kinetically controlled cycloaddition of ylide and aldehyde to give OPA, which decomposes stereospecifically to alkene and phosphine oxide. The TS model for the cycloaddition step of these reactions was initially proposed in order to

[9] OPA equilibration does occur in reactions of benzaldehydes with non-stabilised ethylides, but only at or above the temperature at which alkene formation occurs. Below this temperature, OPAs are formed under kinetic control.

rationalise the observed selectivity in reactions of semi-stabilised ylides with *ortho*-heteroatom substituted benzaldehydes. Based on this, high *Z*-selectivity was predicted in the reactions of non-stabilised, semi-stabilised and stabilised ylides with any aldehyde bearing a suitably oriented β-heteroatom. This prediction was tested and proved to be correct in the reactions that have been described above. The observed results are all entirely consistent with the operation of the cycloaddition mechanism under Li-salt free conditions, and with the existence of a phosphorus-heteroatom bond in the cycloaddition TS leading to *cis*-OPA. Since a cycloaddition mechanism operates in this subset of Wittig reactions that includes examples using all three different types of ylide, it is concluded that all Li-salt free Wittig reactions proceed by the cycloaddition mechanism proposed by Vedejs [6, 16], and modified by Aggarwal, Harvey and co-workers for reactions of stabilised ylides [23–25].

References

1. Harcken C, Martin SF (2001) Org Lett 3:3591
2. Cotter J, Hogan A-ML, O'Shea DF (2007) Org Lett 9:1493–1496
3. Vedejs E, Peterson MJ (1994) In: Eliel EL, Wilen SH (eds) Topics in stereochemistry, vol 21. Wiley, New York
4. Dunne EC, Coyne EJ, Crowley PB, Gilheany DG (2002) Tetrahedron Lett 43:2449
5. Bera R, Dhananjaya G, Singh SN, Kumar R, Mukkanti K, Pal M (2009) Tetrahedron 65:1300–1305
6. Vedejs E, Marth CF, Ruggeri R (1988) J Am Chem Soc 110:3940
7. Maryanoff BE, Reitz AB, Mutter MS, Inners RR, Almond HR Jr, Whittle RR, Olofson RA (1986) J Am Chem Soc 108:7664
8. Reitz AB, Nortey SO, Alfonzo DJ Jr, Mutter MS, Maryanoff BE (1986) J Org Chem 51:3302
9. Ward WJ, McEwen WE (1990) J Org Chem 55:493
10. Appel M, Blaurock S, Berger S (2002) Eur J Org Chem 1143
11. Oh JS, Kim BH, Kim YG (2004) Tetrahedron Lett 45:3925
12. Johnson AW (1993) Ylides and imines of phosphorus. Wiley, New York, chapters 8 and 9 pp 221–305
13. Maryanoff BE, Reitz AB, Mutter MS, Inners RR, Almond HR Jr (1068) J Am Chem Soc 1985:107
14. Vedejs E, Meier GP, Snoble KA (1981) J Am Chem Soc 103:2823
15. Reitz AB, Mutter MS, Maryanoff BE (1873) J Am Chem Soc 1984:106
16. Vedejs E, Fleck T (1989) J Am Chem Soc 111:5861
17. Ward WJ, McEwen WE (1990) J Org Chem 55:493
18. Maryanoff BE, Reitz AB (1989) Chem Rev 89:863
19. Tronchet JMJ, Gentile B (1979) Helv Chim Acta 62:2091
20. Valverde S, Martin-Lomas M, Herradon B, Garcia Ochoa S (1987) Tetrahedron 1987:43
21. Brimacombe JS, Hanna R, Kabir AKMS, Bennett F, Taylor ID (1986) J Chem Soc: Perkin Trans 1:815
22. Brimacombe JS, Kabir AKMS (1986) Carbohydr Res 150:35
23. Robiette R, Richardson J, Aggarwal VK, Harvey JN (2005) J Am Chem Soc 127:13468
24. Robiette R, Richardson J, Aggarwal VK, Harvey JN (2006) J Am Chem Soc 128:2394
25. Harvey JN (2010) Faraday Discuss 145:487

26. Shriver DF, Drezdzon MA (1986) The manipulation of air sensitive compounds, 2nd edn. McGraw-Hill, New York
27. Gilheany DG (1994) Chem Rev 94:1339
28. Vedejs E, Marth CF (1987) Tetrahedron Lett 28:3445
29. Vedejs E, Marth CF (1989) J Am Chem Soc 111:1519
30. Vedejs E, Marth CF (1990) J Am Chem Soc 112:3905
31. Kawashima T, Watanabe K, Okazaki R (1997) Tetrahedron Lett 38:551
32. Matsukawa S, Kojima S, Kajiyama K, Yamamoto Y, Akiba K-y, Re S, Nagase S (2002) J Am Chem Soc 124:13154
33. Kay PB, Trippett S (1986) J Chem Res (S) 6
34. Renard P-Y, Vayron P, Mioskowski C (2003) Org Lett 5:1661
35. Bangerter F, Karpf M, Meier LA, Rys P, Skrabal P (1998) J Am Chem Soc 120:10653

Chapter 3
A Convenient Chromatography-Free Method for the Purification of Alkenes Produced in the Wittig Reaction

3.1 Introduction: Existing Methods for Phosphine Oxide Removal

Phosphines and reagents derived from them are ubiquitous in organic chemistry [1]. Commonly used applications range from the synthesis of alkenes by the Wittig reaction, the synthesis of alkyl halides from alcohols by Appel or Mitsunobu reactions, and the synthesis of amines by the Staudinger reaction. Phosphine ligands are also widely used as ligands in the metal complexes used to catalyse reactions such as the Heck, Suzuki, and Sonagashira reactions among others. The reactions mentioned that involve the use of stoichiometric amounts of the phosphine or phosphine-derived reagent commonly result in the production of stoichiometric amounts of phosphine oxide as by-product. The reactions where phosphines are typically used as ligands in catalytic metal complexes (and are thus present in sub-stoichiometric amounts) often involve the generation of a catalytic complex by in situ reduction of the metal precursor (e.g. $Pd(OAc)_2$) by phosphine, producing phosphine oxide as by-product. The removal of the phosphine oxide by-product of all of these reactions is often not straightforward [2], particularly for large-scale reactions. The currently available methods by which the problem of phosphine oxide removal (and in some cases regeneration) is addressed are exemplified by those used for the purification of alkenes produced in the Wittig reaction. These are reviewed below. Some details of the methods by which the problem is dealt with for crude products from other types of reaction are also briefly discussed.

The Wittig reaction is applied in the synthesis of alkenes on an industrial scale. In the BASF process for carotenoid synthesis [3], the alkene product is separated from the bulk of the phosphine oxide by-product (after acidification using sulphuric acid) by extraction of the gel-like crude product with hydrocarbon solvents, and then scrubbed of the remaining phosphine oxide in a second extractive

P. A. Byrne, *Investigation of Reactions Involving Pentacoordinate Intermediates*, Springer Theses, DOI: 10.1007/978-3-642-32045-3_3,

column with aqueous alcohol. The purification of the alkene is completed using a falling-film evaporator.

The phosphine oxide produced in industrial scale reactions may either be disposed of by conversion to phosphorus pentoxide (P_2O_5) by incineration and hence to calcium phosphate [4], or may be used to regenerate phosphine, which can be used again to form phosphonium salt [3, 4]. Phosphine regeneration is achieved in the BASF process for carotenoid synthesis by reaction of triphenylphosphine oxide with phosgene to give triphenylphosphine dichloride, which is then treated with elemental phosphorus to give triphenylphosphine and phosphorus trichloride (the latter of which is the starting material for the synthesis of triphenylphosphine) [3]. Triphenylphosphine dichloride can also be converted to triphenylphosphine by mixing with finely divided aluminium metal in chlorobenzene. This process forms a triphenylphosphine-aluminium trichloride complex, which is decomposed hydrolytically. Phase separation allows isolation of triphenylphosphine (chlorobenzene phase) from the water-soluble aluminium-containing product, $Al(OH)Cl_2$ [4]. These processes involve the handling of hazardous reagents and products, and are far from straightforward.

Other methods for removal of phosphine oxide from the crude product of a Wittig reaction have also been reported. One involves the addition of a hydrocarbon solvent and a carboxylic acid to the crude product, allowing separation of alkene (in organic phase) and phosphine oxide (in acid phase) by phase separation [5]. In some cases, the alkene can be crystallised, thus facilitating its separation from the other materials in the crude product [6].

On a laboratory scale, phosphine oxide removal from the crude alkene product of a Wittig reaction generally necessitates the use of column chromatography. This presents problems if the product is sensitive to decomposition or isomerisation through contact with the stationary phase or by exposure to light, or if it is sensitive to decomposition in air. For example, certain alkenes produced in Wittig reactions have been shown to undergo isomerisation by light, or by contact with silica or alumina [6]. The products of Appel, Mitsunobu, Staudinger and other reactions may also be sensitive to decomposition if subjected to similar conditions. Thus, there is a strong imperative for the development of general techniques allowing phosphine oxide to be removed from a reaction crude product without having to resort to chromatography. Several methods have been reported that attempt to address this need.

Phosphonium ylides have been developed that are modified in some way relative to the corresponding triphenylphosphine-derived ylide so as to furnish a phosphine oxide by-product that may be removed by aqueous work-up [7–10], or by its filtration [11]. In the latter case, the modified "ion-supported" phosphine oxide was reacted with dimethylsulfate and then $LiAlH_4$ to regenerate phosphine. Alternatively, water-soluble and hydrolytically stable ylides have been utilised to allow the reaction to be conducted in water, giving water-soluble phosphine oxide and insoluble alkene product, which precipitates and can be isolated by filtration [12, 13]. Reactions involving polymer-supported ylides have also been reported. Examples include the studies of Westman [14] and Leung et al. [15] in which the ylide is generated in situ from a polymer supported phosphine. Although the use of polymer supported ylides (and hence the production of polymer-bound phosphine oxide) eliminates the difficulty of phosphine oxide removal, these reactions suffer from the necessity for long reaction times and in some cases poor yield due to problems with polymer swelling, and also from the fact that the resin may have to be synthesised prior to its use. These problems (perhaps with the exception of the long reaction times) have been addressed effectively, at least for one-pot Wittig reactions involving in situ generation of stabilised ylides, by the use of a bifunctional polymer containing phosphine and amine groups, the latter of which ultimately intramolecularly deprotonated the phosphonium salt formed by nucleophilic addition of the phosphine group to an α-halo acetate [16].

Although these methods obviate the difficulty of phosphine oxide removal, they have drawbacks of their own. In all cases the starting phosphine (free or polymer-bound) is either more expensive than triphenylphosphine, or must be synthesised by non-trivial routes, often under conditions requiring the exclusion of air and moisture. Perhaps more serious is that Wittig reactions of the modified phosphonium ylides may not necessarily be selective for the desired isomer of alkene, and the phosphorus substituents cannot easily be tuned to counter this without the undertaking of a much more onerous synthetic procedure to obtain the required phosphonium ylide.

By far the most promising potential solution to the problem of phosphine oxide removal has been the development of processes that are catalytic in the phosphorus-containing entity [2]. The Wittig reaction has for the first time been effected with a catalytic amount of phosphorus reagent (added as phosphine oxide) by using diphenylsilane to reduce the phosphine oxide to phosphine, importantly leaving the carbonyl compound unaffected [17]. Unfortunately, it requires the use of a cyclic phosphine, and does not work with triphenylphosphine. The alkene product was also found to isomerise to the *E*–isomer under the reaction conditions used. However, in as yet unpublished work, the scope of this important development has been further extended to include reactions of semi-stabilised and non-stabilised ylides by O'Brien and co-workers.

The problem of isomerisation of the alkene product is one that is related to that of phosphine oxide removal. Many examples exist where the *Z*-isomer (typically) initially formed by the Wittig reaction is caused to isomerise to the more stable *E*-isomer. Isomerisation can be induced by the presence of acids [18], strong bases [19], the chromatographic stationary phase used [6], the solvent, heat and sunlight. For example, the *Z*-enones synthesised during this project (see Sect. 4.6.2) were found to be extremely sensitive to isomerisation, with samples that had contained only *Z*-enone isomerising to the *E*-isomer when simply allowed to stand. This was more than likely caused by light. For cases such as these, where the alkene cannot be stored or even exposed to chromatographic stationary phase, the development of a quick and easy chromatography-free method of separation of alkene and phosphine oxide is particularly important.

An Appel-type reaction that requires only a catalytic quantity of phosphine oxide has also recently been reported [20]. Phosphine oxide is reacted with oxalyl chloride to give chlorophosphonium salt or dichlorophosphorane,[1] which reacts with alcohol to give alkoxyphosphonium salt and hence giving alkyl chloride and regenerating phosphine oxide. An aza-Wittig reaction that is catalytic in phosphine oxide has also been reported, in which the oxide reacts with an isocyanate to generate aza-ylide with the concomitant formation of CO_2 [21]. In this report, the aza-ylide reacts intramolecularly with a remote carbonyl group on the ylide molecule, resulting in the formation of phenanthridines and benzoxazoles and the regeneration of phosphine oxide. As with the catalytic Wittig reaction discussed above, a cyclic phosphine is also necessary in this example of a catalytic aza-Wittig reaction.

3.2 Chromatography-Free Method for Phosphine Oxide Removal from Wittig Reaction Crude Products

From the foregoing discussion in Sect. 3.1 it can be seen that the development of a chromatography-free process involving phosphine oxide removal from the crude product and the conversion of the phosphine oxide into other entities that can be re-used in the reaction is highly desirable. We now report a convenient method that has been applied to both Wittig and Appel reaction crude products that allows the separation of the reaction product from phosphine oxide, and also can allow the phosphine oxide to be converted back into phosphine, the starting material for the synthesis of the phosphonium salts used to generate phosphonium ylides. In the case of the Wittig reaction, the method is also successful in removing the aldehyde starting material and the conjugate acid of the ylide-generating base from the alkene and phosphine products of the process. The method is based on the generation of chlorophosphonium chloride by reaction of phosphine oxide with oxalyl chloride. Production of such species by this reaction was reported many years ago [22], and has since been put to good use in the reduction and in situ boronation of phosphine oxide [23] and in the development of an Appel reaction catalytic in phosphine oxide [20]. Chlorophosphonium chlorides or dichlorophosphoranes have previously been generated by the reactions of phosphine with chlorine (in various solvents) [24–27], and phosphine oxide with trichloromethyl chloroformate [22]. The structure of the entity produced (i.e. chlorophosphonium salt or phosphorane) seems to depend heavily on the solvent. In particular, it appears to adopt chlorophosphonium salt structure in benzene, acetonitrile and chlorinated solvents, and phosphorane structure in toluene and diethyl ether. It has limited solubility in THF and diethyl ether (what is in solution is phosphorane form), and very limited solubility in alkane solvents.

The method allows the Wittig reaction itself to be carried out using whatever starting materials and reaction conditions are required to give the desired product with the desired stereochemistry. For the reactions used in this study we chose to use the conditions that had become standard in our laboratory for the studies

[1] The exact nature of this intermediate depends heavily on its medium. See Refs. [22–27].

described in Chap. 2, that is in situ generation of ylide from dry precursor phosphonium salt in THF using NaHMDS as base, subsequent addition of aldehyde at −78 °C, and warming to room temperature after stirring at low temperature for a period. The crude product can then be treated in any of the following three ways to results in the separation of phosphine oxide and aldehyde from the alkene product:

(i) The solvent is removed *in vacuo*, oxalyl chloride is added resulting in the evolution of CO and CO_2 gas. When bubbling has ceased, cyclohexane is added, and the upper phase formed is decanted off and then filtered.
(ii) Oxalyl chloride is added directly to the reaction mixture (THF solvent), and stirred, causing the evolution of CO and CO_2. When bubbling has ceased, the solvent is removed *in vacuo*, cyclohexane is added, and the upper phase formed is decanted off and then filtered.
(iii) The solvent is removed *in vacuo*, cyclohexane is added to the crude product, and oxalyl chloride is added, causing CO and CO_2 to be produced. When bubbling has stopped, the upper phase formed is decanted off and then filtered.

In all three methods, the oxalyl chloride reacts with phosphine oxide to form chlorophosphonium salt and release CO and CO_2 [22]. The biphasic mixture that forms in the presence of cyclohexane consists of an upper phase of alkene and residual oxalyl chloride dissolved in cyclohexane (this is typically pale yellow), and a viscous oil that sits at the bottom of the vessel. It has been found for Wittig reaction products that remaining aldehyde starting material usually does not dissolve in the cyclohexane phase,[2] but stays in the residue along with the chlorophosphonium salt. Washing of the cyclohexane solution with aqueous base and then aqueous acid results in the removal of residual oxalyl chloride and also the conjugate acid of the base used to generate the phosphonium ylide (hexamethyldisilamine in this case), furnishing a high yield of alkene that is free of both phosphorus and aldehyde (see Table 3.1). The use of certain other bases to generate the phosphonium ylide (or indeed using a pre-generated stabilised ylide) would obviate the need for the acid wash.

As can be seen from Table 3.1, the procedure was successful for the removal of both triphenylphosphine oxide and methyldiphenylphosphine oxide from Wittig reaction crude products, and has been shown to be applicable to the reactions of all phosphonium ylide types (non-stabilised, semi-stabilised and stabilised) with both aromatic (see Table 3.1 entries 6–16 and 18) and aliphatic aldehydes (see Table 3.1 entries 1–5 and 17). In all cases investigated, comparison of the *Z/E* ratio of the crude alkene product (assigned by integration of characteristic signals in the ^{1}H NMR of the crude product) with that of the product after treatment with oxalyl chloride (again assigned by ^{1}H NMR) showed it to be unaffected by the procedure. This is particularly striking for the syntheses of *Z*-2,2′-difluorostilbene (Table 3.1 entry 15) and Z-1-cyclopentyl-3-methylbut-1-ene (Table 3.1 entry 17). The former compound is known to be sensitive to isomerisation to the *E*-isomer,

[2] The procedure is also effective if cold diethyl ether is employed in place of cyclohexane. See Ref. [28].

Table 3.1 Wittig reactions for which oxalyl chloride treatment of the crude product followed by filtration yields alkene product free of aldehyde and phosphorus

(i) NaHMDS; (ii) R^1CHO; (i) oxalyl chloride; (ii) c-C_6H_{12}, Filtration

Entry	Phosphonium salt		Aldehyde R^1	Alkene yield (%)[a]	Alkene Z/E ratio[b]
	R^3	R^2			
1	Ph	i-Pr		92	100:0
2	Me	$CO_2(t\text{-Bu})$		93	79:21
3	Me	CO_2Me		93	79:21
4	Me	Ph		87	95:5
5	Me	2-BrC_6H_4		96	95:5
6	Me	Ph	Ph	98	15:85
7	Me	2-FC_6H_4	2-MeC_6H_4	90	52:48
8	Me	2-BrC_6H_4	2-MeC_6H_4	94	78:22
9	Me	2-BrC_6H_4	2-BrC_6H_4	90	98:2
10	Me	2-ClC_6H_4	2-MeC_6H_4	78	77:23
11	Me	Ph	2-ClC_6H_4	85	92:8
12	Ph	Ph	2-ClC_6H_4	71	90:10
13	Ph	2-ClC_6H_4	Ph	96	51:49
14[d,e]	Ph	2-ClC_6H_4	2-ClC_6H_4	95	94:6
15	Ph	2-FC_6H_4	2-FC_6H_4	93	94:6[c]
16	Ph	C(O)Me	Ph	87	3:97
17	Ph	i-Pr	c-C_5H_9	N/A[f]	100:0
18[d,g]	Ph	i-Pr	2-BrC_6H_4	90	82:18

[a] Yield based on the mass of isolated alkene

[b] Z/E ratio determined by ^{1}H NMR and ^{19}F NMR (where applicable) analysis of the crude product obtained after aqueous work-up. See Sect. 4.9 for full details of the reaction, work-up and analyses

[c] Z/E ratio determined exclusively by ^{19}F NMR analysis of the crude product obtained after aqueous work-up

[d] Triphenylphosphine was regenerated from the chlorophosphonium salt contained in the residue after filtration of the alkene solution by dissolving the residue in THF and treating with $LiAlH_4$

[e] Ph_3P was isolated in a yield of 50 %

[f] This product could not be completely purified due to decomposition, and was contaminated with a small amount of starting aldehyde

[g] Ph_3P was isolated in a yield of 80 %

especially if subjected to column chromatography [6]. The latter spontaneously degrades under ambient conditions if simply allowed to stand. Thus neither alkene is readily accessible in pure form by methods that demand chromatography. This method has also been successfully applied to the purification of the alkyl chloride products of Appel reactions in the Gilheany laboratory by Kamalraj Rajendran [28].

The oily residue that remains after filtration is shown to contain chlorophosphonium salt by NMR ($\delta_P = 65$ ppm) and also usually a small amount of aldehyde. Repeated NMR observation of the residue over a period of time shows only a single peak in the ^{31}P whose chemical shift becomes lower with the passage of time, until it eventually becomes characteristic of phosphine oxide. It seems likely that phosphine oxide is formed by partial decomposition of the chlorophosphonium salt, and that the two compounds form a complex. The chemical shift of this complex varies depending on its exact composition. There exist several previous reports on the formation of complexes between phosphine oxide and Lewis acids [27, 29, 30].

A method for the net reduction of phosphine oxide to phosphine has been developed in our laboratory. It involves reacting phosphine oxide with oxalyl chloride to give chlorophosphonium salt and treatment of this salt with $LiAlH_4$. This method was applied to the residue containing the chlorophosphonium salt remaining after separation of the cyclohexane solution of alkene. The residue was dissolved in THF and treated with $LiAlH_4$ at low temperature. This resulted in the formation of phosphine for each of the Wittig reactions of Table 3.1 entry 14 (50 % yield of Ph_3P) and entry 18 (80 % yield of Ph_3P). The $LiAlH_4$ was quenched with ethyl acetate.[3] For Wittig reaction crude products, it was best to filter the mixture at this point to remove any remaining phosphonium salt.[4] Typical aqueous work-up of the ethyl acetate solution and solvent removal furnished crude triphenylphosphine product, which could be purified by recrystallisation from methanol.

The use of an alkane solvent for the filtration step after chlorophosphonium salt generation is important for the success of the separation. We have found that the entity generated by treatment of phosphine oxide with oxalyl chloride is at least sparingly soluble in toluene, THF and Et_2O, perhaps due to isomerisation between chlorophosphonium salt and dichlorophosphorane [25].

This procedure is extremely useful preparatively, as in principle it may be applied to the crude product of any Appel-type reaction to give pure alkyl chloride, or to the crude product of any Wittig reaction to give pure alkene whose *Z/E* ratio remains invariant throughout, effectively using only a series of washes to purify the crude product. Thus the troublesome problem of phosphine oxide removal may be easily circumvented without having to resort to modification of the starting materials. We envisage that the method of treating the crude products with oxalyl chloride—i.e. dissolving in cyclohexane and filtering to remove the phosphorus-containing material—can also be applied to other reactions where phosphine oxide removal is a problem e.g. the Mitsunobu and Staudinger reactions.

---

[3] The procedure has since been improved by the use of degassed ammonium chloride solution to quench the $LiAlH_4$, and degassed dichloromethane in place of ethyl acetate. See Ref. [28].

[4] Aqueous hydrolysis of phosphonium salts causes the production of phosphine oxide.

References

1. Cadogan JIG (ed) (1979) Organophosphorus reagents in organic synthesis. Academic Press, London
2. Fairlamb IJS (2009) ChemSusChem 2:1021
3. Pommer H, Nürrenbach A (1975) Pure Appl Chem 43:527
4. Stockburger D (1999) Production-integrated environmental protection and waste management in the chemical industry, In: Christ C (ed) Wiley-VCH, Weinheim, Germany, pp 90–93
5. Yamamoto A, Fukumoto T (1994) Phosphine oxide removal from compounds formed by a Wittig reaction. US Patent 5,292,973, 8 March 1994
6. Dunne EC, Coyne EJ, Crowley PB, Gilheany DG (2002) Tetrahedron Lett 43:2449
7. McNulty J, Keskar K (2008) Tetrahedron Lett 49:7054
8. Wang Q, El Khoury M, Schlosser M (2000) Chem Eur J 6:420
9. Trippet S, Walker DM (1961) J Chem Soc 2130
10. Daniel H, Le Corre M (1987) Tetrahedron Lett. 28:1165
11. Shimojuh N, Imura Y, Moriyama K, Togo H (2011) Tetrahedron 67:951
12. Russell MG, Warren S (1998) Tetrahedron Lett 39:7995
13. McNulty J, Das P (2009) Eur J Org Chem 4031
14. Westman F (2001) Org Lett 3:3745
15. Leung PS-W, Teng Y, Toy PH (2010) Synlett 5:1997
16. Leung PS-W, Teng Y, Toy PH (2010) Org Lett 12:4996
17. O'Brien CJ, Tellez JL, Nixon ZS, Kang LJ, Carter AL, Kunkel SR, Przeworski KC, Chass GC (2009) Angew Chem Int Ed 48:6836
18. Harcken C, Martin SF (2001) Org Lett 3:3591
19. Cotter J, Hogan A-ML, O'Shea DF (2007) Org Lett 9:1493–1496
20. Denton RM, An J, Adeniran B (2010) Chem Commun 46:3025
21. Marsden SP, McGonagle AE, McKeever-Abbas B (2008) Org Lett 10:2589
22. Masaki M, Fukui K (1977) Chem Lett 151
23. Rajendran KV, Gilheany DG (2012) Chem Commun 48:817
24. Godfrey SM, McAuliffe CA, Pritchard RG, Sheffield JM (1996) Chem Commun 2521
25. Godfrey SM, McAuliffe CA, Pritchard RG, Sheffield JM (1998) Chem Commun 921
26. Al-Juboori MAHA, Gates PN, Muir AS (1991) J Chem Soc Chem Commun 1271
27. Gonnella NC, Busacca C, Campbell S, Eriksson M, Grinberg N, Bartholomeyzik T, Ma S, Norwood DL (2009) Magn Reson Chem 47:461
28. Byrne PA, Rajendran, KV, Muldoon J, Gilheany DG (2012) Org Biomol Chem 10:3531
29. Burford N, Spence RE v H, Linden A, Cameron TS (1990) Acta Crystallogr Sect C: Cryst Struct Commun 46, 92
30. Feher FJ, Budzichowski TA, Weller KJ (1993) Polyhedron 12:591

Chapter 4
Experimental

4.1 General Experimental

All chemicals were supplied by Aldrich, with the exception of reagent grade acetone (supplied by Fluka), Zeoprep silica, Merck standardised alumina 90, Merck silica 9385 (partical size 0.040–0.063 mm), and HPLC grade pentane (supplied by Romil). All chemicals were used without further purification except diethyl ether, toluene, and THF, which were processed through an Innovative Technology Inc. Pure Solv-400-3-MD solvent purification (Grubbs still) system and stored in Strauss flasks under a nitrogen atmosphere. The water content of the solvents was checked by Karl Fischer titration and generally found to be less than 10 ppm, and frequently less than 5 ppm. All benzaldehydes (supplied by Aldrich) were checked by NMR for the presence of carboxylic acids, but no trace of acid was found and so the benzaldehydes were used without further purification. Phosphonium salts were dried in a vacuum dessicator over P_2O_5 and, together with KHMDS and NaHMDS, were stored in an mBraun glove box under an atmosphere of argon. Oxalyl chloride was stored under an atmosphere of nitrogen in a Young's flask, and dispensed by nitrogen-flushed syringe. 1,2-O-isopropylidene-3-O-methyl-α-D-xylopentodialdofuranose-(1,4) was dissolved placed in a Schlenk flask under argon gas, and dissolved in dry THF to give a 0.5 mol L^{-1} solution. This solution was stored in the Schlenk flask under nitrogen at -18 °C in a freezer. 2-methylbenzaldehyde (*o*-tolualdehyde) was stored under nitrogen in a sealed Fluka® vessel.

A Harvard Apparatus syringe pump was used for slow additions. For HPLC, a Shimadzu LC 10AT system with autosampler coupled to a Shimadzu SPD 10A UV-vis detector was used. The columns used were an analytical AS-H column of internal diameter 4.6 mm, length 250 mm and with particle size 5 μm along with a guard column of internal diameter 4.6 mm, length 50 mm and particle size 10 μm, and a preparative AS-H column. The pentane solvent used for HPLC (Romil HPLC grade) was filtered through an acrodisc CR 13 mm syringe filter with 0.2 μm PTFE particles prior to injection onto the column.

P. A. Byrne, *Investigation of Reactions Involving Pentacoordinate Intermediates*, Springer Theses, DOI: 10.1007/978-3-642-32045-3_4,

NMR chemical shifts are reported in parts per million (ppm), and coupling constants (J) are reported in hertz (Hz). NMR spectra were obtained on Varian 300, 400, 500 and 600 MHz spectrometers. Double decoupled ^{13}C NMR spectra of the new salts were obtained on the 600 MHz machine (^{1}H and ^{31}P decoupled from ^{13}C signals). ^{1}H and ^{13}C NMR chemical shifts measured relative to tetramethylsilane. ^{31}P NMR chemical shifts were measured relative to an external orthophosphoric acid standard. ^{19}F NMR was carried out on a 400 MHz spectrometer, and the chemical shifts were measured relative to CCl_3F. In the reporting of characterisation details, the NMR signals due to aromatic hydrogen "j" (where j = 1−6) of a benzyl group are referred to as ArH-j and those due to phenyl hydrogens as PhH-j, and likewise the corresponding carbons are referred to as ArC-j and PhC-j respectively. Assignments of signals in ^{1}H and ^{13}C NMR spectra was done by reference to ^{13}C{^{1}H, ^{31}P}, COSY, HSQC, DEPT and HMBC NMR spectra.

High resolution mass spectra were obtained on a LCT electrospray ionisation mass spectrometer or a GCT Premier GC/MS maass spectrometer by electronic or chemical ionisation. Samples were dissolved in acetonitrile or methanol. IR spectra were obtained on a Varian 3100 FTIR spectrometer, and are reported here in units of cm^{-1}. Samples were prepared as thin films between NaCl plates. Melting points were obtained using a Reichert Thermovar melting point apparatus, and are uncorrected. Flash column chromatography was carried out using Aldrich neutral alumina (Brockmann grade I) or Zeoprep silica. TLC was done using Merck pre-coated alumina 90 or silica 60 F-254 plates. Realisation of plates was done by UV irradiation.

All reactions described here were carried out under an atmosphere of nitrogen. The inert atmosphere was established inside a reaction flask by the standard Schlenk pump and fill technique [1], using a Schlenk manifold that allowed each of five silicone rubber tubes to each be open either to vacuum or to the nitrogen supply by means of a three-way tap (third position is closed to both vacuum and nitrogen). The reaction flask was typically flame dried and attached to the Schlenk manifold by one of the silicone rubber tubes, and evacuated by application of a vacuum pump (Edwards RV5 rotary vane pump) while hot. The flask was allowed to cool under vacuum and then filled with nitrogen. It was then evacuated and re-filled a further two times. This technique was also applied to establish a nitrogen atmosphere in the tubing connected to a sealed Schlenk flask that already contained an inert atmosphere (nitrogen or argon gas). The flask could be opened to the nitrogen supply after the connecting tubing had been evacuated and re-filled three times. When not in use, the open end of each length of silicone rubber tubing was fitted with a syringe barrel with an attached needle. The needle was inserted through a rubber septum into a conical flask containing dry KOH pellets, and the tip was embedded in amongst the pellets. In this manner the tubing was kept free of ambient moisture by the hygroscopic KOH.

4.2 Synthesis of Phosphonium Salts

4.2.1 Synthesis of Benzylmethyldiphenylphosphonium Salts

The following procedure was carried out for the synthesis of all benzylmethyldiphenylphosphonium salts unless otherwise indicated.

A Schlenk flask was charged with nitrogen using the standard Schlenk pump and fill technique [1]. Methyldiphenylphosphine (one equivalent) was transferred from a Young's flask to the Schlenk flask by nitrogen-flushed syringe. Dry diethyl ether or toluene was added to give a clear solution of the phosphine (approximately 0.6 mol L^{-1} concentration). The appropriately substituted benzyl bromide (1.05–1.1 equivalents) was then added. The benzyl bromide was added by solid addition funnel if solid and by syringe (neat) if liquid. The solution turned a cloudy white colour upon addition in all cases. The solution was left to stir for 2–3 days, after which time a white solid was observed in the flask. The solid was isolated by filtration under suction. This solid was recrystallised from hot chloroform/ethyl acetate and the crystals isolated by filtration under suction. The filtrate was retained and often yielded further crystals after standing for a period. All syringes and needles used to handle methyldiphenylphosphine were placed in a bleach bath directly after use. The crystals were dried in a vacuum dessicator over P_2O_5 and $CaCl_2$ for several days before being transferred to a glove box under argon. Phosphonium salt dryness was confirmed by the absence of water in the ^{1}H NMR spectrum run under anhydrous conditions using dry $CDCl_3$ as solvent.

^{1}H, ^{13}C, ^{31}P, double decoupled ^{13}C (from ^{1}H to ^{31}P), COSY, HSQC, DEPT and HMBC NMR spectra, melting points and high resolution mass spectra were obtained for salts for which NMR data does not appear to have been reported previously. For phosphonium salts that have previously been reported in the literature, additional characterisation data that has not previously been published has been collected, and is reported below. In the following characterisation details, the NMR signals due to aromatic hydrogen "j" (where j = 1 – 6) of the benzyl aryl group are referred to as ArH-j and those due to phenyl hydrogens as PhH-j, and likewise the corresponding carbons are referred to as ArC-j and PhC-j respectively. Assignments of signals in ^{1}H and ^{13}C NMR spectra was done by reference to $^{13}C\{^1H, ^{31}P\}$, COSY, HSQC, DEPT and HMBC NMR spectra.

1a. Benzylmethyldiphenylphosphonium bromide [2]

Ph₂PMe + PhCH₂Br $\xrightarrow{Et_2O}$ [Ph₂P(Me)CH₂Ph]$^{\oplus}$ Br$^{\ominus}$

To a solution of methyldiphenylphosphine (3.5 ml, 3.73 g, 19 mmol) in dry diethyl ether was added benzyl bromide (2.5 ml, 3.6 g, 21 mmol), which gave benzylmethyldiphenylphosphonium bromide after stirring for 2 days (6.51 g, 94 % yield).

^{1}H NMR (500 MHz, $CDCl_3$): δ 7.91 (m, 4H, ArH-2), 7.70 (m, 2H, ArH-4), 7.60 (m, 4H, PhH-3), 7.27 (m, 2H, ArH-2), 7.20 (m, 1H, ArH-4), 7.13 (m, 2H, ArH-3), 4.91 (d, $^{2}J_{PH} = 15.4$, 2H, CH_2), 2.70 (d, $^{2}J_{PH} = 13.7$, 3H, CH_3),

^{31}P NMR (202 MHz, $CDCl_3$): δ 23.8.

^{13}C NMR (126 MHz, $CDCl_3$): δ 134.5 (d, $J_{PC} = 3.2$, PhC-4), 132.9 (d, $J_{PC} = 9.9$, PhC-2), 130.8 (d, $J_{PC} = 5.6$, ArC-2), 129.8 (d, $J_{PC} = 12.6$, PhC-3), 128.7 (d, $J_{PC} = 3.6$, ArC-3), 128.1 (d, $J_{PC} = 4.2$, ArC-4), 127.4 (d, $J_{PC} = 9.2$, ArC-1), 118.7 (d, $J_{PC} = 85.0$, PhC-1), 30.5 (d, $J_{PC} = 47.6$, CH_2), 7.1 (d, $J_{PC} = 56.4$, CH_3),

HRMS (m/z): Calc. for $[M]^+ = C_{20}H_{20}P$ 291.1303; found 291.1296 (2.4 ppm).

1b. 2-Fluorobenzylmethyldiphenylphosphonium bromide

Ph–P(Ph)–Me + 2-fluorobenzyl bromide —Et_2O→ 2-fluorobenzylmethyldiphenylphosphonium bromide

To a solution of methyldiphenylphosphine (3.8 ml, 4.05 g, 20 mmol) in dry diethyl ether was added 2-fluorobenzyl bromide (2.65 ml, 4.16 g, 22 mmol), which gave 2-fluorobenzylmethyldiphenylphosphonium bromide after stirring for 2 days (7.59 g, 97 % yield).

^{1}H NMR (500 MHz, $CDCl_3$): δ 7.96-7.88 (m, 4H, PhH-2), 7.82-7.77 (m, 1H, ArH-6), 7.76-7.72 (m, 2H, PhH-4), 7.64-7.60 (m, 4H, PhH-3), 7.24 (m, 1H, ArH-4), 7.06 (t, $^{2}J = 7.6$ Hz, 1H, ArH-5), 6.87 (m, 1H, ArH-3), 4.95 (d, $^{3}J_{PH} = 15.4$, 2H, CH_2), 2.79 (d, $^{3}J_{PH} = 13.6$, 3H, CH_3).

^{31}P NMR (202 MHz, $CDCl_3$): δ 22.5.

^{13}C NMR (126 MHz, $CDCl_3$): δ 160.9 (dd, $J_{PC} = 6.1$, $J_{FC} = 247.6$, ArC-2), 135.1 (d, $J_{PC} = 3.1$, PhC-4), 133.4 (dd, $J = 5.1, 2.7$, ArC-6), 133.2 (d, $J = 10.0$, PhC-2), 130.8 (dd, $J_{PC} = 3.9$, $J_{FC} = 8.2$, ArC-4), 130.2 (d, $J_{PC} = 12.6$, PhC-3), 125.2 (app t, $J = 3.5$, ArC-5), 118.8 (d, $J_{PC} = 84.6$, PhC-1), 115.7 (dd, $J_{PC} = 3.2$, $J_{FC} = 21.8$, ArC-3), 115.2 (dd, $J_{PC} = 8.6$, $J_{FC} = 14.9$, ArC-1), 24.7 (d, $J_{PC} = 49.8$, CH_2), 8.0 (d, $J_{PC} = 56.2$, CH_3).

HRMS (m/z): Calc. for $[M]^+ = C_{20}H_{19}PF$ 309.1208; found 309.1209 (0.3 ppm).

MP (crystallised from $CHCl_3$/EtOAc) 202–204 °C.

1c. 2-chlorobenzylmethyldiphenylphosphonium bromide

Ph–P(Ph)–Me + 2-chlorobenzyl bromide —Et_2O→ 2-chlorobenzylmethyldiphenylphosphonium bromide

To a solution of methyldiphenylphosphine (4.5 ml, 5.2 g, 26 mmol) in dry diethyl ether was added 2-chlorobenzyl bromide (3.1 ml, 4.9 g, 24 mmol), which gave 2-chlorobenzylmethyldiphenylphosphonium bromide after stirring for 2 days (6.11 g, 64 % yield).

^{1}H NMR (500 MHz, $CDCl_3$): δ 7.92-7.88 (m, 1H, ArH-3), 7.84-7.79 (m, 4H, PhH-2), 7.79-7.74 (m, 2H, PhH-4), 7.65-7.60 (m, 4H, PhH-3), 7.28-7.24 (m, obscured by C*H*Cl_3, ArH-6), 7.24-7.21 (m, 2H, ArH-4, ArH-5), 5.19 (d, $^2J_{PH} = 15.7$, 2H, CH_2), 2.82 (d, $^2J_{PH} = 13.5$, 3H, CH_3).

^{31}P NMR (202 MHz, $CDCl_3$): δ 23.0.

^{13}C NMR (126 MHz, $CDCl_3$): δ 35.0 (d, $J_{PC} = 3.3$, PhC-4), 134.8 (d, J_{PC} not visible, ArC-2), 133.4 (d, $J_{PC} = 5.1$ Hz, ArC-3), 133.1 (d, $J_{PC} = 9.7$, PhC-2), 130.0 (d, $J_{PC} = 4.2$, ArC-5, signals overlap slightly), 130.0 (d, $J_{PC} = 12.6$, PhC-3), 129.8 (d, $J_{PC} = 3.3$, ArC-4), 127.9 (d, $J_{PC} = 3.8$ Hz, ArC-6), 126.0 (d, $J_{PC} = 8.9$, ArC-1), 118.6 (d, $^1J_{PC} = 84.2$, PhC-1), 28.3 (d, $^1J_{PC} = 48.9$, CH_2), 8.3 (d, $^1J_{PC} = 56.3$ Hz, CH_3).

HRMS (m/z): Calc. for $[M]^+ = C_{20}H_{19}PCl$ 325.0913; found 325.0910 (0.9 ppm).

MP (crystallised from $CHCl_3$/EtOAc) 175–178 °C.

1d. 2-Bromobenzylmethyldiphenylphosphonium bromide

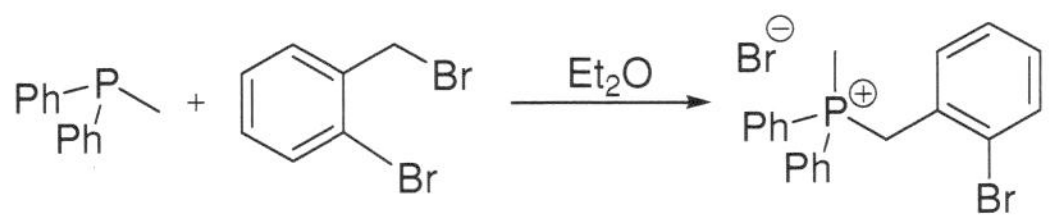

To a solution of methyldiphenylphosphine (3.5 ml, 3.7 g, 19 mmol) in dry diethyl ether was added 2-bromobenzyl bromide (5.18 g, 20 mmol), which gave 2-bromobenzylmethyldiphenylphosphonium bromide after stirring for 2 days (7.40 g, 88 % yield).

^{1}H NMR (500 MHz, $CDCl_3$): δ 7.81 (m, 4H, PhH-2), 7.78 (m, 1H, ArH-6), 7.76 (m, 2H, PhH-4), 7.62 (m, 4H, PhH-3), 7.43 (app d, $J = 8.1$, 1H, ArH-3), 7.27 (partially obscured by C*H*Cl_3, m, 1H, ArH-5), 7.16 (m, 1H, ArH-4), 5.21 (2H, d, $^2J_{PH} = 15.7$, CH_2), 2.82 (d, $^2J_{PH} = 13.5$, 3H, CH_3).

^{31}P NMR (202 MHz, $CDCl_3$): δ 22.8.

^{13}C NMR (126 MHz, $CDCl_3$): δ 135.0 (d, $J_{PC} = 2.8$, PhC-4), 133.3 (d, $J_{PC} = 3.2$, ArC-3), 133.2 (d, PhC-2), 133.2 (d, ArC-6, doublets overlapping, J_{PC} indiscernible), 130.2 (d, $J_{PC} = 4.2$, ArC-4), 130.1 (d, $J_{PC} = 12.5$, PhC-3), 128.4 (d, $J_{PC} = 3.7$, ArC-5), 127.9 (d, $J_{PC} = 8.8$, ArC-1), 125.8 (d, $J_{PC} = 7.0$, ArC-2), 118.4 (d, $^1J_{PC} = 84.7$, PhC-1), 30.9 (d, $^1J_{PC} = 48.9$, CH_2), 8.4 (d, $^1J_{PC} = 36.3$, CH_3),

HRMS: m/z Calc. for $[M]^+ = C_{20}H_{19}PBr$ 369.0408; found 369.0404 (1.1 ppm).

MP (crystallised from $CHCl_3$/EtOAc) 166–167 °C.

1e. 2-iodobenzylmethyldiphenylphosphonium bromide

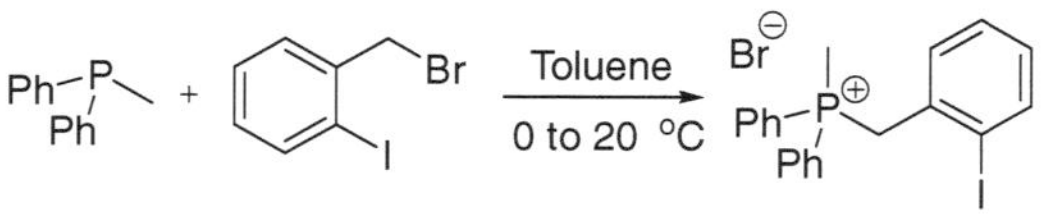

A Schlenk flask was charged with nitrogen using the standard Schlenk pump and fill technique. To this was added 2-iodobenzyl bromide (1.1 g, 3.7 mmol) under a strong flow of nitrogen using a powder funnel, followed by dry toluene (4 ml) by nitrogen-flushed syringe. This was cooled to 0 °C by immersion of the reaction flask in an ice bath, and then methyldiphenylphosphine (0.65 ml, 0.69 g, 3.5 mmol) was added by nitrogen-flushed syringe. This was stirred for 30 minutes at 0 °C, then removed from the cold bath and stirred overnight at room temperature, after which time a white solid had precipitated. This was isolated by cannula filtration of the supernatant into a second Schlenk flask under nitrogen. The white solid was washed with three 12 ml portions of dry toluene and three 12 ml portions of dry diethyl ether. The washings were removed by cannula filtration. The white solid was recrystallised from hot dichloromethane/ethyl acetate, and left in a refrigerator for 2 days, all under a nitrogen atmosphere. The resulting crystals were isolated and washed with dry diethyl ether by cannula filtration, giving 2-iodobenzylmethyldiphenylphosphonium bromide (1.15 g, 66 %), which after drying on high vacuum was stored in a glove box under argon. Characterisation was carried out on a small sample that was left open to air.

^{1}H NMR (500 MHz, $CDCl_3$): δ 7.78 (m, 6H, PhH-3 & PhH-4), 7.72 (m, 2HArH-3 & ArH-6), 7.62 (m, 4H, PhH-2), 7.30 (m, 1H, ArH-5), 6.99 (m, 1H, ArH-4), 5.20 (d, $^{2}J_{PH}$ = 15.5, 2H, CH_2), 2.85 (d, $^{2}J_{PH}$ = 13.5, 3H, CH_3),

^{31}P NMR (202 MHz, $CDCl_3$): δ 21.5.

^{13}C NMR (126 MHz, $CDCl_3$): δ 156.8 (d, J_{PC} = 5.6, ArC-5), 134.5 (d, J_{PC} = 2.8, PhC-4), 133.0 (d, J_{PC} = 10.1, PhC-2), 132.8 (d, J_{PC} = 5.6, ArC-6), 130.0 (d, J_{PC} = 3.9, ArC-4), 129.7 (d, J_{PC} = 12.3, PhC-3), 121.3 (d, J_{PC} = 3.4, ArC-5), 119.7 (d, $^{1}J_{PC}$ = 83.6, PhC-1), 115.9 (d, $^{2}J_{PC}$ = 7.0, ArC-1), 110.3 (d, J_{PC} = 3.3, ArC-3), 54.8 (s, OCH_3), 25.4 (d, $^{1}J_{PC}$ = 48.8, CH_2), 8.0 (d, $^{1}J_{PC}$ = 56.7, CH_3),

HRMS: m/z Calc. for $[M]^+$ = $C_{20}H_{19}PI$ 417.0269; found 417.0265 (1.0 ppm).

MP (crystallised under nitrogen from CH_2Cl_2/EtOAc) >230 °C.

1f. 2-Methoxybenzylmethyldiphenylphosphonium chloride

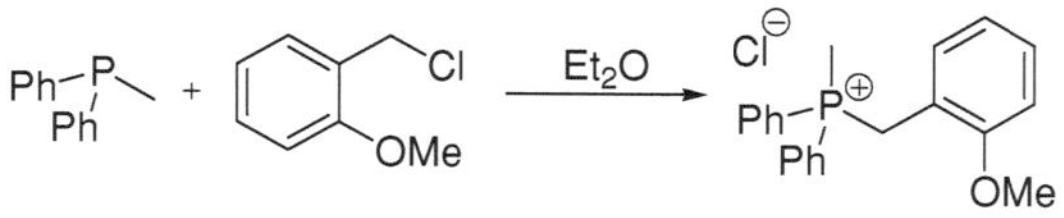

To a solution of methyldiphenylphosphine (4.1 ml, 4.3 g, 21.6 mmol) in dry diethyl ether was added 2-methoxybenzyl chloride (3.3 ml, 3.7 g, 23.8 mmol), which gave 2-chlorobenzylmethyldiphenylphosphonium bromide after stirring for 2 days (7.52 g, 98 % yield).

^{1}H NMR (500 MHz, $CDCl_3$): δ 7.82 (m, 4H, PhH-2), 7.71 (m, 2H, PhH-4), 7.64 (m, 1H, ArH-6), 7.60 (m, 4H, PhH-3), 7.23 (m, 1H, ArH-4), 6.87 (m, 1H, ArH-5), 6.66 (m, 1H, ArH-3), 4.79 (2H, d, $^{2}J_{PH}$ = 15.2, CH_2), 3.20 (s, 3H, OCH_3), 2.75 (d, $^{2}J_{PH}$ = 13.6, 3H, CH_3),

^{31}P NMR (202 MHz, $CDCl_3$): δ 22.9.

^{13}C NMR (126 MHz, $CDCl_3$): δ 156.8 (d, J_{PC} = 5.6, ArC-5), 134.5 (d, J_{PC} = 2.8, PhC-4), 133.0 (d, J_{PC} = 10.1, PhC-2), 132.8 (d, J_{PC} = 5.6, ArC-6), 130.0 (d, J_{PC} = 3.9, ArC-4), 129.7 (d, J_{PC} = 12.3, PhC-3), 121.3 (d, J_{PC} = 3.4, ArC-5), 119.7 (d, $^{1}J_{PC}$ = 83.6, PhC-1), 115.9 (d, $^{2}J_{PC}$ = 7.0, ArC-1), 110.3 (d, J_{PC} = 3.3, ArC-3), 54.8 (s, OCH_3), 25.4 (d, $^{1}J_{PC}$ = 48.8, CH_2), 8.0 (d, $^{1}J_{PC}$ = 56.7, CH_3).

HRMS (m/z): Calc. for $[M]^+$ = $C_{21}H_{22}PO$ 321.1408; found 321.1410 (0.6 ppm).

MP (crystallised from $CHCl_3$/EtOAc) 188–191 °C.

1g. 2-methylbenzylmethyldiphenylphosphonium chloride [3]

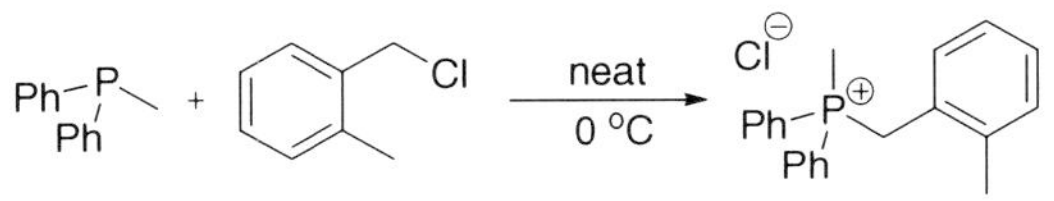

To methyldiphenylphosphine (2.0 ml, 2.1 g, 11 mmol) in a Schlenk flask under a nitrogen atmosphere and cooled to 0 °C in an ice bath was added slowly 2-methylbenzyl chloride (1.45 ml, 1.54 ml, 11.0 mmol). The reaction mixture was allowed to warm slowly to room temperature in the cold bath, and as it did so a white solid formed in the vessel. This was left stirring overnight, after which time only a white solid remained. To this was added dry toluene (5 ml), and the mixture was stirred, then filtered and washed extensively with toluene to yield 2-methylbenzylmethyldiphenylphosphonium chloride, which was recrystallised from chloroform/ethyl acetate (3.5 g, 96 %).

^{1}H NMR (300 MHz, $CDCl_3$): δ 7.7 (m. 6H), 7.6 (m, 4H), 7.4 (m, 1H), 7.1 (m, 3H), 4.9 (d, $^{2}J_{PH}$ = 15.2, 2H, CH_2), 2.9 (3H, d, $^{2}J_{PH}$ = 13.5, P-CH_3), 1.7 (s, 3H, Ar-CH_3) [3].

^{31}P NMR (121 MHz, $CDCl_3$): δ 22.0

4.2.2 Synthesis of Benzyltriphenylphosphonium Salts

Method A: To a solution of triphenylphosphine (1 equivalent) in dry diethyl ether (approximately 0.6 mol L^{-1} concentration) was added the appropriate benzyl halide (1.1 equivalents), which gave benzyltriphenylphosphonium halide salt after stirring for 2 days. The resulting white solid was recrystallised from chloroform/ethyl acetate. The recrystallisation usually occurred very slowly, so batches of crystals were often harvested from the mother liquor, which was then left to stand to allow further crystallisation to occur.

Method B: Triphenylphosphine (one equivalent) was added to a Schlenk flask under N_2. The reaction flask containing the triphenylphosphine was then placed in an oil bath and heated to 100 °C. Triphenylphosphine is a liquid at this temperature,[1] and thus can be used as the reaction solvent. The appropriately substituted benzyl bromide (1.0–1.1 equivalents) was then added. In all cases a white solid formed in the flask virtually instantly after addition of the benzyl bromide, and all of the triphenylphosphine was observed to have disappeared. After approximately 10 minutes, the reaction flask and its contents was removed from the oil bath and allowed to cool to room temperature. The white solid was difficult to remove from the sides of the flasks, so dichloromethane was added to aid removal of the solid from the flask. The suspension of the salt in the dichloromethane was transferred to a round bottomed flask and the solvent was removed using a rotary evaporator. Solvent removal was then completed using the vacuum line. The white salt was recrystallised from hot chloroform/ethyl acetate unless otherwise indicated, and the crystals isolated by filtration under suction.

2a. Benzyltriphenylphosphonium bromide [4]

Ph–P(Ph)–Ph + PhCH2Br —(neat, 100 °C)→ Br⊖ Ph3P⊕CH2Ph

Synthesised by method B from triphenylphosphine (5.878 g, 22.41 mmol) and benzyl bromide (2.7 ml, 3.8 g, 23 mmol). Yield = 8.54 g (88 %) after recrystallisation.

^{1}H NMR (300 MHz, $CDCl_3$): δ 7.80-7.63 (m, 15H), 7.22 (m, 1H), 7.21-7.10 (m, 4H), 5.37 (d, $^2J_{PH}$ = 14.4, 2H) [4].

^{31}P NMR (121 MHz, $CDCl_3$): δ 23.7.

2b. 2-fluorobenzyltriphenylphosphonium chloride [5]

Ph–P(Ph)–Ph + 2-F-C6H4CH2Br —(neat, 100 °C)→ Br⊖ Ph3P⊕CH2(2-F-C6H4)

Synthesised from triphenylphosphine (20.0 g, 76.3 mmo) and 2-fluorobenzyl bromide (14.4 g, 76.3 mmol). Recrystallisation of this from chloroform/petroleum

[1] Melting point is 79–81 °C.

spirits (40–60 °C) gave 2-methylbenzyltriphenylphosphonium bromide (30.1 g, 88 % yield).

^{1}H NMR (300 MHz, $CDCl_3$): δ 7.82-7.69 (m, 19H, aromatic hydrogens), 5.45 (d, $^{2}J_{PH} = 14$, 2H, CH_2) [5].

^{31}P NMR (121 MHz, $CDCl_3$) δ 23.9.

2c. 2-chlorobenzyltriphenylphosphonium bromide [6]

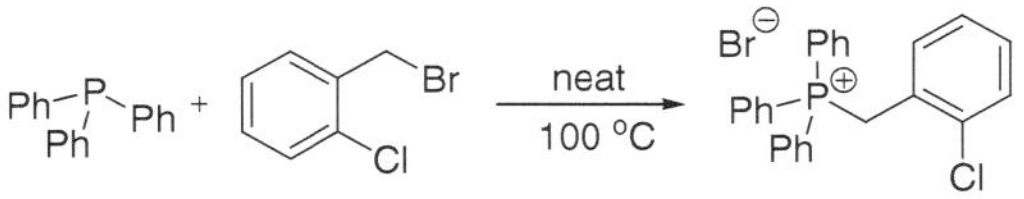

Synthesised by method B from triphenylphosphine (20.0 g, 76.3 mmol) and 2-chlorobenzyl bromide (15.7 g, 76.3 mmol). Recrystallisation of this from chloroform/petroleum spirits (40–60 °C) gave white crystals (31.4 g, 88 %).

^{1}H NMR (300 MHz, $CDCl_3$): δ 7.83-7.10 (m, 19H, ArH), 5.61 (d, $^{2}J_{PH} = 14.0$, 2H, CH_2) [6].

^{31}P NMR (121 MHz, $CDCl_3$) δ 23.7.

2d. 2-Bromobenzyltriphenylphosphonium bromide [7]

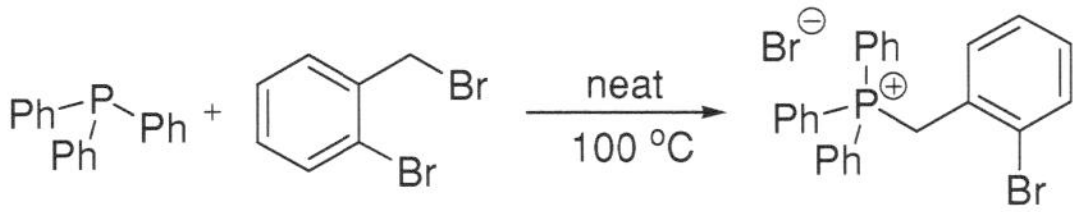

Synthesised by method B from triphenylphosphine (9.50 g, 36.2 mmol) and 2-bromobenzyl bromide (9.71 g, 38.8 mmol). Yield = 15.67 g (85 %) after recrystallisation from chloroform/ethyl acetate.

^{1}H NMR (300 MHz, $CDCl_3$): δ 7.83-7.55 (m, 16H), 7.37 (m, 1H), 7.17 (m, 2H), 5.67 (d, $^{2}J_{PH} = 14.4$, 2H) [7].

^{31}P NMR (121 MHz, $CDCl_3$): δ 23.8.

2e. 2-Iodobenzyltriphenylphosphonium bromide [8]

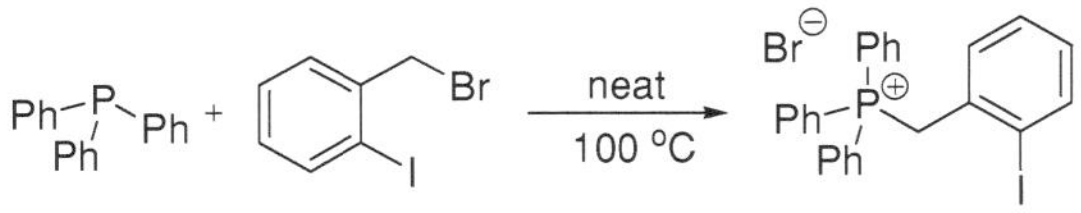

Synthesised by method B from triphenylphosphine (4.46 g, 16.98 mmol) and 2-iodobenzyl bromide (4.99 g, 16.8 mmol). Yield = 9.30 g (99 %) after recrystallisation.

^{1}H NMR (500 MHz, $CDCl_3$): δ 129.3 (d, $^{3}J_{PC}$ = 3.6 Hz, ArC-5), 7.81 (m, 3H, PhH-4), 7.70-7.63 (m, 13H, PhH-3, ArH-3, PhH-2), 7.51 (m, 1H, ArH-6), 7.22 (m, 1H, ArH-5), 6.97 (m, 1H, ArH-4), 5.66 (d, $^{2}J_{PH}$ = 14.1 2H, CH_2),

^{31}P NMR (202 MHz, $CDCl_3$): δ 23.8.

^{13}C NMR (126 MHz, $CDCl_3$): δ 139.7 (d, J_{PC} = 3.1, ArC-3), 135.2 (d, J_{PC} = 3.1, PhC-4), 134.5 (d, J_{PC} = 9.8, PhC-2), 132.4 (d, J_{PC} = 4.8, ArC-6), 131.0 (d, J_{PC} = 9.2, ArC-1), 130.3 (multiplicity indiscernible, ArC-5), 130.3 (d, J_{PC} = 12.7, PhC-3), 117.4 (d, $^{1}J_{PC}$ = 85.3, PhC-1), 104.7 (d, J_{PC} = 7.0, ArC-2), 34.8 (d, $^{1}J_{PC}$ = 48.3, CH_2),

HRMS (m/z): Calc. for $[M]^+$ = $C_{25}H_{21}PI$ 479.0426; found 479.0428 (0.4 ppm).

MP (crystallised from chloroform/ethyl acetate) >230 °C (lit. 265–266 °C) [8].

2f. 2-Methoxybenzyltriphenylphosphonium chloride [9, 10]

Synthesised by method A from triphenylphosphine (34.4 g, 0.131 mol) and methoxybenzyl chloride (21.0 ml, 23.63 g, 0.151 mol). Recrystallisation from chloroform/ethyl acetate gave white crystals of the phosphonium salt (28.6 g, 61 %). The recrystallisation ocurred very slowly.

^{1}H NMR (500 MHz, $CDCl_3$): δ 7.80 (m, 3H, PhH-4), 7.66 (m, 6H, PhH-3), 7.61 (m, 6H, PhH-2), 7.30 (m, 1H, ArH-4), 7.25 (1H, m, ArH-6), 6.80 (m, 1H, ArH-5), 6.62 (d, ^{3}J = 7.6, 1H, ArH-3), 5.05 (d, $^{2}J_{PH}$ = 14.0 Hz, 2H, CH_2), 3.20 (s, 3H, OCH_3).

^{31}P NMR (202 MHz, $CDCl_3$): δ 23.2.

^{13}C NMR (126 MHz, $CDCl_3$): δ 157.0 (d, J_{PC} = 5.5, ArC-5), 135.0 (d, J_{PC} = 3.2, PhC-4), 133.9 (d, J_{PC} = 9.7, PhC-2), 132.2 (d, J_{PC} = 5.6, ArC-6), 130.3 (d, J_{PC} = 3.7, ArC-4), 130.0 (d, J_{PC} = 12.5, PhC-3), 121.0 (d, J_{PC} = 3.8, ArC-5), 117.9 (d, $^{1}J_{PC}$ = 85.6, PhC-1), 115.3 (d, $^{2}J_{PC}$ = 8.9, ArC-1), 110.3 (d, J_{PC} = 3.2, ArC-3), 54.7 (s, OCH_3), 25.1 (d, $^{1}J_{PC}$ = 48.4, CH_2).

HRMS (m/z): Calc. for $[M]^+$ = $C_{26}H_{24}PO$ 383.1565; found 383.1575 (2.6 ppm).

2g. 2-methylbenzyltriphenylphosphonium chloride [11]

Synthesised by method B from triphenylphosphine (20.0 g, 76.3 mmo) and 2-methylbenzyl bromide (14.1 g, 76.3 mmol). Recrystallisation of this from chloroform/petroleum spirits (40–60 °C) gave 2-methylbenzyltriphenylphosphonium bromide (30.4 g, 89 % yield).

^{1}H NMR (300 MHz, $CDCl_3$): δ 7.74-7.05 (m, 19H, aromatic H), 5.30 (d, $^{2}J_{PH}$ = 15 Hz, 2H, CH_2), 1.70 (s, 3H, CH_3) [11].

^{31}P NMR (121 MHz, $CDCl_3$): δ 24.2.

4.2.3 Synthesis of (Alkoxycarbonylmethyl)Methyldiphenylphosphonium Salts

General procedure: A solution of methyldiphenylphosphine (1 equivalent) in dry toluene (a sufficient amount to make a 2 mol L^{-1} solution of phosphine) was prepared in a Schlenk flask under an atmosphere of nitrogen. This was cooled to 0 °C in an ice bath, and the appropriate alkyl haloacetate (1 equivalent) was added slowly by syringe. A white solid was seen to form in the reaction mixture almost instantly. The reaction mixture was stirred for overnight at room temperature, after which time only a white solid was present in the reaction flask. This was isolated by filtration, washed extensively with toluene, and then recrystallised from hot chloroform/ethyl acetate to give white crystals, which were dried using phosphorus pentoxide in a vacuum dessicator. The dry crystals were stored in a glove box under argon gas.

(*tert*-butoxycarbonylmethyl)methyldiphenylphosphonium bromide

From methyldiphenylphosphine (3.7 ml, 3.9 g, 20 mmol) and *tert*-butyl bromoacetate (3.0 ml, 4.0 g, 20 mmol) in a yield of 7.58 g (96 %) after recrystallisation.

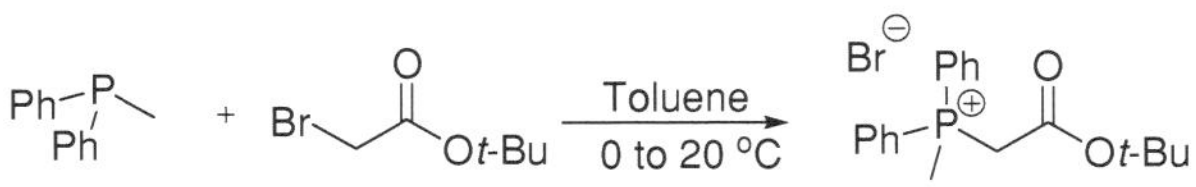

^{1}H NMR (500 MHz, $CDCl_3$) δ 7.95 (m, 4H, PhH-2), 7.75 (m, 2H, PhH-4), 7.70-7.61 (m, 4H, PhH-3), 4.89 (d, $^{2}J_{PH}$ = 13.0, 2H, P-CH_2), 3.04 (d, $^{2}J_{PH}$ = 14.3, 3H, P-CH_3), 1.24 (s, 9H, C(CH_3)$_3$).

^{31}P NMR (162 MHz, $CDCl_3$) δ 21.4.

^{13}C NMR (101 MHz, $CDCl_3$) δ 163.2 (d, $^{2}J_{PC} = 4.3$, $C = O$), 134.8 (d, $J_{PC} = 3.1$, PhC-4), 132.7 (d, $J = 10.6$, PhC-2), 130.1 (d, $J = 12.9$, PhC-3), 119.1 (d, $^{1}J_{PC} = 87.5$, PhC-1), 84.6 (s, C(CH_3)$_3$), 33.0 (d, $^{1}J_{PC} = 55.4$, PCH_2), 27.6 (s,C(CH_3)$_3$), 9.7 (d, $^{1}J_{PC} = 56.3$, PCH_3).

HRMS: Calc. for $[M]^+ = C_{19}H_{24}O_2P$ 315.1514; found 315.1505 (2.8 ppm).

MP (crystallised from $CHCl_3$/EtOAc) 185–188 °C.

(Methoxycarbonylmethyl)methyldiphenylphosphonium chloride

From methyldiphenylphosphine (1.6 ml, 1.7 g, 8.6 mmol) and methyl chloroacetate (0.76 ml, 0.94 g, 8.7 mmol) in a yield of 2.44 g (92 %) after recrystallisation. A sample kept open to air for characterisation purposes was observed to decompose slightly over time. The crystals are hygroscopic, and a doublet was observed to appear in the ^{1}H NMR spectrum at δ 2.92 (d, J_{PH} = 14.1) of a sample left open to air.

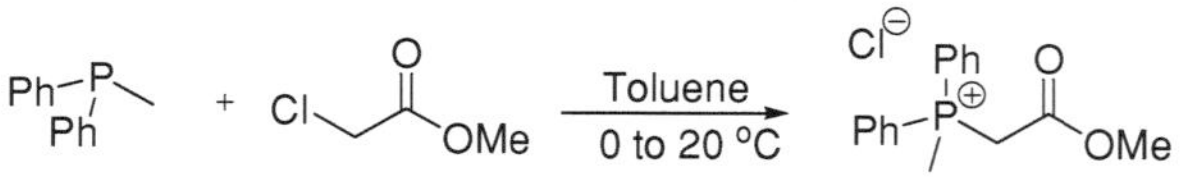

^{1}H NMR (500 MHz, $CDCl_3$) δ 7.99-7.91 (m, 4H, PhH-2), 7.78-7.73 (m, 2H, PhH-4), 7.69-7.63 (m, 4H, PhH-3), 5.30 (d, J = 14.0, 2H, PCH_2), 3.63 (s, 3H, OCH_3), 3.05 (d, J = 14.4, 3H, P-CH_3).

^{31}P NMR (162 MHz, $CDCl_3$) δ 21.6.

^{13}C NMR (101 MHz, $CDCl_3$) δ 165.5 (s, C = O), 134.9 (d, J_{PC} = 3.1, PhC-4), 132.8 (d, $^2J_{PC}$ = 10.6, PhC-2), 130.2 (d, J_{PC} = 13.0, PhC-3), 119.1 (d, $^1J_{PC}$ = 87.7, PhC-1), 53.3 (s, OCH_3), 31.7 (d, $^1J_{PC}$ = 58.6, PCH_2), 9.4 (d, $^1J_{PC}$ = 56.4, PCH_3).

HRMS: Calc. for $[M]^+$ = $C_{16}H_{18}O_2P$ 273.1044; found 273.1052 (2.8 ppm).

MP (crystallised from $CHCl_3$/EtOAc) 145–147 °C.

(Ethoxycarbonylmethyl)methyldiphenylphosphonium bromide [12]

From methyldiphenylphosphine (0.60 ml, 0.64 g, 3.2 mmol) and ethyl bromoacetate (0.36 ml, 1.54 g, 3.2 mmol) in a yield of 1.01 g (86 %) after recrystallisation.

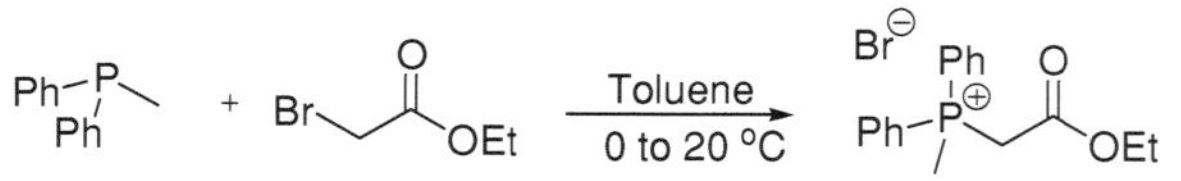

^{1}H NMR (300 MHz, $CDCl_3$) δ 8.05-7.87 (m, 4H, PhH-2), 7.84-7.72 (m, 2H, PhH-4), 7.71-7.59 (m, 4H, PhH-3), 5.11 (d, J = 13.9, 2H, PCH_2), 4.07 (d, J = 7.1, 2H, OCH_2), 3.06 (d, J = 14.3, 3H, PCH_3), 1.10 (t, J = 7.1, 3H, OCH_2CH_3) [12].

^{31}P NMR (121 MHz, $CDCl_3$) δ 21.3 (lit. 22) [12].

^{13}C NMR (75 MHz, $CDCl_3$) δ 164.6 (d, J = 4.0, C = O), 134.9 (d, J = 3.0, PhC-4), 132.8 (d, J = 10.6, PhC-2), 130.1 (d, J = 13.1, PhC-3), 118.9 (d, J = 87.9, PhC-1), 62.7 (s, OCH_2), 32.0 (d, J = 57.6, PCH_2), 13.8 (s, OCH_2CH_3), 9.7 (d, J = 56.3, PCH_3).

4.2.4 *Synthesis of Acetonylmethyldiphenylphosphonium Salts and Derived Ylides*

Acetonylmethyldiphenylphosphonium chloride

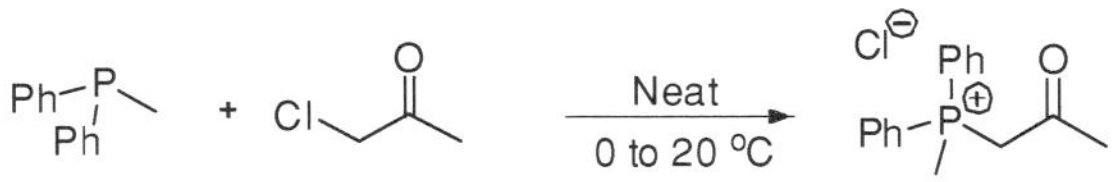

Methyldiphenylphosphine (0.85 ml, 0.91 g, 4.5 mmol) was added to a flame dried Schlenk flask under a nitrogen atmosphere, and cooled to 0 °C in an ice/water bath, and chloroacetone (0.36 ml, 0.42 g, 4.5 mmol) was added slowly by syringe. The solution was allowed to warm slowly to room temperature in the cooling bath. The initially transparent solution turned into a white solid as it was allowed to warm, and was stirred overnight. Toluene was added to the flask and the white solid was isolated by filtration, washed with toluene and recrystallised from chloroform/ethyl acetate gave colourless crystals, which were dried over phosphorus pentoxide in a vacuum dessicator. The dry crystals (1.25 g, 95 %) were stored in a glove box under argon gas.

^{1}H NMR ($CDCl_3$, 500 MHz): δ 7.93 (m, 4H, PhH-2), 7.73 (m, 2H, PhH-4), 7.65 (m, 4H, PhH-3), 5.61 (d, $J = 12.3$, 2H, CH_2), 2.83 (d, $J = 14.3$, 3H, P-CH_3), 2.46 (s, 3H, C(O)CH_3).

^{31}P NMR ($CDCl_3$, 202 MHz): δ 19.6.

^{13}C NMR ($CDCl_3$, 126 MHz): δ 201.2 (d, $J = 7.0$, $C = $ O), 134.5 (d, $J = 3.0$, PhC-4), 132.7 (d, $J = 10.6$, PhC-2), 130.1 (d, $J = 12.9$, PhC-3), 119.6 (d, $J = 87.8$, PhC-1), 39.7 (d, $J = 59.1$, PCH$_2$), 32.1 (d, $J = 6.1$, C(O)CH$_3$), 9.1 (d, $J = 56.7$, P-CH$_3$).

HRMS (m/z): Calc. for $[M]^+ = C_{16}H_{18}OP$ 257.1095; found 257.1106 (4.3 ppm).

MP (crystallised from $CHCl_3$/EtOAc) 175-178 °C (lit. 170–171 °C) [2].

3-chloroacetonylmethyldiphenylphosphonium chloride

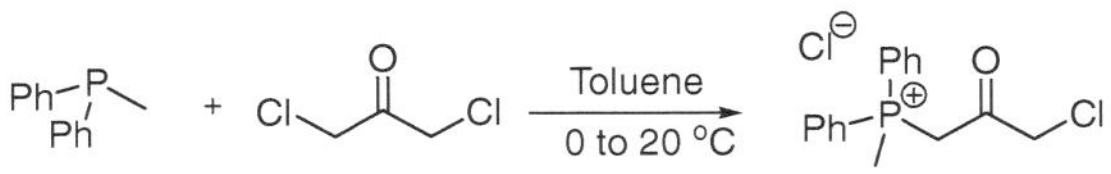

Methyldiphenylphosphine (2.9 ml, 3.1 g, 16 mmol) was added to a flame dried Schlenk flask under a nitrogen atmosphere, and dry toluene (20 ml) was added. The resulting solution was cooled to 0 °C in an ice/water bath, and 1,3-dichloroacetone (2.29 g, 18 mmol) was added by solid addition funnel. The solution was allowed to warm slowly to room temperature in the cooling bath. A precipitate gradually formed in the solution, which was left to stir for 3 days. After this time, the supernatant liquid was removed by cannula filtration under an inert atmosphere.

The isolated white solid was washed extensively with toluene, and the washings were removed by cannula filtration. The white solid was then recrystallised from hot dry chloroform/ethyl acetate. The crystals were isolated by cannula filtration under an inert atmosphere and washed with ethyl acetate as before. The dry phosphonium salt was stored in a glove box under an argon atmosphere. Yield = 4.56 g (87 %). This compound was found to degrade in methanol and DMSO, and if left open to air, and thus no melting point could be obtained.

^{1}H NMR (600 MHz, DMSO-*d6*): δ 7.95-7.87 (4H, m, PhH-2), 7.83-7.77 (2H, m, PhH-4), 7.72-7.67 (4H, m, PhH-3), 5.07 (2H, d, $^{2}J_{PH}$ = 13.2, P-CH_2), 4.76 (2H, s, CH_2Cl), 2.73 (3H, d, $^{2}J_{PH}$ = 14.6, P-CH_3).

^{31}P NMR (243 MHz, DMSO-*d6*): δ 20.0.

^{13}C NMR (151 MHz, DMSO-*d6*): δ 195.5 (d, $^{2}J_{PC}$ = 6.6, *C* = O), 135.0 (d, $^{4}J_{PC}$ = 2.6 Hz, PhC-4), 132.6 (d, $^{3}J_{PC}$ = 10.9, PhC-2), 130.3 (d, $^{2}J_{PC}$ = 12.8, PhC-3), 120.5 (d, $^{1}J_{PC}$ = 86.1, PhC-1), 50.1 (d, $^{1}J_{PC}$ = 9.5, *C*H_2Cl), 35.1 (d, $^{1}J_{PC}$ = 58.7, P-*C*H_2), 7.0 (d, $^{1}J_{PC}$ = 54.2, P-*C*H_3).

HRMS (m/z): Calc. for $[M]^+$ = $C_{16}H_{17}OPCl$ 291.0706; found 291.0715 (3.1 ppm).

(3,3-dimethylbutan-2-on-1-yl)methyldiphenylphosphonium bromide

Et$_2$O, 0 to 20 °C

A solution of methyldiphenylphosphine (1.90 ml, 2.03 g, 10 mmol) in dry diethyl ether (5 ml) was prepared in a Schlenk flask under an atmosphere of nitrogen. This was cooled to 0 °C in an ice bath, and 1-bromo-3,3-dimethylbutan-2-one (1.38 ml, 1.83 g, 10.2 mmol) was added slowly by syringe. A white solid was seen to form in the reaction mixture almost instantly. The reaction mixture was stirred for 4 h at room temperature. The white solid was then isolated by filtration. The solid was washed extensively with diethyl ether, and then recrystallised from hot chloroform/ethyl acetate to give white crystals, which were dried over phophorus pentoxide in a vacuum dessicator. The dry crystals (2.98 g, 77 %) were stored in a glove box under argon gas.

^{1}H NMR (600 MHz, $CDCl_3$): δ 7.96-7.85 (m, 4H, PhH-2), 7.75-7.66 (m, 2H, PhH-4), 7.65-7.56 (m, 4H, PhH-3), 5.52 (d, *J* = 12.4, 2H, PCH_2), 2.90 (d, *J* = 14.2, 3H, PCH_3), 1.24-1.14 (m, 9H, C(CH_3)$_3$).

^{31}P NMR (243 MHz, $CDCl_3$): δ 21.4.

^{13}C NMR (151 MHz, $CDCl_3$): δ 208.8 (d, $^{2}J_{PC}$ = 7.7, *C* = O), 134.5 (d, J_{PC} = 3.4, PhC-4), 132.8 (d, J_{PC} = 10.7, PhC-2), 130.0 (d, J_{PC} = 12.9, PhC-3), 120.0 (d, $^{1}J_{PC}$ = 88.1, PhC-1), 45.5 (d, *J* = 4.0, *C*(CH$_3$)$_3$), 36.2 (d, *J* = 60.3, CH_2), 26.2 (s, C(*C*H$_3$)$_3$), 10.2 (d, *J* = 56.8, P*C*H$_3$).

HRMS (m/z): Calc. for $[M]^+$ = $C_{19}H_{24}OP$ 299.1565; found 299.1553 (4.0 ppm).

3-methoxyacetonylidenetriphenylphosphorane [13, 14]

(i) Et_2O
(ii) Filtration

(i) $Ph_3P{=}CH_2$
(ii) Aqueous HCl
(iii) Aqueous K_2CO_3

Imidazole (2.10 g, 0.031 mol) was added to a nitrogen flushed Schlenk flask. Sufficient dry diethyl ether/dry THF (1:1) was added to give a 0.2 mol l^{-1} solution of imidazole, which was then cooled to 0 °C in an ice bath. Methoxyacetyl chloride (1.71 g, 0.015 mol) was added slowly over 30 min to the solution at 0 °C. A white precipitate of imidazole hydrochloride formed. The ethereal solution of methoxyacetyl imidazolide was isolated by cannula filtration into a second Schlenk flask using a cannula that had been dried for 1 h at 100 °C, leaving the white precipitate in the first vessel. The residue was washed with two 3 ml aliquots of dry diethyl ether, and the washings were cannula filtered into the second Schlenk flask.

Methyltriphenylphosphonium bromide (5.37 g, 0.015 mol) was added to a nitrogen flushed Schlenk flask surrounded with aluminium foil (to prevent ylide decomposition by carbene formation) and dry diethyl ether was added to give a cloudy white suspension of the salt (0.1 mol L^{-1}). A 1 mol L^{-1} solution of NaHMDS in THF (15 ml, 0.015 mol) was added, giving a yellow solution of ylide. A precipitate of NaBr was also formed. The solution left to stir for 1 h. After this time the solution was transferred by cannula filtration to a Schlenk flask wrapped in aluminium foil, giving a transparent yellow coloured solution of salt-free ylide.

The salt free ylide solution was cooled to −78 °C in a dry ice/acetone bath. The solution of methoxyacetyl imidazolide was added from a syringe to the salt free ylide solution at −78 °C over 4 h using a syringe pump (0.25 equivalents added per hour) while stirring at a high rate. By slow addition of the methoxyacetyl imidazolide, it was hoped that Claisen-type self condensation of the imidazolide would be inhibited. After this time, the reaction vessel was removed from the cold bath. Inside was a white/orange suspension (containing solid imidazole). The solid dissolved upon warming of the reaction mixture to room temperature, leaving a yellow/orange solid on the sides of the reaction vessel.

For aqueous work-up a solution of HCl (1 mol L^{-1}, 200 ml) was added to the reaction mixture. Diethyl ether (50 ml) was added to the reaction mixture, and the resulting mixture was shaken in a separatory funnel. The ethereal phase was removed, and solid K_2CO_3 was added to the aqueous phase until its pH changed to 10 from its initial value of 1. A pale brown solid settled out of solution. This was isolated by filtration, dissolved in dichloromethane and dried over $MgSO_4$. The ylide was then recovered by filtration and solvent removal in a yield of 12 % (0.62 g).

^{1}H NMR (400 MHz, $CDCl_3$): δ 3.45 (s, 3H, OCH_3), 3.95 (s, 2H, CH_2Cl), 4.05-4.15 (broad, 1H, PCH), 7.50 (m, 6H). 7.60 (m, 3H), 7.70 (m, 6H) [14].

^{31}P NMR (162 MHz, $CDCl_3$): δ 17.3

4.2.5 Synthesis of Phosphonium Salt Precursors of Non-stabilised Ylides

Ethyltriphenylphosphonium bromide [15]

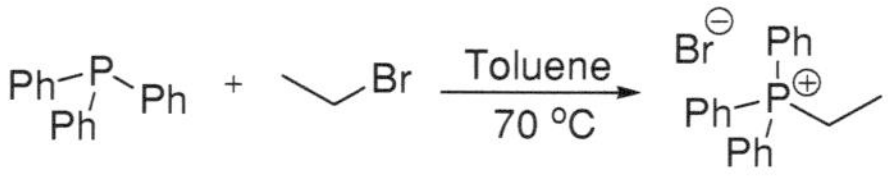

Triphenylphosphine (4.04 g, 15.4 mmol), ethyl bromide (1.69 ml, 2.51 g, 23.0 mmol), and toluene (5 ml) were mixed together in a Schlenk flask under a nitrogen atmosphere to give a clear solution, and this solution was heated at 70 °C for a day resulting in the formation of a white precipitate. This was isolated by filtration, and recrystallised from chloroform/ethyl acetate in a yield of 3.73 g (79 %). The white crystals thus obtained were placed in a dessicator under nitrogen and dried using phosphorus pentoxide and calcium chloride, and then placed in a glove box under an atmosphere of argon.

^{1}H NMR (400 MHz, $CDCl_3$) δ 7.90-7.76 (m, 9H), 7.73-7.67 (m, 6H), 3.93 (dq, J = 12.5, 7.4, 2H, P-CH_2), 1.40 (dt, J = 20.1, 7.4, 3H, CH_3) [15].

^{31}P NMR (162 MHz, $CDCl_3$): δ 26.5 (lit. 26.8) [15].

^{13}C NMR (101 MHz, $CDCl_3$): δ 135.0 (d, $^{4}J_{PC}$ = 3.1 Hz, PhC-4), 133.8 (d, J_{PC} = 10.1 Hz), 130.5 (d, J_{PC} = 12.5), 118.3 (d, $^{1}J_{PC}$ = 85.7, PhC-1), 17.3 (d, $^{1}J_{PC}$ = 51.4, PCH_2), 6.9 (d, $^{2}J_{PC}$ = 4.7 Hz, CH_3) [15].

Diethyldiphenylphosphonium bromide [16]

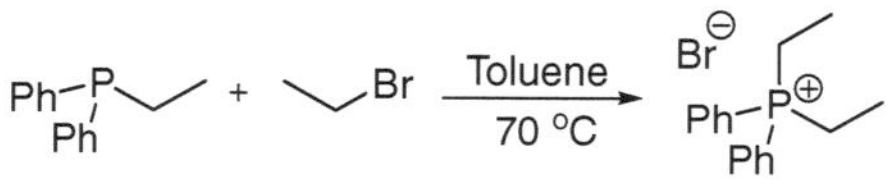

Ethyldiphenylphosphine (2.0 ml, 2.1 g, 9.9 mmol) and ethyl bromide (0.80 ml, 1.2 g, 10.8 mmol) were mixed together in a Schlenk flask at 0 °C under a nitrogen atmosphere to give a clear solution. The solution was stirred at room temperature 5 days resulting in the formation of a white paste. This was isolated by filtration, and recrystallised from chloroform/ethyl acetate in a yield of 2.64 g (83 %). The white crystals thus obtained were placed in a dessicator under nitrogen and dried using phosphorus pentoxide and calcium chloride, and then placed in a glove box under an atmosphere of argon.

^{1}H NMR (400 MHz, $CDCl_3$) δ 7.93-7.88 (m, 4H), 7.82-7.77 (m, 2H, PhH-4), 7.73-7.68 (m, 4H), 3.39 (dq, J = 12.7, 7.6, 4H, P-CH_2), 1.23 (dt, J = 19.7, 7.6, 6H, CH_3) [16].

^{31}P NMR (162 MHz, $CDCl_3$): δ 32.7.

^{13}C NMR (101 MHz, $CDCl_3$): δ 134.8 (d, J_{PC} = 3.1 Hz, PhC-4), 133.3 (d, J_{PC} = 8.6 Hz), 130.4 (d, J_{PC} = 11.7), 117.2 (d, $^1J_{PC}$ = 81.8, PhC-1), 16.0 (d, $^1J_{PC}$ = 50.6, PCH_2), 6.3 (d, $^2J_{PC}$ = 5.4 Hz, CH_3).

***P*-(*isobutyl*)triphenylphosphonium bromide** [17]

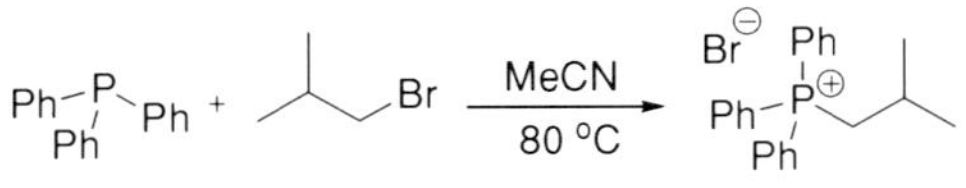

Triphenylphosphine (4.26 g, 16.2 mmol), dry acetonitrile (10 ml), and 1-bromo-2-methylpropane (2.0 ml, 2.2 g, 18.4 mmol) were added to a flame dried Schlenk flask underan atmosphere of nitrogen. Stirring gave a clear solution. The flask was fitted with a reflux condenser (flame dried and connected to nitrogen supply) and the reaction mixture was heated to 80 °C for 3 days under a gentle flow of nitrogen. The progress of the reaction was periodically checked by taking a ^{31}P NMR spectrum of a small sample of the reaction mixture diluted with $CDCl_3$. After 3 days, the acetonitrile solvent was removed, and the residue was recrystallised from hot chloroform/ethyl acetate to give white crystals of the product in a yield of 5.5 g (13.8 mmol, 85 %). The white crystals thus obtained were placed in a dessicator under nitrogen and dried using phosphorus pentoxide and calcium chloride, and then placed in a glove box under an atmosphere of argon.

^{1}H NMR (400 MHz, $CDCl_3$) δ 7.90 (m, 6H, PhH-2), 7.79 (m, 3H, PhH-4), 7.71 (m, 6H, PhH-3), 3.80 (dd, J = 12.9, 6.2, 2H, PCH_2), 2.18-1.99 (m, 1H, CHMe$_2$), 1.08 (d, J = 6.7, 6H, CH(CH_3)$_2$) [17].

^{31}P NMR (162 MHz, $CDCl_3$): δ 23.0.

^{13}C NMR (101 MHz, $CDCl_3$) δ 134.9 (d, J = 3.0, PhC-4), 133.7 (d, J = 10.0, PhC-2), 130.4 (d, J = 12.5, PhC-3), 119.0 (d, J = 85.3, PhC-1), 30.5 (d, J = 47.4, P-CH_2), 24.6 (d, J = 4.3, CHMe$_2$), 24.34 (d, J = 9.2, CH(CH_3)$_2$) [17].

***P*-phenyl-5*H*-dibenzophosphole** [18, 19]

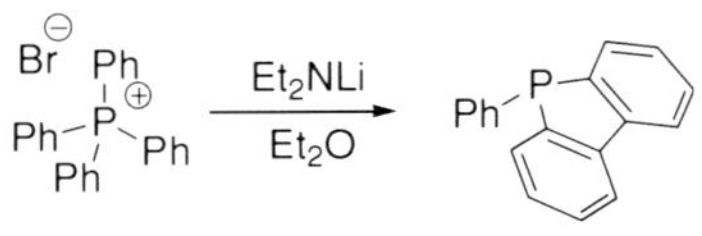

Tetraphenylphosphonium bromide (3.654 g, 8.715 mmol) was weighed into a flame-dried 100 ml Schlenk flask (with a male joint) under an atmosphere of argon in a glove box. This was then fitted with a U-bend tube and sealed on the other end of the U-bend tube with a small round-bottom flask under argon atmosphere. The sealed apparatus was removed from the glove box and attached to a Schlenk manifold using standard Schlenk pump and fill technique.

From this point, all manipulations were carried out under a nitrogen atmosphere. Diethylamine was distilled from calcium hydride into a Schlenk flask. The freshly distilled diethylamine (3.0 ml, 29 mmol) was dissolved in dry diethyl ether (18 ml) in a 100 ml Schlenk flask, and to this was added dropwise a 1.6 mol l^{-1} solution of *n*-BuLi in hexanes (19.5 ml, 31 mmol) at room temperature with gentle stirring to give lithium diethylamide. The addition was slightly exothermic. After stirring this solution for 10 min, further dry ether (30 ml) was added.

At this point the 100 ml Schlenk flask containing the tetraphenylphosphonium bromide was fitted via the U-bend tube to the 250 ml flask containing the lithium diethylamide solution under a flow of nitrogen. The tetraphenylphosphonium bromide was then slowly tipped into the solution of lithium diethylamide with stirring at room temperature. The reaction mixture instantly became red in colour. The volume of the mixture was increased by the addition of further dry ether (19 ml) [18]. This mixture was stirred overnight, during which time it became dark red in colour.

After this time, the reaction mixture was cooled to 0 °C, and degassed aqueous HCl (3.3 mol L^{-1}, 25 ml, 83 mmol) was slowly added, causing the formation of a biphasic mixture in which the ether layer was pale yellow and the aqueous layer was darker after stirring for an hour. The reaction mixture was quickly poured into a separatory funnel (itself under nitrogen flow attachment to Schlenk manifold). The layers were separated, and the ether layer was washed a further two times with degassed saturated aqueous $NaHCO_3$ solution. The isolated ether phase (after removal of water phases) was dried over Na_2SO_4. The drying agent was removed by quickly pouring the mixture into a flame-dried Schlenk flask through a frit under a flow of nitrogen. Removal of the solvent under vacuum (attached to Schlenk manifold) gave a slightly orange solid, which was recrystallised from hot dry methanol under nitrogen to give the phosphine (colourless crystals, 1.65 g, 73 %), which was stored under argon in the glove box.

^{1}H NMR (300 MHz, $CDCl_3$) δ 7.95 (d, J = 7.7, 2H), 7.74-7.66 (m, 2H), 7.47 (td, J = 7.6, 1.1, 2H), 7.37-7.18 (m, 7H, contains $CHCl_3$ signal) [20].

^{31}P NMR (121 MHz, $CDCl_3$) δ -10.1 [20].

P-(isobutyl)-P-phenyldibenzophospholium bromide

Ph–P + Br MeCN 80 °C P⊕ Ph Br⊖

P-Phenyldibenzophosphole (488 mg, 1.87 mmol) was added to a flame dried Schlenk flask under argon in a glove box. This flask was removed from the glove box and attached to a Schlenk manifold using standard Schlenk pump and fill technique. To this was added dry acetonitrile (5 ml) and then 1-bromo-2-methylpropane (1.00 ml, 1.26 g, 9.20 mmol). The Schlenk flask was then fitted with a flame-dried reflux condenser (fitted to Schlenk manifold by nitrogen adaptor) under a flow of nitrogen. The resulting solution was stirred under nitrogen atmosphere at 80 °C for 4 days. Reaction progress was monitored periodically by obtaining the ^{31}P NMR of a sample of the reaction mixture under inert atmosphere using $CDCl_3$ as solvent. When conversion of phosphine to phosphonium salt was observed to be complete by ^{31}P NMR, the solvent was removed *in vacuo* and the resulting white solid was recrystallised from hot chloroform/ethyl acetate in a yield of 0.50 g (84 %), dried in a vacuum dessicator containing P_2O_5 and stored in a glove box under argon.

^{1}H NMR (500 MHz, $CDCl_3$): δ 8.68-8.56 (m, 2H, DBP H-4 & H-6), 8.37-8.23 (m, 2H, PhH-2), 8.02 (dd, J = 7.8, 2.9, 2H, DBP H-1 & H-9), 7.84 (m, 2H, DBP H-2 & H-8), 7.71-7.65 (m, 3H, contains PhH-4 and DBP H-3 & H-7), 7.64-7.59 (m, 2H, PhH-3), 3.95 (dd, J = 13.9, 6.6, 2H, PCH_2), 2.06-1.89 (m, 1H, CHMe$_2$), 0.96 (dt, J = 17.4, 8.7, 6H, CH_3).

^{31}P NMR (162 MHz, $CDCl_3$): δ 28.2.

^{13}C NMR (126 MHz, $CDCl_3$): δ 144.0 (d, J = 17.7, DBP C-1a & C-9a), 135.7 (d, J = 2.4, DBP C-2 & C-8), 134.9 (d, J = 3.3, PhC-4), 133.6 (d, J = 10.5, DBP C-4 & C-6), 133.3 (d, J = 11.4, PhC-2), 131.1 (d, J = 11.8, DBP C-3 & C-7), 130.4 (d, J = 13.2, PhC-3), 122.9 (d, J = 9.4, DBP C-1 & C-9), 121.7 (d, J = 89.1, DBP C-4a & C-5a), 117.8 (d, J = 83.4, PhC-1), 30.6 (d, J = 42.7, CH_2), 25.0 (d, J = 4.2, CHMe$_2$), 24.0 (d, J = 10.4, CH_3).

HRMS (m/z): Calc. for $[M]^+$= $C_{22}H_{22}P$ 317.1459; found 317.1461 (0.6 ppm).

MP (crystallised from chloroform/ethyl acetate) >230 °C.

***P*-Ethyl-*P*-phenyldibenzophospholium bromide** [21]

P-Phenyldibenzophosphole (0.823 g, 3.16 mmol) was added to a flame-dried Schlenk flask under argon in a glove box. This flask was removed from the glove box and attached to a Schlenk manifold using standard Schlenk pump and fill technique. To this was added dry acetonitrile (10 ml) and then ethyl bromide (0.93 ml, 1.4 g, 13 mmol). The Schlenk flask was then fitted with a flame-dried reflux condenser (fitted to Schlenk manifold by nitrogen adaptor) under a flow of nitrogen. The resulting solution was stirred under nitrogen atmosphere at 70 °C for 3 days, with periodic monitoring of reaction progress by ^{31}P NMR of a smple of the reaction mixture. The solvent was removed *in vacuo* on completion of the reaction, and the resulting white solid was recrystallised from hot chloroform/ethyl acetate (yield 1.12 g, 96 %) and dried in a vacuum dessicator containing P_2O_5 and stored in a glove box under argon.

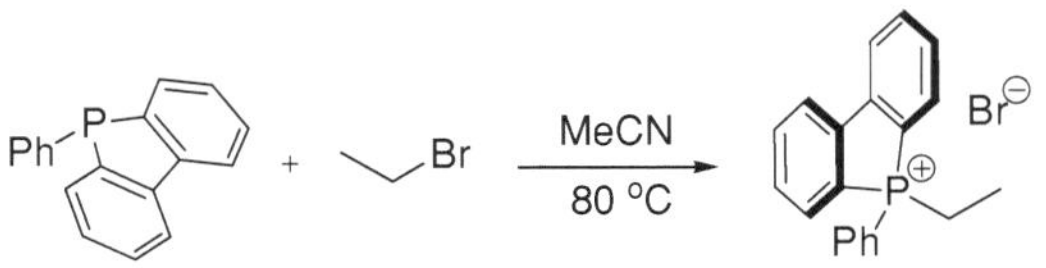

^{1}H NMR (300 MHz, $CDCl_3$): δ 8.60-8.48 (m, 2H), 8.36-8.23 (m, 2H), 8.01 (dd, J = 7.8, 2.9, 2H), 7.84 (t, J = 7.7, 2H), 7.77-7.58 (m, 5H), 4.01 (dq, J = 14.8, 7.4, 2H), 1.16 (dt, J = 22.5, 7.4, 3H) [21].

^{31}P NMR (121 MHz, $CDCl_3$): δ 32.7 [21].

4.3 General Procedures for Wittig Reactions and Assignment of *Z/E* Ratios

The procedures employed for the Wittig reactions differ only in the method by which ylide formation was realised [22].

4.3.1 Procedure A: Ylide Generated Using NaHMDS Solution

The appropriately substituted phosphonium salt (one equivalent) was added to a Schlenk flask under a nitrogen atmosphere. Dry THF was added to make a transparent suspension of the salt (approximately 0.2 mol L^{-1}). A solution of NaHMDS in THF (1 mol L^{-1}, containing 0.95 equivalents of base) was added. Mixing of the NaHMDS and phosphonium salt in THF solution caused the formation of a solution of ylide (orange or yellow coloured for non-stabilised and semi-stabilised ylides, pale yellow or colourless for ester-stabilised ylides) and a precipitate of sodium bromide or chloride. The ylide solution was stirred for one hour for semi-stabilised and ester-stabilised ylides, 15–20 min for non-stabilised ylides. After this time, the solution was cooled to −78 °C in a dry ice/acetone bath. A solution of the appropriately substituted benzaldehyde[2] (neat or as solution in dry THF, 1.0 equivalent) was added dropwise by syringe, causing the ylidic colour to fade where it had been present. The resulting solution was left to stir at −78 °C for 15 min, and then either removed from the cold bath and allowed to warm slowly to room temperature (for non-stabilised and semi-stabilised ylides) or quenched at low temperature by the addition of ammonium chloride solution (for ester-stabilised ylides).

4.3.2 Procedure B: Ylide Generated from "Instant Ylide Mix" Using Solid NaHMDS or KHMDS

Dry phosphonium salt (1 equivalent) and NaHMDS or KHMDS (0.95 equivalents) were placed in a flame-dried Schlenk flask in a glove box under an atmosphere of argon. The flask was sealed, removed from the glove box and charged with an atmosphere of nitrogen by the pump and fill technique using a Schlenk manifold. Dry THF was added, giving a solution of ylide (orange or yellow coloured for non-stabilised and semi-stabilised ylides, maroon for *P*-(alkylidene)-*P*-phenyldibenzophophoslanes, and pale yellow or colourless for ester-stabilised ylides)

[2] It was ensured that the aldehydes to be used were free of carboxylic acid by NMR analysis. Diagnostic singlets were observed at approximately 10.4 ppm for the 2-halobenzaldehydes, and at 10 ppm for benzaldehyde itself. No signals pertaining to carboxylic acid were observed.

and a precipitate of alkali metal halide upon stirring. The ylide solution was stirred for 1 h for semi-stabilised and ester-stabilised ylides, 15–20 min for non-stabilised ylides. This was cooled to −78 °C in a dry ice/acetone bath, and then the aldehyde (1.0 equivalent, neat or as a solution in dry THF) was added dropwise, causing the ylidic colour to fade where it had been present. The resulting solution was left to stir at −78 °C for 15 min, and then either removed from the cold bath and allowed to warm slowly to room temperature (for non-stabilised and semi-stabilised ylides) or quenched at low temperature by the addition of ammonium chloride solution (for ester-stabilised ylides).

4.3.3 Work-Up

Different work-up procedures were employed for reactions of different ylide types. The work-up procedures used for reactions of each of semi-stabilised, ester-stabilised and non-stabilised ylides are described in Sects. 4.4, 4.5 and 4.7, respectively.

4.3.4 Assignment of *Z*/*E* Ratios

The *Z*/*E* ratios for the various mixture of alkene isomers from reactions of all ylide types was obtained by comparing the integrations of signals in the ^{1}H NMR spectrum of the crude product that could unambiguously be assigned to the *E* and *Z* isomers respectively. The assignments were assisted by reference to ^{1}H NMR spectra of the individual alkene isomers, some of which had been synthesised and purified in this project, and some of which were available in the literature. For all the alkenes synthesised, details of the specific ^{1}H NMR signals used to determine the assigned *Z*/*E* ratios are listed below—the data is listed in Sect. 4.4 for stilbenes, in Sect. 4.5 for enoates, in Sect. 4.6 for enones, in Sect. 4.7 for alkenes derived from reactions of non-stabilised ylides, and in Sect. 4.8 for alkenes derived from 1,2-O-isopropylidene-3-O-methyl-α-D-xylopentodialdofuranose-(1,4).

For all reactions of semi-stabilised ylides and benzaldehydes to give stilbenes (see Sect. 4.4), the alkene hydrogen signals for the *Z*-isomer were clearly discernible in the spectrum of the crude product. However, for many of the stilbenes, none of the signals assigned to the *E*-isomer had sufficient baseline separation from other signals to allow an accurate appraisal of the integrations of the signals. In these cases, it was necessary to carry out some further purification of the crude product in order to determine the *Z*/*E* ratio. For all other alkenes, it was relatively straightforward to assign baseline-separated signals to each isomer and thus to obtain the *Z*/*E* ratio for the compound, but some of these were purified in the course of developing a new technique for phosphine oxide removal. The two methods of purification that were employed are described below.

The first method was chromatographic, and was only applied to crude products of reactions of semi-stabilised ylides. A small sample of the crude product was eluted through a neutral alumina plug using pentane as the eluting solvent to remove the remaining aldehyde, salts and phosphine oxide.[3] The pentane was then removed to give a mixture of the isomers of the stilbene that was free of impurities. This purified product was characterised by NMR, and assignment of the *Z/E* ratio was made as described above. Unless otherwise indicated, the integrations of all signals quoted for each compound are consistent with the assigned *Z/E* ratio. For 2-bromo-2′-fluorostilbene the same purification procedure was followed but with a silica plug. The non-isomerisation of the stilbenes purified in this way was demonstrated by a series of isomerisation tests—for further details see Sect. 4.4.1. This technique was applied for 2-fluorostilbene, 2,2′-dichlorostilbene, 2-bromo-2′-fluorostilbene, 2,2′-dimethoxystilbene, 2-bromo-2′-methoxystilbene, 2-bromostilbene (for the samples synthesised by reactions of benzaldehyde with 2-bromobenzylidene-methyldiphenylphosphorane and 2-bromobenzylidene-triphenylphosphorane, and of 2-bromobenzaldehyde with benzylidene-triphenylphosphorane), 2,2′-diiodostilbene (from 2-iodobenzylidene-triphenylphosphorane and 2-iodobenzaldehyde), and 2-methoxystilbene (for the samples synthesised by reactions of benzaldehyde with 2-methoxybenzylidenemethyldiphenylphosphorane and 2-methoxybenzylidene-triphenylphosphorane, and of 2-methoxybenzaldehyde with benzylidene-methyldiphenylphosphorane).

The second method was more general, and was applied to representative examples of all the types of alkene produced in this project. It involved the addition of neat oxalyl chloride (1.1 equivalents) to the crude product to form chlorophosphonium salt, followed by six cyclohexane washes of the resulting yellow oil. The cyclohexane washings were decanted, combined and filtered through a cannula with a Whatman® grade GFD glass microfibre filter secured over the end with PTFE tape. The clear filtrate was washed with saturated $NaHCO_3$ solution (2 × 2 ml) and then HCl solution (1 mol L^{-1}). The isolated organic phase was then dried over $MgSO_4$, filtered and concentrated *in vacuo* to give the purified alkene whose *Z/E* ratio matched that of the crude product, as judged by ^{1}H NMR. See Chap. 3 and Sect. 4.9 for further details on this method, and on the compounds to which the method was applied.

An example of an assignment of the *Z/E* ratio of an alkene produced in a Wittig reaction is shown below. The reaction in question is the production of 2-iodostilbene from benzylidenemethyldiphenylphosphorane and 2-iodobenzaldehyde (see relevant part if Sect. 4.4.2 for details on signal assignments). The reaction was subjected to the aqueous work-up conditions described above, and the section of the ^{1}H NMR of the crude product that contains all of the alkene signals is shown in Fig. 4.1. The spectrum also contains phosphine oxide. Comparison of the chemical

[3] An alumina plug was used here rather than a silica plug, as was done in Ref. [22], as it has previously been demonstrated by work in our group that some alkenes are more prone to isomerisation in contact with silica than with alumina (see Ref. [23]).

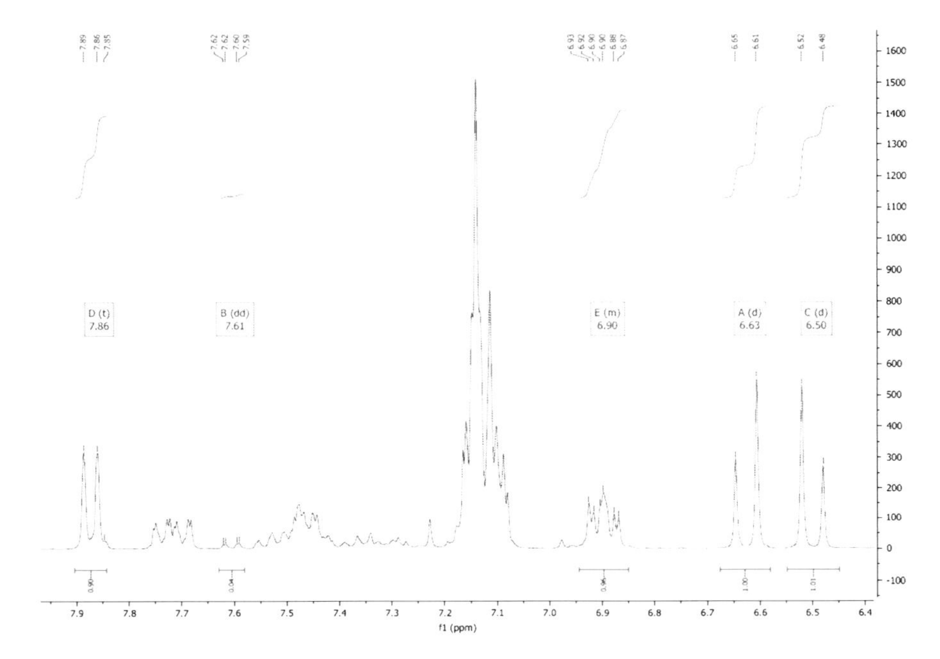

Fig. 4.1 Alkene region of the ^{1}H NMR spectrum of the crude 2-iodostilbene produced in a Wittig reaction

shifts, relative integrations and multiplicities of the signals present with those of authentic samples of the alkene isomers allows the signals at 7.86 (not baseline-separated from an *E*-alkene signal), 6.90, 6.63 (doublet, $J = 12.1$ Hz, characteristic of *Z*-alkene) and 6.50 ($J = 12.1$ Hz, characteristic of *Z*-alkene) to be assigned to the *Z*-isomer, and the double-doublet at 7.61 ($J = 7.8$ and 1.2 Hz) to be assigned to the *E*-isomer. Comparison of the integrations of the baseline-separated signals allows the *Z/E* ratio to be assigned to be 96:4.

4.4 Wittig Reactions of Semi-stabilised Ylides with Benzaldehdyes

4.4.1 Work-Up Procedure and Stilbene Isomerisation Tests

Work-Up

All reactions were carried out by one of the general procedures A or B described in Sect. 4.3. For work-up, distilled water (1–2 ml aliquot) was added to dissolve salt and quench remaining base. The resulting mixture was added to a separating funnel, and the aqueous phase was then extracted of product three times with pentane or cyclohexane (each aliqout being approximately twice the volume of THF used in the reaction). The separated organic phases were combined and dried over $MgSO_4$ or Na_2SO_4. After removal of drying agent by filtration, the solvent was removed *in vacuo*. An oil was typically obtained, which in some cases crystallised upon standing. The crude product (which typically contained significant quantities of phosphine oxide and some aldehyde) was then characterised by NMR using $CDCl_3$ as solvent. ^1H, ^{31}P and COSY NMR techniques were typically employed. Conversion was observed to be high in all cases by the large (relative) amount of alkene present in the crude product. In a large number of selected examples, this high conversion was confirmed by isolation of the alkene product. Yields no lower than 70 %, and in many cases greater than 90 % were observed in these examples. Isolation of the alkene product was done by column chromatography on neutral alumina using cyclohexane as the eluting solvent, or by making use of the oxalyl chloride treatment and filtration technique that is described in Sect. 4.3 above, and discussed in full in Chap. 3. The high conversion was substantiated by NMR analysis of the *unquenched* reaction mixture. The THF solvent was removed *in vacuo* and $CDCl_3$ was added, causing the dissolution of everything present but the inorganic salt. Filtration of this mixture into an NMR tube gave a transparent solution of the crude product. ^1H and ^{31}P NMR of this generally showed high consumption of phosphonium salt and aldehyde, and the expected large proportions of alkene and phosphine oxide. The *Z/E* ratio of the alkene in the crude product was determined by comparison of the integrations of all of the baseline-separated signals belonging to the *E* and *Z* isomers. The signals used to determine the *Z/E* ratio for each alkene synthesised are listed below.

Isomerisation Tests

To ensure that no isomerisation of the *Z*-stilbenes was occuring due to contact with the neutral alumina, samples of 2-bromostilbene, 2,2′-dibromostilbene, 2-iodostilbene, 2-methoxystilbene, 2,2′-dimethoxystilbene and 2-bromo-2′-fluorostilbene containing a high proportion of the *Z*-isomer were subjected to elution through a neutral alumina column using pentane as solvent. A ^1H NMR spectrum of the pure stilbene obtained as a mixture of isomers after solvent removal was taken, and the *Z/E* ratio of the stilbene sample was determined. This sample was then subjected to a second alumina column using pentane as the eluting solvent, a ^1H NMR spectrum of the resulting stilbene mixture was taken, and the *Z/E* ratio was established. In all cases mentioned above the *Z/E* ratio was found to be unchanged by the process of elution through th alumina plug. It had previously been shown that isomerisation under these conditions does not occur for 2-fluorostilbene or 2-chlorostilbene, but does occur for 2,2′-difluorostilbene [23]. Partial isomerisation of a sample *Z*-2, 2′-diiodostilbene dissolved in $CDCl_3$ in an NMR tube when left exposed to light was observed. The samples of stilbene that were purified by oxalyl chloride treatment and filtration (see below, and Sect. 4.9) were also shown not to undergo isomerisation during this process by comparison of the *Z/E* ratios of the crude and purified products, both of which were established by ^1H NMR.

4.4.2 Synthesis of Stilbenes from Benzylides

Stilbene [24, 25]

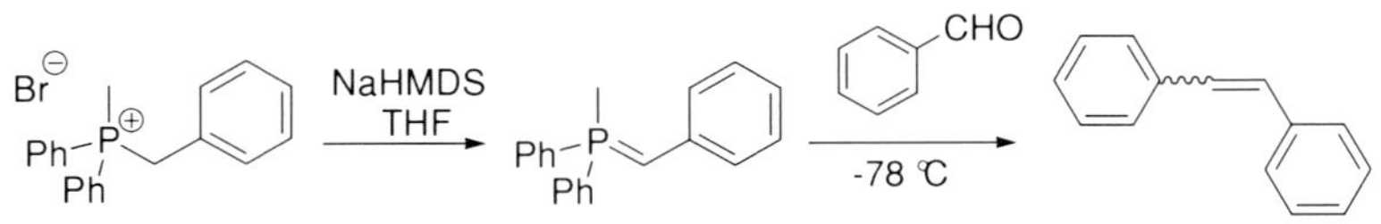

From benzylmethyldiphenylphosphonium bromide (986 mg, 2.66 mmol), NaHMDS (488 mg, 2.66 mmol), and benzaldehyde (0.27 ml, 0.28 g, 2.6 mmol) by procedure B. The crude product was treated with oxalyl chloride (0.29 ml, 0.44 g, 3.4 mmol) to give the product as a clear oil that crystallised on standing (0.46 g, 98 %). See Table 2.1 entry 6.
^{1}H spectrum of product after treatment with oxalyl chloride (300 MHz, $CDCl_3$):

Assigned to *E*-isomer: δ 7.56-7.46 (m, 4H), 7.41-7.31 (m, 4H), 7.11 (s, 2H, alkene H) [24].

Assigned to *Z*-isomer: δ 6.60 (s, 2H, alkene H) [25].

Integrations indicate a *Z/E* ratio of 15:85.

2-fluorostilbene [26, 27]

(a) From benzylmethyldiphenylphosphonium bromide (0.895 g, 2.41 mmol), NaHMDS (2.40 ml of a 1 mol L^{-1} solution in THF, 2.40 mmol) and 2-fluorobenzaldehyde (0.301 g, 2.43 mmol) by procedure A. See Table 2.1, entry 4.

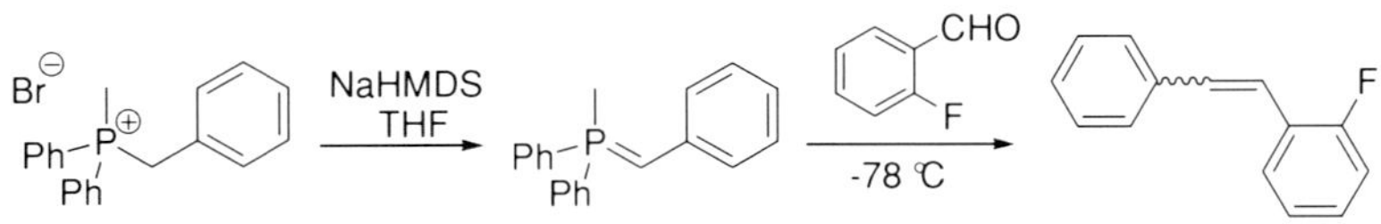

^{1}H NMR (500 MHz, $CDCl_3$) of the alumina plug purified product:

Assigned to *E*-isomer: δ 7.65-7.61 (m, 1H), 7.58-7.54 (m, 2H), 7.41-7.38 (m, 2H), 7.12-7.09 (m, 1H) [26].

Assigned to *Z*-isomer: δ 7.08-7.03 (m, 1H), 6.95 (m, 1H), 6.74 (d, ^{3}J = 12.2, 1H), 6.63 (d, ^{3}J = 12.2, 1H) [27].

^{19}F NMR (376 MHz, $CDCl_3$): −115.36 (m, *Z*-alkene), −118.37 (m, *E*-alkene). Integrations in both ^{1}H and ^{19}F NMR spectra give *Z/E* ratio of 84:16.

(b) From 2-fluorobenzylmethyldiphenylphosphonium bromide (0.933 g, 2.40 mmol), NaHMDS (2.40 ml of a 1 mol L^{-1} solution in THF, 2.40 mmol) and benzaldehyde (0.255 g, 2.40 mmol) by procedure A. See Table 2.1 entry 10.

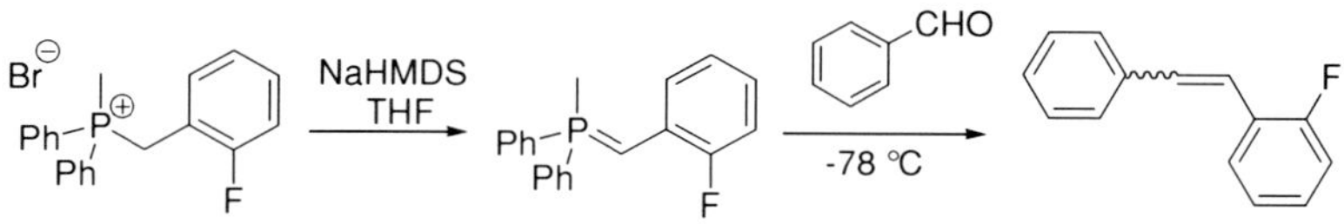

^{1}H NMR (400 MHz, $CDCl_3$) of the alumina plug purified product:

Assigned to *E*-isomer: δ 7.65-7.61 (app td, J = 7.7, 1.7, 1H), 7.41-7.37 (m, 2H) [26].

Assigned to *Z*-isomer: δ 7.08-7.03 (m, 1H), 6.97-6.93 (m, 1H), 6.74 (d, ^3J = 12.2, 1H), 6.63 (d, ^3J = 12.9, 1H) [27].

^{19}F spectrum (376 MHz, $CDCl_3$): δ -115.36 (m, *Z*-alkene), −118.37 (m, *E*-alkene). Integrations in both ^{1}H and ^{19}F NMR spectra give a *Z/E* ratio of 41:59.

2-chlorostilbene [28, 29]

(a) From benzylmethyldiphenylphosphonium bromide (160 mg, 0.431 mmol), KHMDS (82 mg, 0.41 mmol) and 2-chlorobenzaldehyde (0.046 ml, 57 mg, 0.41 mmol using 100 μl syrninge) by procedure B. See Table 2.1 entry 1.

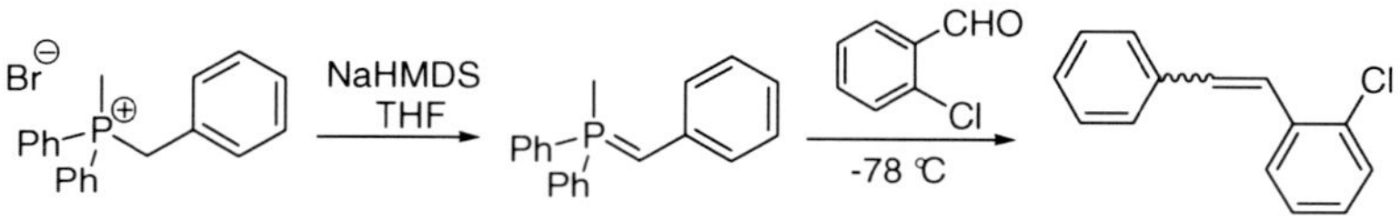

The *Z/E* ratio was assigned based on the ^{1}H NMR spectrum (500 MHz, $CDCl_3$) of crude product:

Assigned to *E*-isomer: δ 7.08 (d, ^3J = 16.4, *trans* alkene H) [28].

Assigned to *Z*-isomer: δ 7.03 (td, J = 7.5, 0.7, 1H), 6.73 (d, ^3J = 12.2, *cis* alkene H, 1H), 6.69 (d, ^3J = 12.2, *cis* alkene H, 1H) [29].

Integrations give *Z/E* ratio of 92:8.

The crude product was purified by treatment with oxalyl chloride (0.04 ml, 0.06 g, 0.5 mmol, see Sect. 4.3) giving the alkene as a clear oil (77 mg, 85 %).

(b) From 2-chlorobenzylmethyldiphenylphosphonium bromide (0.959 g, 2.36 mmol), NaHMDS (2.36 ml of a 1 mol L^{-1} solution in THF, 2.40 mmol) and benzaldehyde (0.26 ml, 0.27 g, 2.5 mmol) by procedure A. See Table 2.1 entry 7.

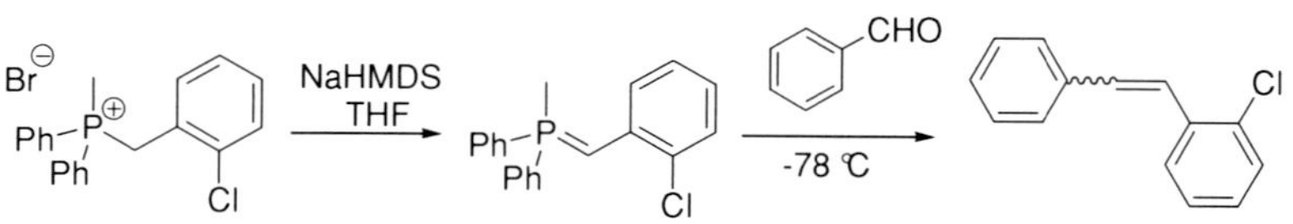

^{1}H NMR of crude stilbene mixture (500 MHz, $CDCl_3$):

Assigned to *E*-isomer: δ 7.70 (m, partially obscured by phosphine oxide signal), 7.08 (d, ^3J = 16.2, *trans* alkene H, 1H) [28].

Assigned to Z-isomer: δ 7.03 (td, J = 7.5, 0.7, 1H), 6.73 (d, ^3J = 12.2, *cis* alkene H, 1H), 6.69 (d, ^3J = 12.2, *cis* alkene H, 1H) [29].

Integrations give *Z/E* ratio of 34:66.

2-bromostilbene [30]

(a) From benzylmethyldiphenylphosphonium bromide (117 mg, 0.315 mmol), KHMDS (60 mg, 0.30 mmol) and 2-bromobenzaldehyde (0.035 ml, 5.5 mg, 0.3 mmol, using 100 µl syrninge) by procedure B. See Table 2.1 entry 2.

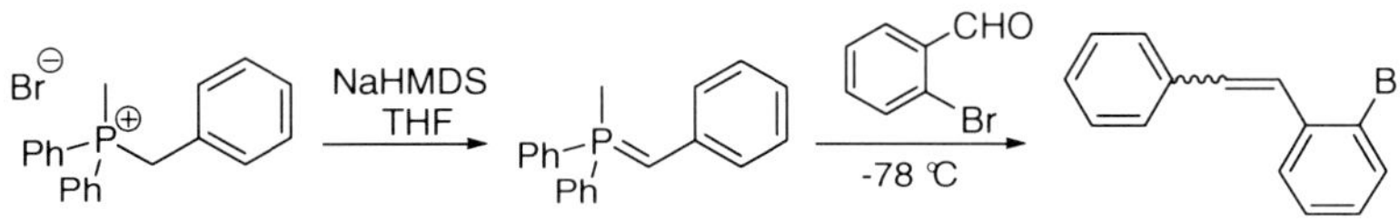

The crude product was purified by treatment with oxalyl chloride (0.03 ml, 0.4 mmol, see Sect. 4.3) giving the alkene as a clear oil (90 mg, 90 %). ^{1}H NMR (500 MHz, $CDCl_3$) of stilbene mixture obtained after oxalyl chloride treatment:

Assigned to *E*-isomer: δ 7.70 (dd, ^3J = 7.83 Hz, ^4J = 1.51 Hz, 1H), 7.51 (d, ^3J = 16.2, *trans* alkene H, 1H) [30].

Assigned to Z-isomer: δ 7.60 (m, 1H), 7.24-7.16 (m, 6H), 6.73 (d, ^3J = 12.1, *cis* alkene H, 1H), 6.66 (d, ^3J = 12.1, *cis* alkene H, 1H) [30].

Integrations give *Z/E* ratio of 94:6.

(b) From benzyltriphenylphosphonium bromide (1.03 g, 2.38 mmol)), NaHMDS (2.40 ml of a 1 mol L^{-1} solution in THF, 2.40 mmol) and 2-bromobenzaldehyde (0.28 ml, 0.44 g, 2.6 mmol) by procedure A. See Table 2.2 entry 1.

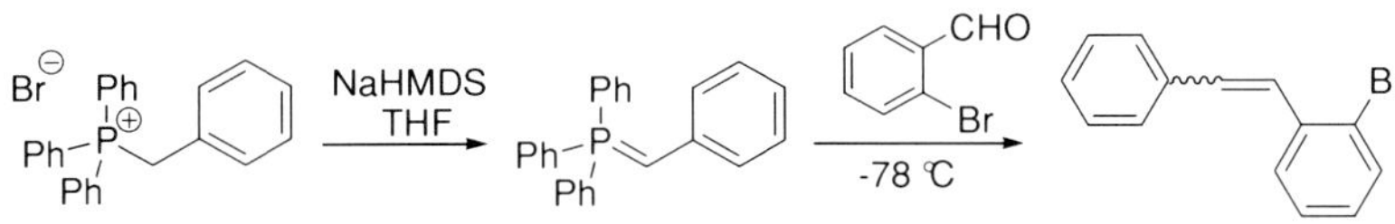

^{1}H NMR (500 MHz, $CDCl_3$) of alumina plug-purified stilbene mixture:

Assigned to *E*-isomer: δ 7.07 (d, 6.64 ^3J = 16.2, *trans* alkene H, 1H) [30].

Assigned to Z-isomer: δ 6.72 (1H, d, ^3J = 12.1, *cis* alkene H, 1H), 6.64 (1H, d, ^3J = 12.1, *cis* alkene H, 1H) [30].

Integrations give *Z/E* ratio of 87:13.

(c) From 2-bromobenzylmethyldiphenylphosphonium bromide (270 mg, 0.60 mmol), NaHMDS (114 mg, 0.57 mmol) and benzaldehyde (0.06 ml, 0.1 g, 0.6 mmol). See Table 2.1 entry 8.

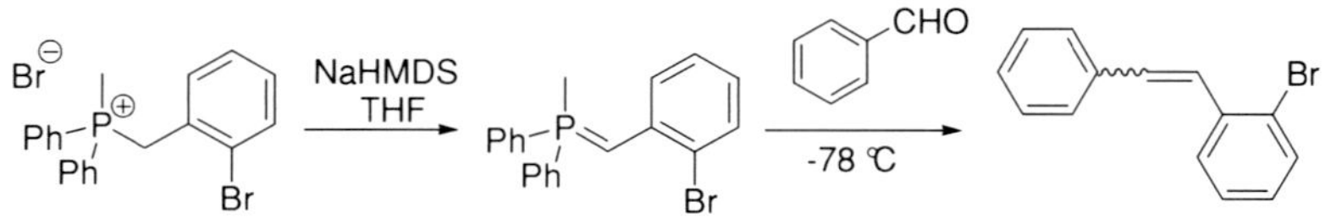

^{1}H NMR (500 MHz, $CDCl_3$) of alumina plug-purified stilbene mixture:
Assigned to *E*-isomer: δ 7.69 (dd, $^3J = 7.81$, $^4J = 1.7$, 1H), 7.60 (dd, slightly obscured by other signals, 2H), 7.57 (m, poorly resolved, 2H), 7.49 (d, $J = 16.2$), 7.42-7.38 (m, 2H), 7.35-7.29 (m, 2H), 7.06 (d, $^3J = 16.3$, *trans* alkene H, 1H) [30].

Assigned to *Z*-isomer: δ 6.71 (1H, d, $^3J = 12.2$, *cis* alkene, 1H), 6.64 (d, $^3J = 12.2$, *cis* alkene, 1H). Other signals overlap with those of *E*-isomer. [30]. Integrations give *Z/E* ratio of 33:67.

The crude product was purified by treatment with oxalyl chloride (0.06 ml, 0.7 mmol, see Sect. 4.3) giving the alkene as a clear oil (0.14 g, 74 %).

(d) From 2-bromobenzyltriphenylphosphonium bromide (1.24 g, 2.41 mmol), NaHMDS (2.40 ml of a 1 mol L^{-1} solution in THF, 2.40 mmol) and benzaldehyde (0.24 ml, 0.25 g, 2.4 mmol). See Table 2.2 entry 8.

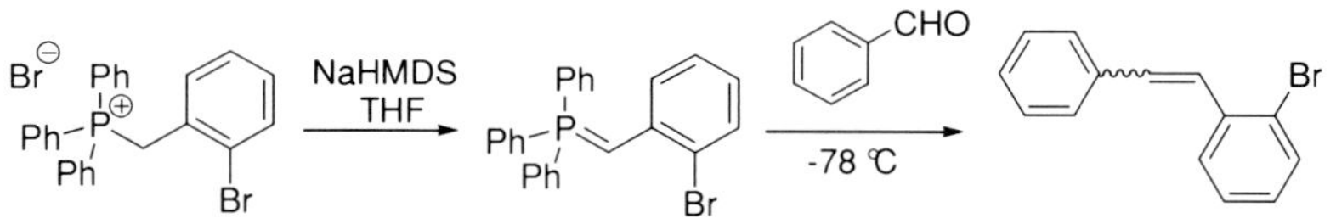

^{1}H NMR (500 MHz, $CDCl_3$) of alumina plug-purified stilbene mixture:

Assigned to *E*-isomer: δ 7.07 (d, $^3J = 16.2$, 1H) [30].

Assigned to *Z*-isomer: δ 6.72 (d, $^3J = 12.2$, 1H), 6.64 (d, $^3J = 12.2$, 1H) [30].

Integrations give *Z/E* ratio of 42:58.

2-iodostilbene

(a) From benzylmethyldiphenylphosphonium bromide (1.55 g, 4.16 mmol), KHMDS (0.836 g, 4.0 mmol) and 2-iodobenzaldehyde (0.934 g, 4.03 mmol as a solution in 1 ml dry THF) using procedure B. See Table 2.1 entry 3.

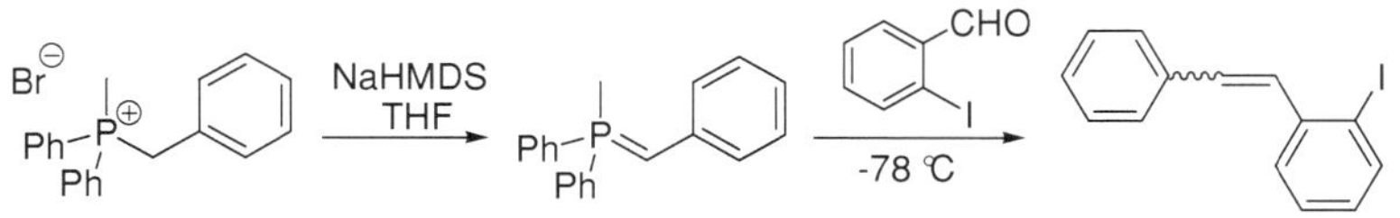

^{1}H NMR (500 MHz, $CDCl_3$) of crude stilbene mixture:

Assigned to *E*-isomer: δ 7.64 (dd, J = 7.8, 1.3, 1H, aromatic H).[4]
Assigned to *Z*-isomer: δ 6.63 (d, ^{3}J = 12.1, 1H, alkene H), 6.50 (d, ^{3}J = 12.1, 1H, alkene H).

Integrations indicate a *Z/E* ratio of 96:4.

After normal work-up and NMR analysis of the crude product, the *Z*-alkene was isolated by column chromatography on neutral alumina (Brockmann grade I) using cyclohexane as the eluting solvent (R_f=0.54). 1.09 g (89 %) of the *Z*-isomer was obtained as a clear oil after solvent removal.

^{1}H NMR (500 MHz, $CDCl_3$) δ 7.86 (d, J = 7.9, 1H, I ring H-3), 7.17-7.07 (m, 7H, contains Ph H-2, H-3, H-4, H-5 & H-6; I ring H-5 & H-6), 6.88 (ddd, J = 8.4, 4.5, 0.9, 1H, I ring H-4), 6.61 (d, J = 12.1, 1H, PhC*H*=C), 6.50 (d, J = 12.0, 1H, 2-IC_6H_4C*H*=C).

^{13}C NMR (126 MHz, $CDCl_3$) δ 141.8 (I ring C-1), 139.1 (I ring C-3), 136.3 (Ph C-1), 133.7 (2-IC_6H_4C*H*=C), 131.1 (PhC*H*=C), 130.2 (I ring C-6), 129.1 (Ph C-2), 128.7 (I ring C-4), 128.2 (Ph C-3), 128.0 (I ring C-5), 127.3 (Ph C-4), 99.8 (I ring C-2).

HRMS (TOF MS EI+): 305.9906 (Calculated for $[M]^+$ = $C_{14}H_{11}I$); found 305.9909 (1.0 ppm), 179.0861 (Calculated for $[M\text{-}I]^+$ = $C_{14}H_{11}$) found 179.0873 (1.7 ppm), 178.0782 (Calculated for $[M\text{-}H\text{–}I]^+$ = $C_{14}H_{10}$); found 178.0781 (0.6 ppm).

IR: 3,060-3,020 (w, alkene and aromatic C–H stretch), 1,630 (w, alkene C=C stretch), 1,584 (m), 1,556 (m), 1,493 (m), 1,464 (s), 1437 (s) (aromatic C–C stretch), 1,024 (s), 1,012 (s) (aromatic C–H in plane bend), 778 (s, C–H out of plane bend, 1,2-disubstituted arene), 755 (s, C–H out of plane bend, monosubstituted arene), 736 (s), 696 (s).

(b) From benzyltriphenylphosphonium bromide (1.034 g, 2.386 mmol), NaHMDS (2.36 ml of a 1 mol L^{-1} solution in THF, 2.36 mmol) and 2-iodobenzaldehyde (0.554 g, 2.39 mmol as a solution in 1.11 ml dry THF). See Table 2.2 entry 2.

[4] Characterisation details for *E*-2-iodostilbene is given above in this section.

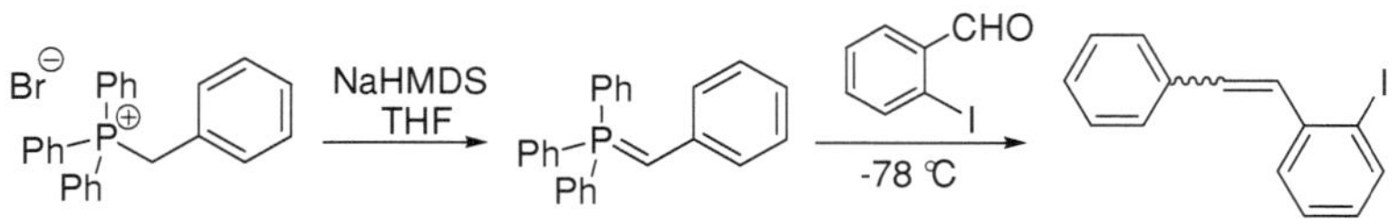

^{1}H NMR (500 MHz, $CDCl_3$) of crude stilbene mixture:

Assigned to *E*-isomer: δ 7.64 (dd, J = 7.9, 1.5, 1H).[4]

Assigned to *Z*-isomer: δ 6.66 (d, ^3J = 12.2, 1H), 6.53 (d, ^3J = 12.2, 1H).[5]

Integrations indicate a *Z/E* ratio of 88:12.

(c) From 2-iodobenzylmethyldiphenylphosphonium bromide (456 mg, 0.92 mmol), NaHMDS (167 mg, 0.91 mmol) and benzaldehyde (0.093 ml, 97 mg, 0.92 mmol by 100 μl syringe). See Table 2.1 entry 9.

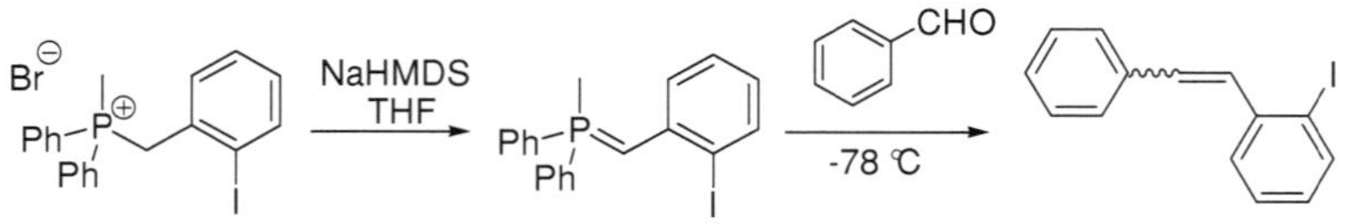

^{1}H NMR (500 MHz, $CDCl_3$) of crude stilbene mixture:

Assigned to *E*-isomer: δ 7.64 (dd, ^3J = 7.8, ^4J = 1.4, 1H).

Assigned to *Z*-isomer: δ 6.67 (d, ^3J = 12.5, 1H), 6.54 (d, ^3J = 12.9, 1H).[5]

Integrations indicate a *Z/E* ratio of 28:72.

The crude product was then purified by column chromatography on neutral alumina (Brockmann grade I) using cyclohexane as the eluting solvent (R_f = 0.43), yielding the *Z*-isomer as the first fraction, and the *E*-isomer as the second. Combination of the fractions containing only one isomer and solvent removal *in vacuo* gave samples of each of the Z and the *E* isomers in respective yields of 58 mg (21 %) & 168 mg (60 %). The former of these was confirmed as being identical to the product obtained in reaction (a) above by NMR.

E-isomer:

^{1}H NMR (500 MHz, $CDCl_3$) δ 7.87 (dd, J = 11.6, 4.4, 1H, I ring H-3), 7.61 (dd, J = 7.8, 1.0, 1H, I ring H-6), 7.55 (d, J = 7.4, 2H, PhH-2), 7.40-7.26 (m, 5H, contains 2-IC_6H_4CH=C; Ph H-3 & H-4; I ring H-4), 6.99-6.91 (m, 2H, contains PhC*H*=C & I ring H-5).

[5] Characterisation data for for Z-2-iodostilbene is given above in this section.

^{13}C NMR (126 MHz, $CDCl_3$) δ 140.4 (I ring C-1), 139.6 (I ring C-3), 137.0 (Ph C-1), 132.5 (2-IC_6H_4CH=C), 131.6 (Ph*C*H=C), 129.0 (I ring C-5), 128.8 (Ph C-3), 128.4 (I ring C-4), 128.1 (Ph C-4), 126.8 (Ph C-2), 126.3 (I ring C-6), 100.4 (I ring C-2). HRMS (TOF MS EI + m/z): Calc. for $[M]^+$= $C_{14}H_{11}I$ 305.9906; found 305.9920 (4.6 ppm).

IR: 3,080-3,000 (m, alkene and aromatic C–H stretch), 1,495 (s), 1,630 (w, C = C stretch), 1,462 (s), 1,448 (s), 1,432 (s), 1,264 (s), 1,011 (s), 959 (s, *E*-alkene C–H bend), 780-760 (two strong bands, 1,2-disubstituted arene C–H bend), 706 (s), 690 (s). MP 52-54 °C (lit. 54-56 °C) [31].

2-methoxystilbene [32]

(a) From benzylmethyldiphenylphosphonium chloride (0.896 g, 2.41 mmol), NaHMDS (2.40 ml of a 1 mol L^{-1} solution in THF, 2.40 mmol) and 2-methoxybenzaldehyde (0.29 ml, 0.33 g, 2.41 mmol) by procedure A. See Table 2.1 entry 5.

^{1}H NMR (500 MHz, $CDCl_3$) of alumina-plug purified product:

Assigned to *E*-isomer: δ 7.61 (dd, J = 7.7, 1.5), 7.55 (app d, J = 7.2, 2H), 7.50 (d, J = 165, 1H, alkene H), 7.36 (app t, J = 7.7, 2H), 7.13 (d, J = 16.5, 1H, alkene H), 6.99 (app t, J = 8.2, 1H), 3.91 (s, 3H, CH_3).

Assigned to *Z*-isomer: δ 6.76 (m, 1H), 6.70 (d, J = 12.3, 1H, alkene H), 6.63 (d, J = 12.3, 1H, alkene H), 3.84 (s, 3H, CH_3).

Integrations of these signals indicate a *Z/E* ratio of 88:12. Signal of *E*-isomer at δ 6.92 (app d, J = 8.2) overlaps with signal of *Z*-isomer. Total integral of the resulting multiplet is 1.14H relative to 1H of *E*-isomer.

This reaction was also carried out using a strictly salt-free ylide by the following procedure:

Benzylmethyldiphenylphosphonium chloride (0.903 g, 2.43 mmol) was added to a Schlenk flask under a nitrogen atmosphere. Dry THF (12.5 ml) was added to make a transparent suspension of the salt (approximately 0.2 mol/l). A 1 mol l^{-1} solution of NaHMDS in THF was added (2.40 ml, 2.40 mmol). A precipitate of NaCl was immediately observed upon addition of the NaHMDS. A deep orange solution of the ylide resulted, and was stirred for 1 h. The stirring of the reaction mixture was then stopped to allow the precipitate to settle. The coloured solution containing the ylide was then isolated by cannula filtration into a second nitrogen flushed Schlenk flask (using a cannula that had been in the oven at 100 °C for 1 h) leaving a white precipitate of NaCl in the original Schlenk flask. This was washed with two 1 ml aliquots of dry THF which were then cannula filtered into the second Schlenk flask.

After this time, the salt free solution of the ylide was cooled to −78 °C in a dry ice/acetone bath, or a liquid nitrogen/ethanol bath. 2-methoxybenzaldehyde[6] (0.29 ml, 0.33 g, 2.4 mmol) was syringed from a sealed, nitrogen flushed crimped vessel into the Schlenk flask. The resulting solution was left to stir at −78 °C for 5 min, then removed from the cold bath and allowed to warm slowly back to room temperature. The solution was left to stir at room temperature for 12 h.

Non-aqueous work up: The THF reaction solvent was removed on the vacuum line from the reaction mixture, and pentane (ca. 20 ml) was added to dissolve stilbenes. The resulting mixture was then filtered through a cotton wool plug into a round-bottom flask of known mass to remove most of the phosphine oxide, giving a clear or faintly yellow solution. The pentane was then removed from the filtrate on a rotary evaporator, and final removal of solvent was done using the vacuum line in a fume cupboard. An oil was typically obtained, which in some cases crystallised upon standing. A ^{1}H NMR spectrum was taken of the crude product on a 500 MHz spectrometer. Significant amounts of phosphine oxide remained at this stage. Assignment of the *Z/E* ratio for the sample of 2-methoxystilbene synthesised by this method was done as described in the general procedure.

^{1}H NMR (500 MHz, $CDCl_3$) of crude stilbene mixture (assigned by comparison with spectrum of mixture of isomers after alumina plug filtration):

Assigned to *E*-isomer: δ 7.62 (dd, $J = 7.5, 1.5$, 1H), 3.91 (s, 3H) [32].

Assigned to *Z*-isomer: δ 6.77 (m, 1H), 6.71 (d, $^3J = 12.3$, 1H), 6.64 (d, $^3J = 12.3$, 1H), 3.84 (s, 3H) [32].

Integrations indicate a *Z/E* ratio of 88:12. Signals due to 2-methoxybenzaldehyde were identified by spiking the NMR sample with that compound.

(b) From benzyltriphenylphosphonium bromide (1.06 g, 2.45 mmol), NaHMDS (2.4 ml of a 1 mol L^{-1} solution in THF, 2.4 mmol) and 2-methoxybenzaldehyde (0.30 ml, 0.34 g, 2.5 mmol) by procedure A. See Table 2.2 entry 4.

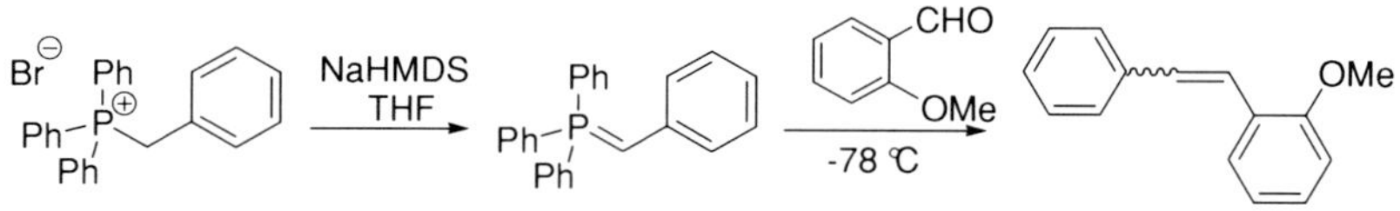

^{1}H NMR (500 MHz, $CDCl_3$) of crude stilbene mixture:

[6] It was ensured that the 2-methoxybenzaldehyde used was free of 2-methoxybenzoic acid by dissolving the compound in dichloromethane, and washing the dichloromethane solution with two aliquots of saturated aqueous sodium hydrogencarbonate solution. The dichloromethane phase was then dried using $MgSO_4$, and the drying agent was removed by filtration. Evaporation of the dichloromethane first on rotary evaporator and then on the vacuum line gave the 2-methoxybenzaldehyde.

Assigned to *E*-isomer: δ 7.62 (J = 7.5, 1.5, 1H, ArH) 3.91 (s, 3H), 7.50 (d, ^3J = 16.5, *trans* alkene H, 1H partially obscured) [32].

Assigned to *Z*-isomer: δ 6.77 (m, 1H), 6.71 (d, ^3J = 12.3, 1H, *cis* alkene H), 6.64 (d, ^3J = 12.3, *cis* alkene H, 1H), 3.84 (s, 3H) [32].

Integrations indicate a *Z/E* ratio of 90:10. Signals due to 2-methoxybenzaldehyde were identified by spiking the NMR sample with that compound.

(c) From 2-methoxybenzylmethyldiphenylphosphonium chloride (0.845 g, 2.37 mmol) NaHMDS (2.25 ml of a 1 mol L^{-1} solution in THF, 2.25 mmol) and benzaldehyde (0.24 ml, 0.25 g, 2.4 mmol) as a solution in 0.5 ml dry THF by procedure A. See Table 2.1 entry 12. The *Z/E* ratio was assigned based on the ^{1}H NMR spectrum of the alumina plug purified product.

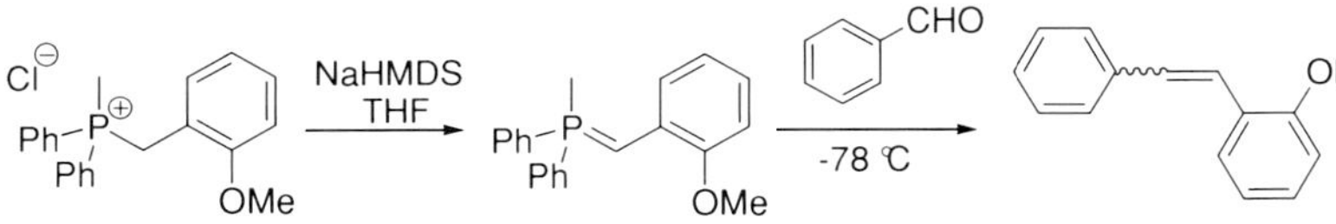

^{1}H NMR (500 MHz, $CDCl_3$) of stilbene mixture after alumina plug purification: Assigned to *E*-isomer: δ 7.62 (dd, J = 7.5, 1.5 Hz, 1H), 7.50 (d, ^3J = 16.5 Hz, 1H, *trans* alkene H), 7.36 (app t, 2H, J = 7.7), 7.13 (^3J = 16.5 Hz, 1H, *trans* alkene H, partially obscured), 3.91 (s, 3H) [32].

Assigned to *Z*-isomer: δ 6.77 (m, 1H), 6.71 (d, ^3J = 12.3 Hz, 1H, *cis* alkene H), 6.64 (d, ^3J = 12.3 Hz, 1H, *cis* alkene H), 3.84 (s, 3H) [32].

Integrations indicate a *Z/E* ratio of 12:88. Two further ^{1}H NMR spectra were obtained for this sample, the first one after spiking the sample with a small quantity of 2-methoxybenzaldehyde, and the second after spiking with 2-methylanisole. Analysis of these spectra allowed the signals due to these two compounds in the original ^{1}H spectrum of the crude product to be identified.

(d) From 2-methoxybenzyltriphenylphosphonium chloride (1.000 g, 2.387 mmol), NaHMDS (2.25 ml of a 1 mol L^{-1} solution in THF, 2.25 mmol) and benzaldehyde (0.25 ml, 0.26 g, 2.5 mmol) by procedure A. See Table 2.2 entry 6. *Z/E* ratio was assigned based on the ^{1}H NMR spectrum of the alumina plug purified product.

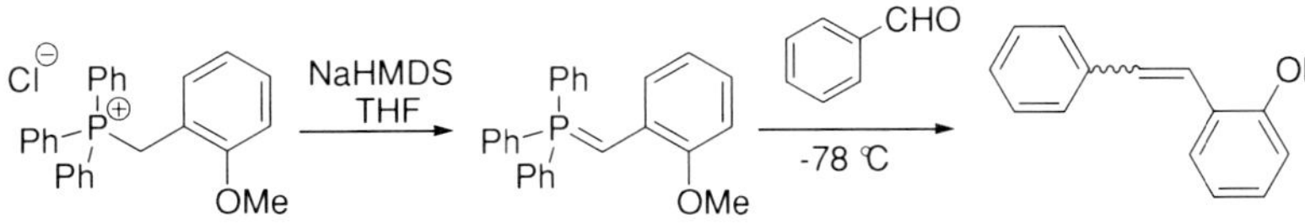

^{1}H NMR (500 MHz, $CDCl_3$) of alumina plug-purified stilbene mixture:

Assigned to *E*-isomer: δ 7.62 (dd, J = 7.5, 1.5), 7.50 (d, ^3J = 16.5, *trans* alkene H, 1H), 7.14 (d, ^3J = 16.5, *trans* alkene H, baseline obscured), 3.91 (s, 3H) [32].

Assigned to Z-isomer: δ 6.77 (m, 1H), 6.72 (d, ^3J = 12.3, *cis* alkene H, 1H), 6.66 (d, ^3J = 12.3, *cis* alkene H, 1H), 3.84 (s, 3H) [32].

Integrations indicate a *Z/E* ratio of 42:58. Two further ^{1}H NMR spectra were obtained for this sample, the first one after spiking the sample with a small quantity of 2-methoxybenzaldehyde, and the second after spiking with 2-methylanisole. Analysis of these spectra allowed the signals due to these two compounds in the original ^{1}H spectrum of the crude product to be identified.

2-methylstilbene [29, 33]

(a) From benzylmethyldiphenylphosphonium bromide (0.996 g, 2.68 mmol), NaHMDS (2.6 ml of a 1 mol L^{-1} solution in THF, 2.6 mmol) and 2-methylbenzaldehyde (0.32 ml, 0.33 g, 2.7 mmol) by procedure A. See Table 2.1, entry 26. *Z/E* ratio was assigned based on the ^{1}H NMR spectrum crude product.

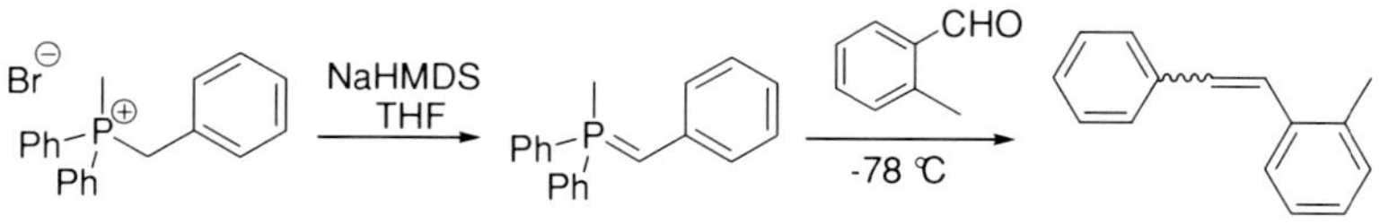

^{1}H NMR (400 MHz, $CDCl_3$) of crude stilbene mixture:

Assigned to *E*-isomer: δ 2.43 (s, 3H, CH_3) [33].

Assigned to Z-isomer: δ 6.65, 6.60 (pair of roofed doublets, ^3J = 12.3, 2H), 2.26 (s, 3H, CH_3) [29, 33].

Integrations indicate a *Z/E* ratio of 33:67.

(b) From 2-methylbenzylmethyldiphenylphosphonium chloride (1.00 g, 2.93 mmol), NaHMDS (2.8 ml of a 1 mol L^{-1} solution in THF, 2.8 mmol) and benzaldehyde (0.29 ml, .30 g, 2.9 mmol), by procedure A. See Table 2.1 entry 11.

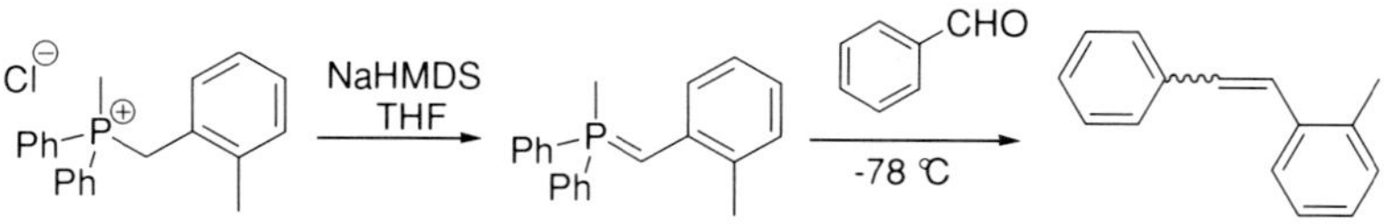

^{1}H NMR (400 MHz, $CDCl_3$) of crude stilbene mixture:

Assigned to *E*-isomer: δ 7.00 (d, $^3J_{HH}$ = 16.2, 1H), 2.43 (s, 3H, CH_3) [33].

Assigned to Z-isomer: δ 6.65, 6.60 (pair of roofed doublets, $^3J_{HH}$ = 12.3, 2H), 2.26 (s, 3H, CH_3) [29, 33].

Integrations indicate a *Z/E* ratio of 6:94.

2,2′-dichlorostilbene [34, 35]

(a) From 2-chlorobenzylmethyldiphenylphosphonium bromide (0.940 g, 2.317 mmol), NaHMDS (2.25 ml of a 1 mol L^{-1} solution in THF, 2.25 mmol) and

2-chlorobenzaldehyde (0.33 g, 2.34 mmol as a solution in 0.5 ml dry THF) by procedure A. See Table 2.1 entry 13.

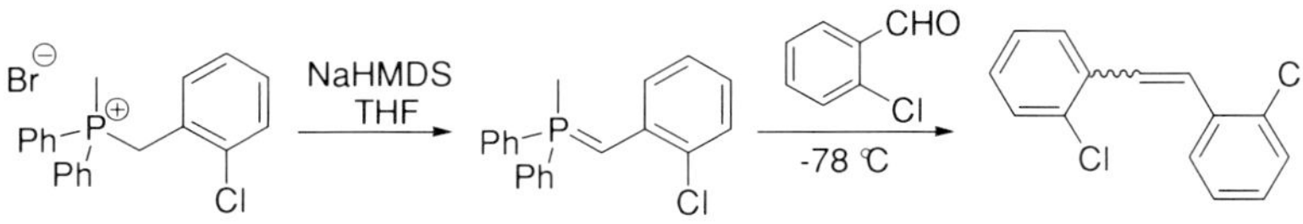

^{1}H NMR (500 MHz, $CDCl_3$) of alumina plug-purified stilbene mixture:

Assigned to *E*-isomer: δ 7.48 (s, *trans* alkene H, 2H) [34, 35].

Assigned to *Z*-isomer: δ 7.39 (dd, $J = 8.2$, 1.0, 2H), 7.15 (m, 2H), 7.01 (m, 4H), 6.87 (s, *cis* alkene H, 2H) [35].

Integrations give *Z/E* ratio of 97:3.

2, 2′dibromostilbene [36, 37]

(a) From 2-bromobenzylmethyldiphenylphosphonium bromide (2.47 g, 5.49 mmol), NaHMDS (5.0 ml of a 1 mol L^{-1} solution in THF, 5.0 mmol) and 2-bromobenzaldehyde (0.64 ml, 1.01 g, 5.49 mmol) by procedure A. See Table 2.1 entry 14. The crude product was treated with neat oxalyl chloride (0.50 ml, 0.75 g, 6.0 mmol), and the alkene was obtained as a clear oil (1.52 g, 90 %).

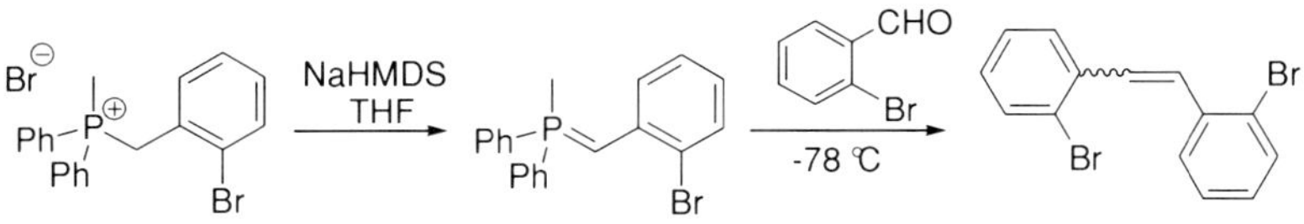

^{1}H NMR (500 MHz, $CDCl_3$) of purified stilbene mixture:

Assigned to *E*-isomer: δ 7.71 (dd, $J = 7.9$, 1.5, 2H), 7.59 (d, $J = 1.1$, partially obscured by signal at 7.55), 7.39 (s, 2H, *H*C=C*H*), 7.35-7.30 (m, 2H), 7.15-7.11 (m, 2H) [36, 37].

Assigned to *Z*-isomer: δ 7.55 (dd, $J = 7.8$, 1.2, 2H), 7.06-6.94 (m, 6H), 6.77 (s, 2H, *H*C=C*H*) [36, 37].

Integrations indicate a *Z/E* ratio of 98:2.

(b) From 2-bromobenzyltriphenylphosphonium bromide (1.182 g, 2.308 mmol), NaHMDS (2.30 ml of a 1 mol L^{-1} solution in THF, 2.30 mmol) and 2-bromobenzaldehyde (0.27 ml, 2.308 mmol). See Table 2.2 entry 7. The *Z/E* ratio was assigned based on the ^{1}H NMR spectrum of the alumina plug purified product.

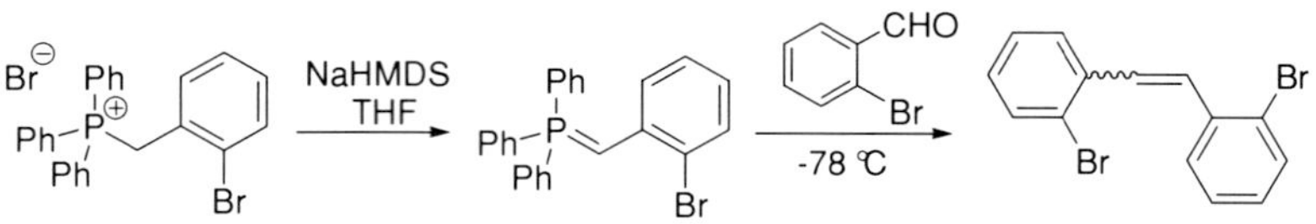

^{1}H NMR (500 MHz, $CDCl_3$) alumina plug-purified stilbene mixture:

Assigned to *E*-isomer: δ 7.73 (dd, $J = 7.8, 1.2$, 2H, ArH-6,) [36, 37].

Assigned to *Z*-isomer: δ 7.08-6.97 (m, 6H, ArH-3, ArH-4, ArH-5), 6.80 (s, *cis* alkene H, 2H), [36, 37].

Integrations give *Z/E* ratio of 94:6.

2,2′-diiodostilbene [38]

(a) From 2-iodobenzylmethyldiphenylphosphonium bromide (0.400 g, 0.805 mmol), NaHMDS (0.73 of a 1 mol L^{-1} solution in THF, 0.73 mmol) and 2-iodobenzaldehyde (1.6 ml of a 0.45 mol L^{-1} solution in dry THF, 0.73 mmol) by method A. See Table 2.1 entry 15.

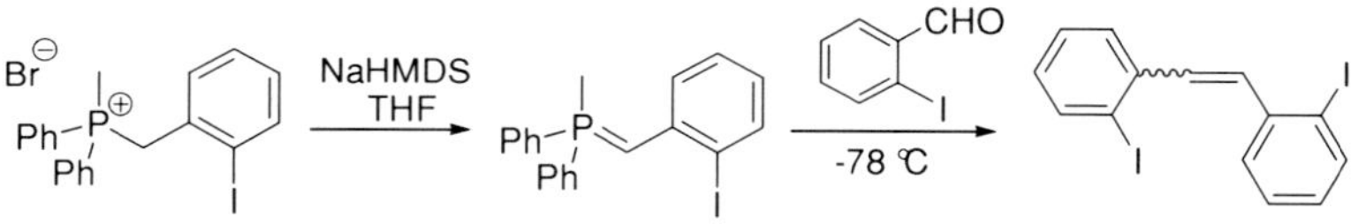

^{1}H NMR (300 MHz, $CDCl_3$) of the product after elution through an alumina plug:

Assigned to *E*-isomer: δ 7.67 (dd, $J = 7.8, 1.4$, 2H), 7.36 (t, $J = 8.0$, 2H), 7.18 (s, 2H, *trans* alkene H) [38].

Assigned to *Z*-isomer: δ 7.83 (dd, $J = 7.9, 0.9$, 2H), 7.03 (td, $J = 7.5, 1.0$, poor resolution on middle doublet of td, 2H), 6.62 (s, 2H, *cis* alkene H).

Integrations indicate a *Z/E* ratio of >99:1.

The crude product was eluted with cyclohexane through an alumina plug to remove aldehyde and phosphine oxide. The resulting mixture of isomers was recrystallised by dissolving in hot cyclohexane, allowing the solution thus obtained to cool to room temperature and then placing in a refrigerator overnight. The white crystals that resulted were isolated by filtration of the mixture, yielding a pure sample of *Z*-2, 2′-diiodostilbene (0.27 g, 78 % yield). Note that it was observed for a sample of *Z*-2, 2′-diiodostilbene (pure by NMR) subjected to GC–MS that two peaks were detected on the GC, each having an identical mass (by HRMS). It is suggested that this alkene undergoes partial isomerisation on the GC column and that the *Z* and *E* isomers have different retention times, thus giving two peaks of the same mass.

^{1}H NMR (500 MHz, $CDCl_3$): δ 7.84 (dd, J = 7.9, 1.0, 2H, ArH-6), 7.05 (td, J = 7.6, 1.1, 2H, ArH-4), 6.91 (dd, J = 7.7, 1.6, 2H, ArH-3), 6.86 (td, J = 7.7, 1.7, 2H, ArH-5), 6.62 (s, 2H, HC=CH).

^{13}C NMR (126 MHz, $CDCl_3$): δ 140.5 (*C*–I), 139.0 (ArC-6), 134.8 (H*C*=*C*H), 130.4 (ArC-3), 128.7 (ArC-5), 127.8 (ArC-4), 99.8 (ArC-1).

HRMS (m/z, TOF MS EI +): Calc. for $[M]^+ = C_{14}H_{10}I_2$ 431.8872, found 431.8878 (1.4 ppm).

Elemental Analysis: $C_{14}H_{10}I_2$ requires: C, 38.92; H, 2.33; Found: C, 38.94; H, 2.26;

IR: 3,049-3,100 (w, aromatic C–H stretch), 1,580 (w), 1,553 (w), 1,455 (m), 1,427 (m), 1,262 (w), 1,157 (s), 1,038 (w), 1,012 (s, aromatic C–H in-plane bend), 946 (w), 768-739 (4 strong bands, 1,2-disubstituted arene C–H bend).

MP (crystallised from cyclohexane) 65–67 °C.

(b) From 2-iodobenzyltriphenylphosphonium bromide (1.085 g, 1.94 mmol), NaHMDS (1.75 ml of a 1 mol L^{-1} solution in THF, 1.75 mmol) and 2-iodobenzaldehyde (0.41 g, 1.75 mmol, as a solution in 0.82 ml dry THF) by method A. See Table 2.1 entry 8.

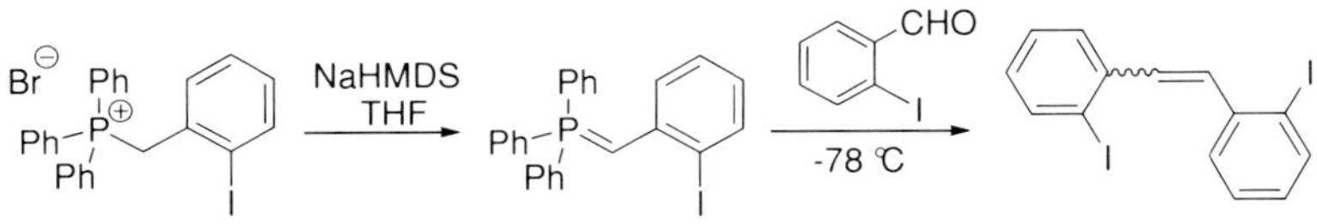

^{1}H NMR (500 MHz, $CDCl_3$) of crude stilbene mixture:

Assigned to *E*-isomer: δ 7.18 (s, 2H, alkene H) [38].

Assigned to *Z*-isomer: δ 6.94-6.85 (m, 4H), 6.63 (s, 2H, alkene H).[7]

Integrations indicate a *Z/E* ratio of 94:6.

2,2′-dimethoxystilbene [36]

(a) From 2-methoxybenzylmethyldiphenylphosphonium chloride and 2-methoxybenzaldehyde (Table 2.1 entry 20).

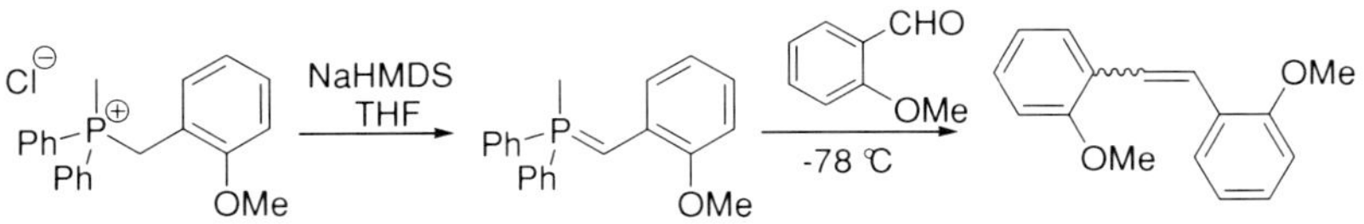

Attempts at synthesising 2,2′-bismethoxystilbene using non-dry 2-methoxybenzylmethyldiphenylphosphonium chloride (this salt is hygroscopic) were

[7] Characterisation details for *Z*-2,2′-diiodostilbene are given in this section.

unsuccessful due to decomposition of the ylide; significant loss of the ylidic colour was invariably noted if left to stir for a period. The use of dry phosphonium salt (stored in a glove box under argon) ameliorated this problem. The alkene was synthesised by procedure B using dry 2-methoxybenzyltriphenylphosphonium chloride (66 mg, 0.18 mmol), NaHMDS (32 mg, 0.17 mmol), and 2-methoxybenzaldehyde[2] (0.21 ml of a 0.83 mol L^{-1} solution in dry THF, 0.95 equivalents).

^{1}H NMR (500 MHz, $CDCl_3$) of stilbene mixture obtained after alumina plug treatment:

Assigned to *E*-isomer: δ 7.68 (dd, $J = 7.7$, 1.5, 2H, ArH-6), 7.50 (s, 2H, *trans* alkene H), 3.90 (s, 6H) [36].

Assigned to *Z*-isomer: δ 6.80 (s, 2H, *cis* alkene H), 6.77 (2H, dt, $J = 7.5$, 1.0 Hz, 1H, ArH-5), 3.84 (s, overlapping with 2-methylanisole signal) [36].

Integrations indicate a *Z/E* ratio of 48:52. Signals due to 2-methoxybenzaldehyde and 2-methylanisole (product of ylide hydrolysis from aqueous work-up) were identified by spiking the NMR sample with those compounds.

(b) From 2-methoxybenzyltriphenylphosphonium chloride (1.022 g, 2.44 mmol), NaHMDS (2.35 ml of a 1 mol L^{-1} solution in THF, 2.35 mmol) and 2-methoxybenzaldehyde (0.30 ml, 0.34 g, 2.5 mmol) by procedure A. See Table 2.2 entry 10. *Z/E* ratio was assigned based on the ^{1}H NMR spectrum of the alumina plug purified product, which was a clear oil that crystallised to give a white solid.

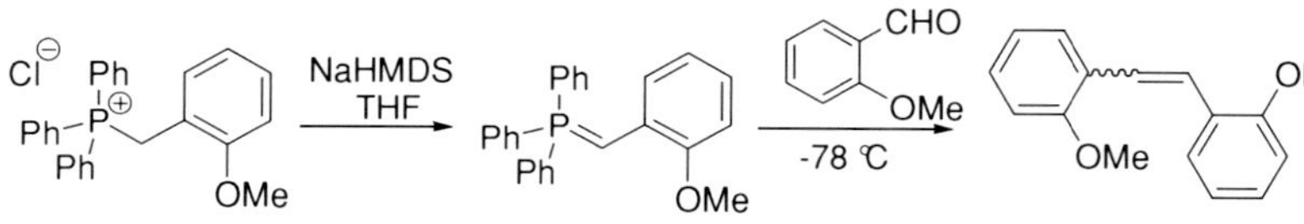

^{1}H NMR (500 MHz, $CDCl_3$) of sample purified by alumina plug:

Assigned to *E*-isomer: δ 7.70 (dd, 2H, ArH-6), 7.54 (s, 2H, *trans* alkene H), 3.92 (s, 6H, OCH_3) [36].

Assigned to *Z*-isomer: δ 6.83 (s, 2H, *cis* alkene H), 6.77-6.73 (m, 2H), 3.86 (s, 6H, OCH_3). [36].

Integrations indicate a *Z/E* ratio of 90:10. Signals due to 2-methoxybenzaldehyde and 2-methylanisole (product of ylide hydrolysis) identified by spiking the NMR sample with these compounds.

2-bromo-2′-methoxystilbene [39]

From 2-bromobenzylmethyldiphenylphosphonium bromide (113 mg, 0.25 mmol), NaHMDS (44 mg, 0.24 mmol) and 2-methoxybenzaldehyde (0.29 ml of a 0.83 mol L^{-1} solution in dry THF, 0.24 mmol) by procedure B. The crude mixture was eluted through an alumina plug to remove phosphine oxide and aldehyde. See Table 2.1 entry 21.

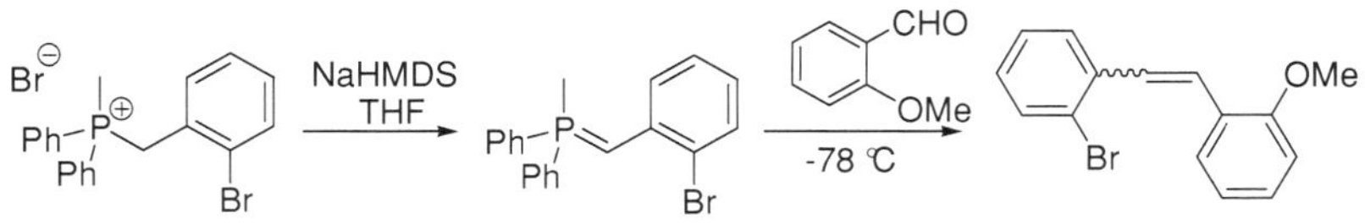

^{1}H NMR (500 MHz, $CDCl_3$) of sample purified by alumina plug:

Assigned to *Z*-isomer: δ 7.59-7.53 (m, 1H), 7.19-7.13 (m, 1H), 7.13-7.09 (m, 1H), 7.05-6.98 (m, 2H), 6.99-6.94 (m, 1H), 6.87-6.81 (m, 2H), 6.71-6.64 (m, 2H), 3.80 (s, 3H, OCH_3) [39].

Assigned to *E*-isomer: δ 7.71 (dd, $J = 7.9, 1.5$, 1H), 7.63 (dd, $J = 7.6, 1.5$, 1H), 7.48 (d, $J = 16.4$, 0H), 7.41 (d, $J = 16.4$, 0H), 3.88 (s, 3H, OCH_3).

Integrations indicate a *Z/E* ratio of 95:5.

2-bromo-2′-fluorostilbene

(a) From 2-bromobenzylmethyldiphenylphosphonium bromide (1.010 g, 2.24 mmol), NaHMDS (2.4 ml of a 1 mol L^{-1} solution in THF, 2.4 mmol) and 2-fluorobenzaldehyde (0.304 g, 2.45 mmol as solution in 0.5 ml dry THF) by procedure A. See Table 2.1 entry 18. The crude product was eluted through a silica plug to remove phosphine oxide and aldehyde, giving a pale yellow oil (0.51 g, 75 %).

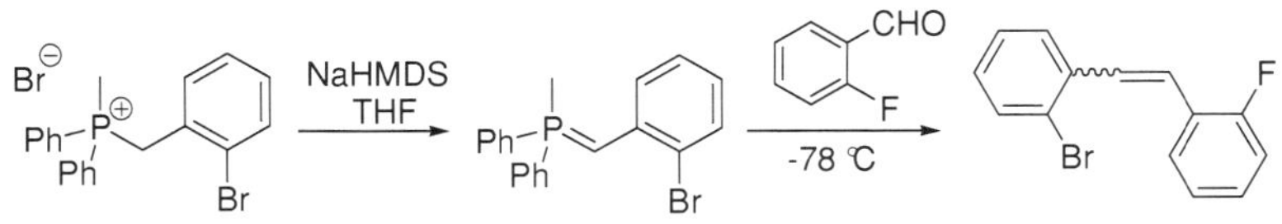

^{1}H NMR (400 MHz, $CDCl_3$) of silica plug-purified stilbene mixture:

Assigned to *E*-isomer: δ 7.72 (dd, $J = 8.1, 1.5$, 1H), 7.69 (td, $J = 7.8, 1.5$, 1H), 7.55 (d, $^3J = 16.4$, 1H, 2-FC_6H_4CH), 7.37-7.32 (m, 1H).

Assigned to *Z*-isomer: δ 7.05-7.00 (m, 2H), 6.88 (td, $J = 7.6, 1.2$, 1H), 6.79 (s, 2H, *HC*=*CH*).

Integrations indicate *Z/E* ratio of 84:16. The alkene isomers were separated by column chromatography on Merck standardised alumina 90 using cyclohexane as the eluting solvent. The *Z*-isomer eluted first, the *E*-isomer second.

Z-isomer:

^{1}H NMR (400 MHz, $CDCl_3$) δ 7.59 (m, 1H), 7.20-7.04 (m, 4H), 7.01 (app t, $J = 8.9$, 2H), 6.85 (app t, $J = 7.6$, 1H), 6.77 (s, 2H, *HC*=*CH*).

^{13}C NMR (101 MHz, $CDCl_3$) δ 160.4 (d, J = 248.4, C–F), 137.5 (s), 132.7 (s), 131.4 (d, J = 1.5), 130.5 (s), 130.4 (d, J = 3.3), 129.1 (d, J = 8.3), 128.8 (s), 127.0 (s), 124.2 (d, J = 13.8), 123.9 (s), 123.9 (d, J = 3.9), 123.5 (d, J = 3.5), 115.5 (d, J = 21.9).

^{19}F NMR (376 MHz, $CDCl_3$) δ −115.3-115.4 (m).

IR: 3,100-300 (w, alkene & aromatic C–H stretch, obscured by broad water peak), 1,638 (m, broad), 1,575 (m), 1,488 (m), 1,453 (m), 1,435 (m), 1,238 (m), 1,095 (m), 1,026 (m), 750 (s, 1,2-disubstituted arene C– H bend), 654 (m).

E-isomer:

^{1}H NMR (400 MHz, $CDCl_3$) δ 7.73-7.64 (m, 2H), 7.61-7.57 (m, 1H), 7.53 (d, J = 16.4, 1H, alkene H), 7.32 (t, J = 7.6, 1H), 7.30-7.24 (m, obscured by C*H*Cl_3), 7.22 (d, J = 16.4, 1H, alkene H), 7.19-7.05 (m, 3H).

^{13}C NMR (101 MHz, $CDCl_3$) δ 160.5 (d, J = 250.1, *C*-F), 137.0 (s, 0H), 133.1 (s), 129.5 (d, J = 4.5, 1), 129.3 (d, J = 8.4), 129.1 (s), 127.6 (s), 127.2 (d, J = 3.4), 126.9 (s), 124.9 (d, J = 12.0), 124.3 (d, J = 3.6), 124.2 (s), 123.5 (d, J = 4.1), 115.8 (d, J = 22.1).

^{19}F NMR (376 MHz, $CDCl_3$) δ −117.8-117.9 (m).

IR: 3,056-3,000 (w, alkene & aromatic C–H stretch), 1,578 (m), 1,488 (s), 1,456 (s), 1,438 (m), 1,259 (w), 1,232 (s), 1,210 (w), 1,115 (w), 1,094 (m), 1,024 (s), 963 (m, *E*-alkene C–H bend), 909 (w), 760 (s, 1,2-disubstituted arene C–H bend), 670 (m).

(b) From 2-fluorobenzylmethyldiphenylphosphonium bromide (0.936 g, 2.40 mmol), NaHMDS (2.4 ml of a 1 mol L^{-1} solution in THF, 2.4 mmol) and 2-bromobenzaldehyde (0.445 g, 0.28 ml, 2.4 mmol) by procedure A. See Table 2.1 entry 17. The crude product was eluted through a silica plug using pentane as the eluting solvent, giving a pale yellow oil (0.50 g, 76 %).

^{1}H NMR (400 MHz, $CDCl_3$) of alumina plug-purified stilbene mixture:

Assigned to *E*-isomer: δ 7.55 (d, ^{3}J = 16.4, 1H, 2-FC_6H_4C*H*).

Assigned to *Z*-isomer: δ 7.05-7.00 (m, 2H), 6.88 (td, ^{3}J = 7.6, 1.2, 1H), 6.79 (s, 2H). Integrations indicate *Z/E* ratio of 94:6.

2-fluoro-2′methylstilbene

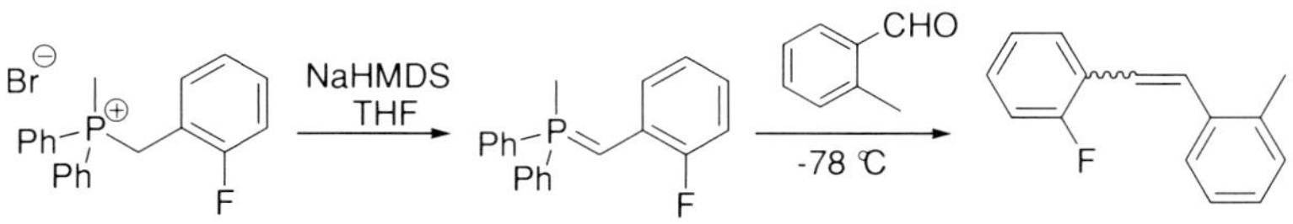

From 2-fluorobenzylmethyldiphenylphosphonium bromide (1.069 g, 2.75 mmol), NaHMDS (2.6 ml of a 1 mol L^{-1} solution in THF, 2.6 mmol) and 2-methylbenzaldehyde (0.32 ml, 0.33 g, 2.8 mmol) in dry THF (11 ml) by procedure A. See Table 2.1 entry 25.

^{1}H NMR ((400 MHz, $CDCl_3$) of crude product:

Assigned to *E*-isomer: δ 2.43 (s, 3H).

Assigned to *Z*-isomer: δ 2.28 (s, 3H).

^{19}F NMR (376 MHz, $CDCl_3$) of crude product:

Assigned to *E*-isomer: δ -117.8-118.0 (m).

Assigned to *Z*-isomer: δ -116.1-116.3 (m).

Integrations indicate a *Z/E* ratio of 51:49.

The crude product was treated with oxalyl chloride (0.23 ml, 0.35 g, 92.8 mmol) to give the product as a clear oil that crystallised on standing (0.51 g, 90 %). The alkene isomers were separated by column chromatography on neutral alumina 90 (Brockmann grade I) using cyclohexane as the elution solvent. The *Z*-isomer was first to elute, followed by some fractions containing both *Z* and *E* alkene, and then *E*-alkene only was last to elute. Combination of fractions containing the *E* or *Z* isomer respectively followed by solvent removal yielded pure samples of each isomer, each of which solidified on standing.

Z-isomer:

^{1}H NMR (600 MHz, $CDCl_3$) δ 7.18 (d, J = 7.6, 1H, ArH-3 Me ring), 7.16-7.09 (m, 2H, contains ArH-4 of Me ring & ArH-4 of F ring), 7.08 (d, J = 7.5, 1H, ArH-6 Me ring), 7.03-6.93 (m, 3H, ArH-5 or Me ring, ArH-3 & ArH-6 of F ring), 6.80 (td, J = 7.9, 1.0, 1H, ArH-5 of F ring), 6.78 (d, J = 12.2, 1H, 2-MeC_6H_4C*H* = C), 6.73 (d, J = 12.3, 1H, 2-FC_6H_4C*H* = C), 2.28 (s, 3H, CH_3). Underlined signals are not baseline separated.

^{13}C NMR (151 MHz, $CDCl_3$) δ 160.4 (d, J = 248.1, *C*–F), 136.6 (s, Me ring ArC-1), 136.1 (s, Me ring ArC-2), 131.3 (d, J = 1.5, 2-MeC_6H_4*C*H=C), 130.1 (d, J = 3.4, F ring ArC-6), 130.0 (s, Me ring ArC-3), 128.8 (s, Me ring ArC-6), 128.7

(d, J = 8.3, F ring ArC-4), 127.4 (s, Me ring ArC-4), 125.6 (s, Me ring ArC-5), 124.8 (d, J = 13.4, F ring ArC-1), 123.4 (d, J = 3.6, F ring ArC-5), 122.6 (d, J = 4.4, 2-FC_6H_4CH=C), 115.3 (F ring ArC-3), 19.8 (s, CH_3).

^{19}F NMR (376 MHz, $CDCl_3$) δ −116.1-116.3 (m).

HRMS (TOF MS EI +): Calc. for $[M]^+$= $C_{15}H_{13}F$ 212.1001, found 212.1004 (1.4 ppm).

IR: 3,056-3,000 (w, alkene & aromatic C–H stretch), 2,991-2,850 (w, methyl C–H stretch), 1,608 (w), 1,572 (m), 1,489 (m), 1,453 (s), 1,225 (m), 1,191 (m), 1,102 (m), 1,031 (w), 950 (m), 886 (w), 854 (w), 829 (m), 770 (s, 1,2-disubstituted arene C–H bend).

MP: 38–39 °C.

E-isomer:

^{1}H NMR (500 MHz, $CDCl_3$) δ 7.63-7.56 (m, 2H, contains F-ring H-6 & Me ring H-4), 7.40 (d, J = 16.4, 1H, 2-FC_6H_4CH=C), 7.25-7.10 (m, 6H, contains F ring H-4 & H-5, Me ring H-3, H-5 & H-6, 2-MeC_6H_4CH=C), 7.06 (ddd, J = 10.8, 8.2, 1.2, 1H, F ring H-3), 2.42 (s, 3H, CH_3).

^{13}C NMR (126 MHz, $CDCl_3$) δ 160.5 (d, J = 249.4, F ring C-2), 136.4 (s, Me ring C-1), 135.9 (s, Me ring C-2), 130.4 (s, Me ring C-3), 128.9 (d, J = 5.0, 2-FC_6H_4CH=C), 128.7 (d, J = 8.4, F ring C-4), 127.8 (s, Me ring C-6), 127.3 (d, J = 3.7, F ring C-6), 126.3 (s, Me ring C-5), 125.6 (d, J = 12.1, F ring C-1), 125.5 (s, Me ring C-4), 124.2 (d, J = 3.6, F ring C-5), 122.3 (d, J = 3.5, 2-MeC_6H_4CH=C), 115.8 (d, J = 22.2, F ring C-3), 19.9 (s, CH_3).

^{19}F NMR (282 MHz, $CDCl_3$) δ −117.85-117.93 (m).

HRMS (TOF MS EI +): Calc. for $[M]^+$= $C_{15}H_{13}F$ 212.1001, found 212.1004 (1.4 ppm).

IR: 3,057-3,000 (w, alkene & aromatic C–H stretch), 3000-2850 (w, methyl C–H stretch), 1,490 (m), 1,456 (m), 1,237 (m), 1,093 (m), 980 (w), 961 (s, *E*-alkene C–H bend), 806 (w), 758 (s, s, 1,2-disubstituted arene C–H bend), 725 (m).

MP: 41–42 °C.

2-chloro-2′-methylstilbene [40]

(a) From 2-methylbenzylmethyldiphenylphosphonium chloride (0.45 g, 1.3 mmol), NaHMDS (1.18 of a 1 mol L^{-1} solution in THF, 1.18 mmol) and 2-chlorobenzaldehyde (0.15 ml, 0.19 g, 1.3 mmol) by procedure A. See Table 2.1 entry 19. *Z/E* ratio was assigned based on the ^{1}H NMR spectrum of the crude product.

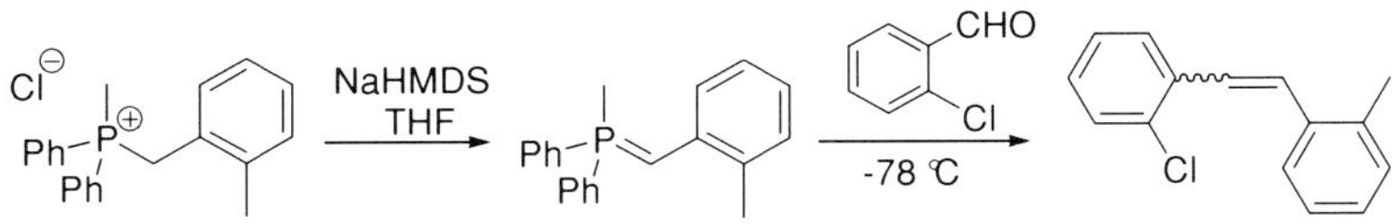

^{1}H NMR (400 MHz, $CDCl_3$) of crude product:

Assigned to *E*-isomer: δ 2.43 (s, 3H, Ar-CH_3) [40].

Assigned to *Z*-isomer: δ 6.79 (s, 2H, *H*C=C*H*), 2.27 (s, 3H, Ar-CH_3) [40].

Integrations indicate a *Z/E* ratio of 94:6.

(b) From 2-methylbenzyltriphenylphosphonium chloride (1.05 g, 2.61 mmol), NaHMDS (2.5 ml of a 1 mol L^{-1} solution in THF, 2.5 mmol) and 2-chlorobenzaldehyde (0.31 ml, 0.39 g, 2.8 mmol) by procedure A. See Table 2.2 entry 9. *Z/E* ratio was assigned based on the ^{1}H NMR spectrum of the crude product.

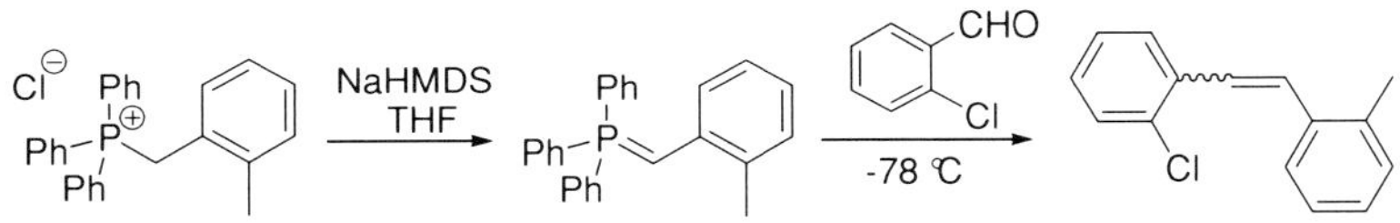

^{1}H NMR (400 MHz, $CDCl_3$) of crude product:

Assigned to *E*-isomer: δ 2.44 (s, 3H, Ar-CH_3) [40].

Assigned to *Z*-isomer: δ 6.80 (s, 2H, *H*C=C*H*), 2.27 (s, 3H, Ar-CH_3) [40].

Integrations indicate a *Z/E* ratio of 95:5

(c) From 2-chlorobenzylmethyldiphenylphosphonium chloride (1.93 g, 4.76 mmol), NaHMDS (4.6 ml of a 1 mol L^{-1} solution in THF, 4.6 mmol) and 2-methylbenzaldehyde (0.55 ml, 0.57 g, 4.8 mmol) by procedure A. See Table 2.1 entry 24.

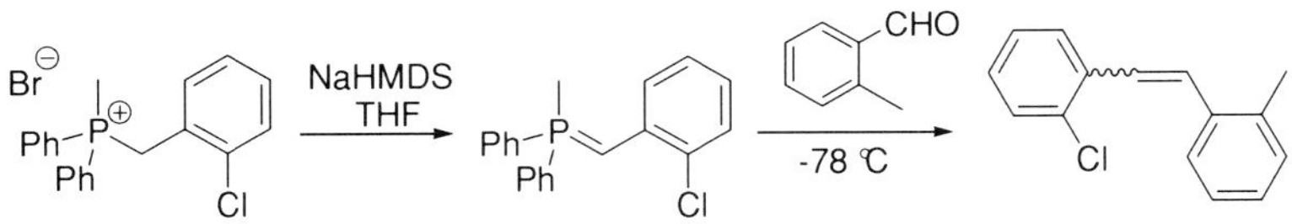

^{1}H NMR (400 MHz, $CDCl_3$) of crude product:

Assigned to *E*-isomer: δ 2.43 (s, 3H, Ar-CH_3) [40].

Assigned to *Z*-isomer: δ 6.79 (app s, 2H, *H*C=C*H*), 2.27 (s, 3H, Ar-CH_3) [40]. Integrations indicate a *Z/E* ratio of 77:23. The crude product was purified by treatment with oxalyl chloride (0.44 ml, 0.66 g, 5.2 mmol) by the procedure described in Sect. 4.3, giving a pure sample of the alkene as a clear oil (0.85 g, 78 %).

2-bromo-2′methylstilbene [30]

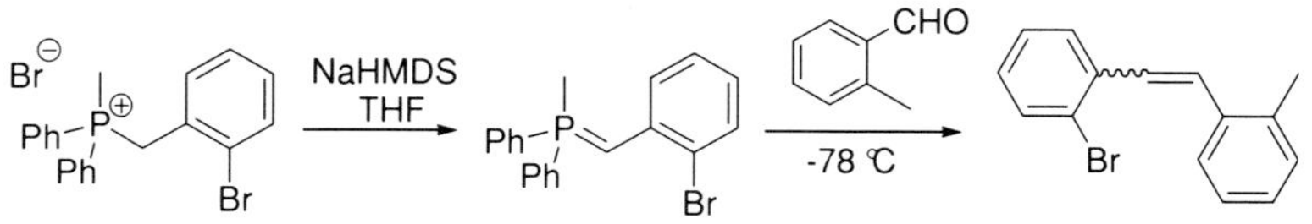

From 2-bromobenzylmethyldiphenylphosphonium bromide (1.115 g, 2.48 mmol), NaHMDS (2.3 ml of a 1 mol L^{-1} solution in THF, 2.3 mmol) and 2-methylbenzaldehyde (0.30 ml, 0.31 g, 2.6 mmol) by procedure A. The crude product was treated with oxalyl chloride (0.22 ml, 0.33 g, 2.6 mmol) to give the product as a clear oil that crystallised on standing (0.59 g, 94 %). See Table 2.1 entry 23.

^{1}H NMR (300 MHz, $CDCl_3$) of oxalyl chloride-treated product:

Assigned to *E*-isomer: δ 2.42 (s, 3H) [30].

Assigned to *Z*-isomer: δ 6.81-6.69 (pair of roofed doublets, AB system, 2H), 2.26 (s, 3H). 400 MHz ^{1}H spectrum of crude product (before oxalyl chloride treatment) shows two doublets at δ 6.81-6.69, each with J = 12.0.

Integrations indicate a *Z/E* ratio of 75:25. The alkene isomers were separated by column chromatography on neutral alumina (Brockmann grade I) using cyclohexane as the eluting solvent. Pure samples of each isomer were obtained after combination in each case of the column fractions containing the pure alkene isomer and solvent removal. Each sample solidified on standing.

Z-isomer:

^{1}H NMR (500 MHz, $CDCl_3$) δ 7.55 (dd, J = 7.8, 1.2, 1H, Br ring H-3), 7.15 (d, J = 7.6, 1H, Me ring H-3), 7.10 (td, J = 7.1, 1.8, 1H, Me ring H-4), 7.03-6.91 (m, 5H, contains Me ring H-5 & H-6, Br ring H-4, H-5 & H-6), 6.78 (d, J = 12.0, 1H, 2-MeC_6H_4CH=C), 6.74 (d, J = 12.0, 1H, 2-BrC_6H_4CH=C), 2.27 (s, CH_3).

^{13}C NMR (151 MHz, $CDCl_3$) δ 137.5 (Br ring C-1), 136.2 (Me ring C-2), 135.9 (Me ring C-1), 132.6 (Br ring C-3), 130.8 (Br ring C-6), 130.7 (2-MeC_6H_4CH=C), 130.0 (Me ring C-3), 129.9 (2-BrC_6H_4CH=C), 129.2 (Me ring C-6), 128.4 (Br ring C-4), 127.3 (Me ring C-4), 126.7 (Br ring C-5), 125.5 (Me ring C-5), 124.1 (Br ring C-2), 19.8 (CH_3).

HRMS (TOF MS EI+): Calc. for $[M]^+$= $C_{15}H_{13}\,^{79}Br$ 272.0201, found 272.0211 (100 %, 3.7 ppm); Calc. for $[M]^+$= $C_{15}H_{13}\,^{81}Br$ 274.0180, found 274.0168 (97 %, 4.4 ppm).

IR: 3,064-3,000 (w, alkene & aromatic C–H stretch), 2,965-2,856 (w, methyl C–H stretch), 1,590 (w), 1,466 (m), 1,434 (m), 1,261 (w), 1,102 (m), 1,044 (m), 1,024 (s), 964 (w), 777 & 755 (s, 1,2-disubstituted arene C–H bend), 736 (s).

MP: 54–55 °C.

E-isomer [30]:

^{1}H NMR (500 MHz, $CDCl_3$) δ 7.66 (dd, J = 7.8, 1.3, 1H, Br ring H-6), 7.63 (d, J = 7.0, 1H, Me ring H-6), 7.59 (dd, J = 8.0, 1.0, 1H, Br ring H-3), 7.37-7.29 (m, 2H, contains 2-BrC_6H_4CH=C, d, J = 16.0 & Br ring H-5), 7.28–7.16 (m, 4H, contains 2-MeC_6H_4CH=C, and Me ring H-3, H-4 & H-5), 7.12 (td, J = 8.0, 1.6, 1H, Br ring H-4), 2.44 (s, 3H, CH_3) [30].

^{13}C NMR (101 MHz, $CDCl_3$) δ 137.5 (Br ring C-1), 136.1 (Me ring C-1), 136.0 (Me ring C-2), 133.1 (Br ring C-3), 130.4 (Me ring C-3), 129.4 (2-MeC_6H_4CH=C), 128.8 (2-BrC_6H_4CH=C), 128.7 (Br ring C-4), 128.0 (Me ring C-4), 127.5 (Br ring C-5), 126.9 (Br ring C-6), 126.3 (Me ring C-5), 125.9 (Me ring C-6), 124.1 (Br ring C-2), 19.9 (CH_3).

HRMS (TOF MS EI +): Calc. for $[M]^+$= $C_{15}H_{13}^{79}$ Br 272.0201, found 272.0213 (100 %, 4.4 ppm); Calc. for $[M]^+$= $C_{15}H_{13}^{81}$ Br 274.0180, found 274.0207 (97 %, 9.8 ppm).

IR: 3,061-3,000 (w, alkene & aromatic C–H stretch), 2,964-2,849(w, methyl C–H stretch), 1,629 (w, alkene C = C?), 1,486 (m), 1,458 (m), 1,436 (m), 1,258 (w), 1,225 (w), 1,101 (w), 1,024 (m), 961 (s, *E*-alkene C–H bend), 865 (w), 800 (w), 756 (s, 1,2-disubstituted arene C–H bend), 721.

2-methoxy-2′methylstilbene [40]

From 2-methoxybenzylmethyldiphenylphosphonium chloride (0.700 g, 1.96 mmol), NaHMDS (1.9 ml of a 1 mol L^{-1} solution in THF, 1.9 mmol) and 2-methylbenzaldehyde (0.23 ml, 0.24 g, 1.98 mmol) by procedure A. See Table 2.1 entry 27.

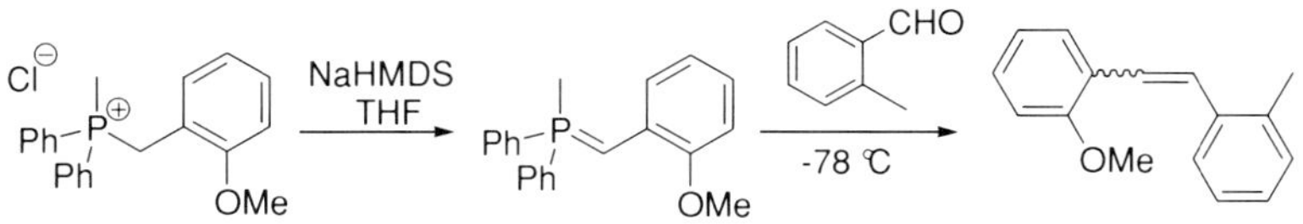

^{1}H NMR (400 MHz, $CDCl_3$) of crude product:.

Assigned to *E*-isomer: δ 3.88, (s. 3H, lies over another small signal), 2.42 (s, 3H) [40].

Assigned to *Z*-isomer: δ 3.82 (s. 3H), 2.28 (s. 3H) [40].

Integrations indicate a *Z/E* ratio of 66:34.

2,2′-dimethylstilbene [36]

From 2-methylbenzylmethyldiphenylphosphonium chloride (0.416 g, 1.16 mmol), NaHMDS (1.06 ml of a 1 mol L^{-1} solution in THF, 1.06 ml) and 2-methylbenzaldehyde (0.13 ml, 0.13 g, 1.1 mmol) by procedure A. See Table 2.1 entry 22.

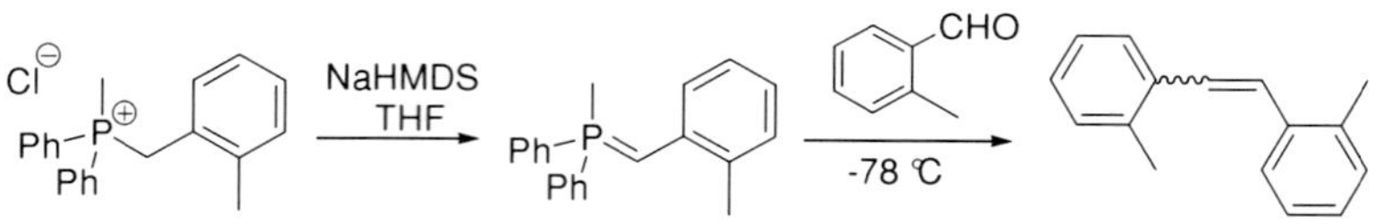

^{1}H NMR (300 MHz, $CDCl_3$) of crude product:.

Assigned to *E*-isomer: δ 2.42 (s, 3H, Ar-CH_3) [36].

Assigned to *Z*-isomer: δ 6.71 (s, 2H, C=C*H*), 2.28, (s, 3H, Ar-CH_3) [36].

Integrations indicate a *Z/E* ratio of 31:69.

4.5 Wittig Reactions of Ester-Stabilised Ylides with Benzaldehydes

4.5.1 Work-Up Procedure

All reactions in this section were carried out by procedure B from Sect. 4.3. For work-up, each reaction was quenched at −78 °C by addition of saturated aqueous ammonium chloride solution to ensure that the reaction had actually occurred at low temperature. The quenched reaction mixture was allowed to warm to room temperature, poured into a separatory funnel and extracted three times with diethyl ether (10 ml). The combined ether phases were dried over Na_2SO_4, filtered to remove the drying agent, and the solvent removed *in vacuo*. This gave an oil that was analysed by ^{1}H and COSY NMR. The *Z/E* ratio was determined by comparison of the integrations of the all of the signals of the *E* and *Z* isomers of enoate that showed sufficient baseline separation to alow for accurate integration. The signals used to assign the *Z/E* ratio are listed below for each compound.

Conversion was observed to be high in all cases by the large (relative) amount of alkene present in the crude product. This high conversion was confirmed in the case of *tert*-butyl 3-(2-bromophenyl)prop-2-enoate by isolation of the alkene product in a yield of 84 % after column chromatography.

4.5.2 Synthesis of Alkyl Cinnamates from Ester-Stabilised Ylides

Methyl 3-phenylprop-2-enoate [41]

From (methoxycarbonylmethyl)methyldiphenylphosphonium chloride (78 mg, 0.25 mmol), NaHMDS (44 mg, 0.24 mmol) and benzaldehyde (0.025 ml, 26 mg, 0.25 mmol using 100 μl syringe) by procedure B.

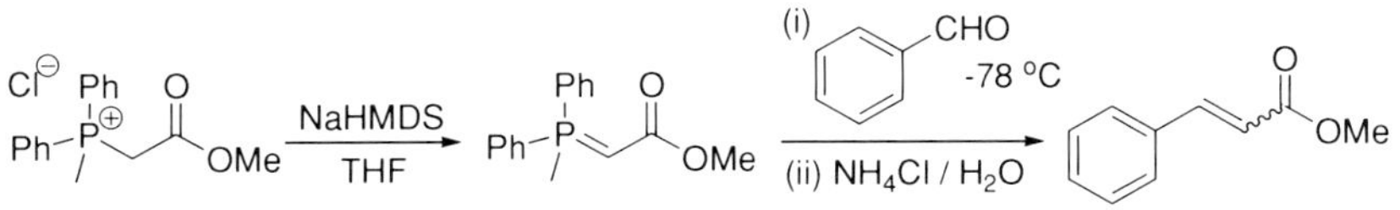

^{1}H NMR (500 MHz, $CDCl_3$) of crude product:

Assigned to E-isomer: δ 6.44 (d, J = 16.0, 1H, C(O)CH=C), 3.80 (s, 3H, OMe) [41].

Assigned to Z-isomer: δ 6.95 (d, J = 12.6, 1H, ArCH=C), 5.95 (d, J = 12.6, 1H, C(O)CH=C), 3.70 (s, 3H, OMe) [41].

Integrations show a *Z/E* ratio of 36:64.

Ethyl 3-phenylprop-2-enoate [42, 43]

From (ethoxycarbonylmethyl)methyldiphenylphosphonium bromide (53 mg, 0.14 mmol), KHMDS (27 mg, 0.13 mmol) and benzaldehyde (0.013 ml, 14 mg, 0.13 mmol) by procedure B.

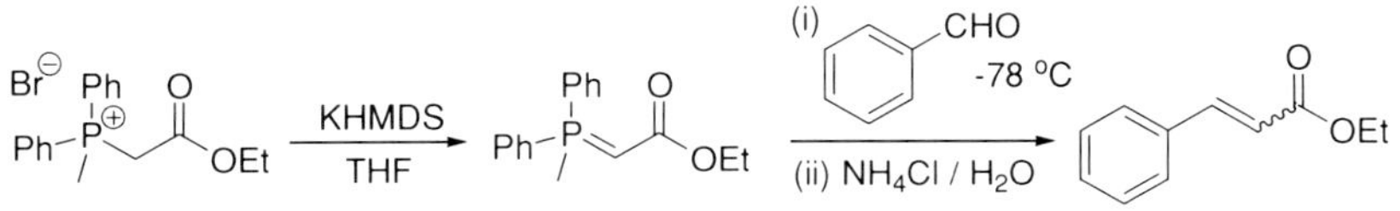

^{1}H NMR (300 MHz, $CDCl_3$) of crude product:

Assigned to E-isomer: δ 6.43 (d, J = 16.0, 1H, ArCH=C), 4.17 (q, J = 7.1, 2H, OCH_2), 1.23 (t, J = 7.1, 3H, CH_3) [42].

Assigned to Z-isomer: δ 6.93 (d, J = 12.6, 1H, ArCH=C), 5.94 (d, J = 12.6, 1H, C(O)CH=C), 4.26 (q, J = 7.1, 2H, OCH_2), 1.33 (t, J = 7.1, 3H, CH_3) [43].

Integrations show a *Z/E* ratio of 36:64.

***tert*-butyl 3-phenylprop-2-enoate** [44, 45]

From (*tert*-butoxycarbonylmethyl)methyldiphenylphosphonium bromide (116 mg, 0.29 mmol), NaHMDS (52 mg, 0.28 mmol) and benzaldehyde (0.028 ml, 29 mg, 0.28 mmol) by procedure B.

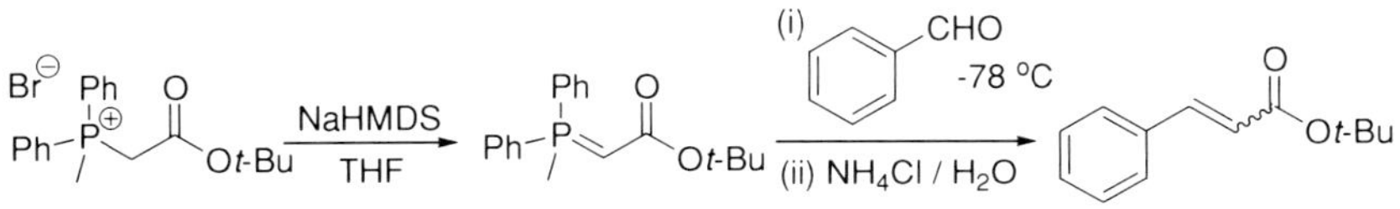

^{1}H NMR (500 MHz, $CDCl_3$) of crude product:

Assigned to E-isomer: δ 6.37 (d, J = 16.0, 1H, COCH), 1.53 (s, baseline separation not perfect due to presence of small signals, C(CH_3)$_3$) [44].

Assigned to Z-isomer: δ 6.85 (d, J = 12.6, 1H, ArCH), 5.87 (d, J = 12.6, 1H, COCH), 1.42 (s, baseline separation not perfect due to presence of small signals, $C(CH_3)_3$) [45].

Integrations show a Z/E ratio of 40:60.

Ethyl 3-(2-methoxyphenyl)prop-2-enoate [46]

From (ethoxycarbonylmethyl)methyldiphenylphosphonium bromide (50 mg, 0.14 mmol), KHMDS (27 mg, 0.13 mmol) and 2-methoxybenzaldehyde (0.13 ml of a 1 mol L^{-1} solution in dry THF, 0.13 mmol) by procedure B [46].

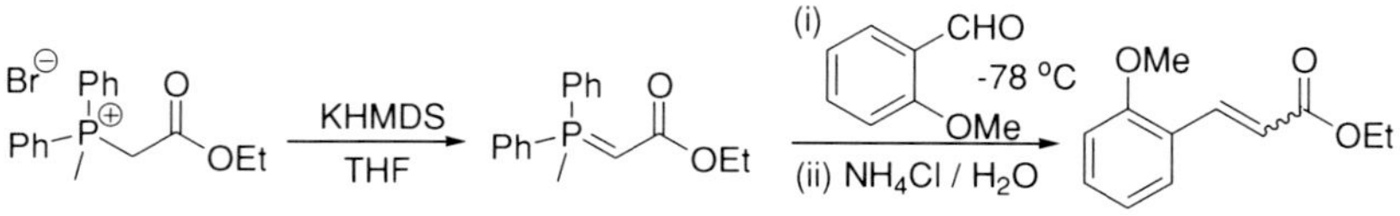

KHMDS, THF; (i) 2-methoxybenzaldehyde (CHO, OMe), -78 °C; (ii) NH_4Cl / H_2O

^{1}H NMR (400 MHz, $CDCl_3$) of crude product:

Assigned to E-isomer: δ 7.99 (d, J = 16.2, 1H, ArCH=C), 6.53 (d, J = 16.2, 1H, C=CHC(O)), 4.26 (q, J = 7.1, 2H, OCH_2), 3.89 (s, 3H, OCH_3), 1.34 (t, J = 7.0, baseline separation not sufficient for integration, CH_2CH_3) [46].

Assigned to Z-isomer: δ 7.16 (d, J = 12.5, 1H, ArCH=C), 5.97 (d, J = 12.5, 1H, C=CHC(O)), 4.13 (q, J = 7.0, 2H, OCH_2), 3.83 (s, 3H, OCH_3), 1.20 (t, J = 7.1, baseline separation not sufficient for integration, CH_2CH_3) [46].

Integrations indicate a Z/E ratio of 66:34.

***tert*-butyl 3-(2-methoxyphenyl)prop-2-enoate** [47]

From (*tert*-butoxycarbonylmethyl)methyldiphenylphosphonium bromide (122 mg, 0.301 mmol), KHMDS (58 mg, 0.29 mmol) and 2-methoxybenzaldehyde (0.29 ml of a 1 mol L^{-1} solution in dry THF, 0.29 mmol) by procedure B.

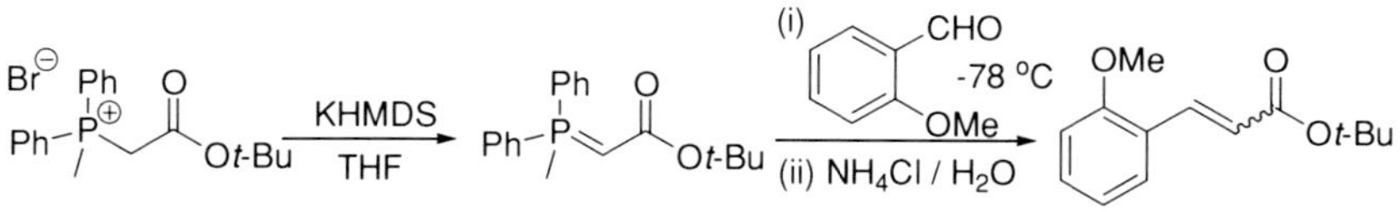

KHMDS, THF; (i) 2-methoxybenzaldehyde (CHO, OMe), -78 °C; (ii) NH_4Cl / H_2O

^{1}H NMR (400 MHz, $CDCl_3$) of crude product:

Assigned to E-isomer: δ 7.83 (d, J = 16.2, 1H, ArCH=C), 6.36 (d, J = 16.2, 1H, C=CHC(O)), 3.75 (s, 3H, OCH_3), 1.44 (s, 9H, $(CH_3)_3$) [47].

Assigned to Z-isomer: δ 5.81 (d, J = 12.5, 1H, C=CHC(O)), 3.72 (s, 3H, OCH_3), 1.29 (t, J = 7.1, 3H, $CH_3)_3$) [47].

Integrations indicate a Z/E ratio of 77:23.

Methyl 3-(2-chlorophenyl)prop-2-enoate [48, 49]

From (methoxycarbonylmethyl)methyldiphenylphosphonium chloride (15 mg, 0.049 mmol), KHMDS (9 mg, 0.049 mmol) and 2-chlorobenzaldehyde (0.006 ml, 7 mg, 0.05 mmol) by procedure B. This reaction was quenched with a solution of HCl in methanol (1.25 mol L^{-1}, 1 ml, 1.25 mmol) rather than ammonium chloride solution.

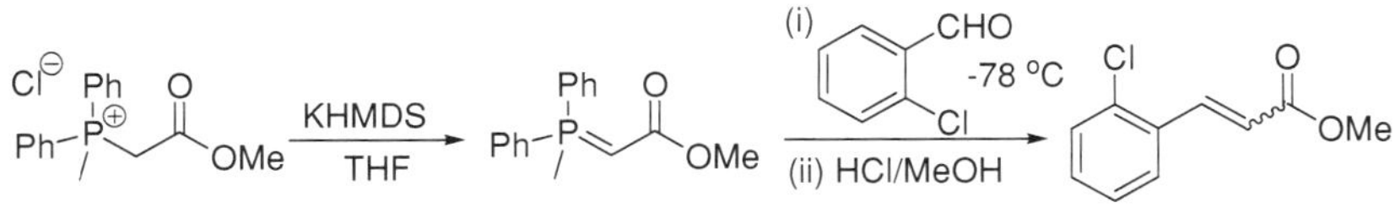

^{1}H NMR (300 MHz, $CDCl_3$) of crude product:

Assigned to *E*-isomer: δ 8.10 (d, J = 16.0, 1H, ArC*H*=C), 6.43 (d, J = 16.0, 1H, C(O)C*H*=C), 3.82 (s, 3H, OCH_3) [48].

Assigned to Z-isomer: δ 7.15 (d, J = 12.2, 1H, ArC*H*=C), 6.09 (d, J = 12.2, 1H, C(O)C*H*=C) [49].

Integrations show a *Z/E* ratio of 79:21.

Ethyl 3-(2-chlorophenyl)prop-2-enoate [50]

From (ethoxycarbonylmethyl)methyldiphenylphosphonium bromide (52 mg, 0.14 mmol), KHMDS (26 mg, 0.13 mmol) and 2-chlorobenzaldehyde (0.015 ml, 19 mg, 0.14 mmol) by procedure B.

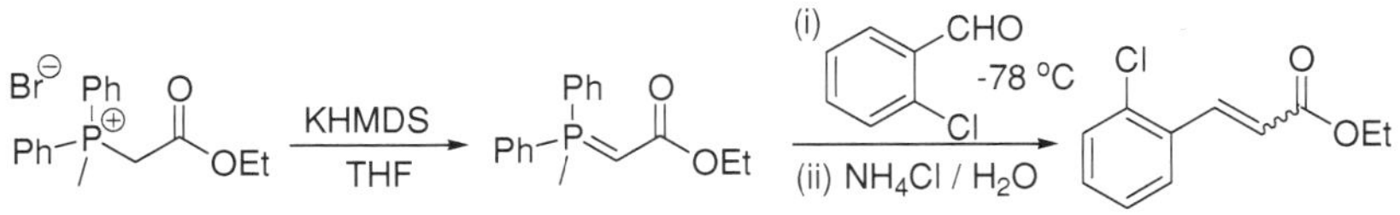

^{1}H NMR (300 MHz, $CDCl_3$) of crude product:

Assigned to *E*-isomer: δ 8.01 (d, J = 16.0, 1H, ArC*H*=C), 6.35 (d, J = 16.0, 1H, C(O)C*H*=C), 4.20 (q, J = 7.1, 2H, CH_2), 1.34 (t, J = 7.1, 3H, CH_3) [50].

Assigned to Z-isomer: δ 7.05 (d, J = 12.2, 1H, ArC*H*=C), 6.00 (d, J = 12.2, 1H, C(O)C*H*=C), 4.03 (q, J = 7.1, 2H, CH_2), 1.26 (t, J = 7.1, 3H, CH_3) [50].

Integrations show a *Z/E* ratio of 77:23.

Methyl 3-(2-bromophenyl)prop-2-enoate [51, 52]

From (methoxycarbonylmethyl)methyldiphenylphosphonium chloride (108 mg, 0.35 mmol), NaHMDS (62 mg, 0.34 mmol) and 2-bromobenzaldehyde (0.04 ml, 63 mg, 0.34 mmol) by procedure B.

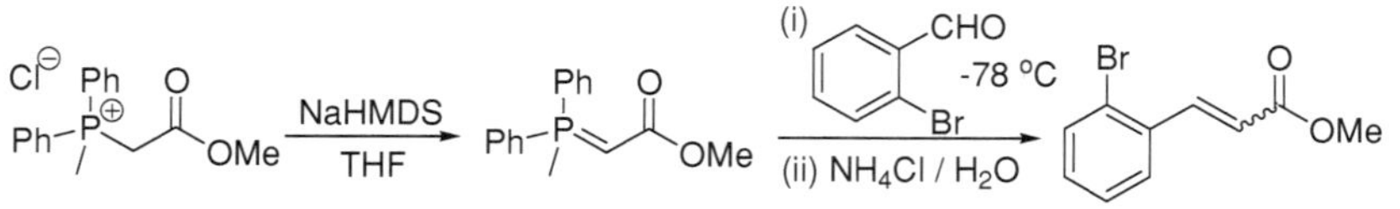

^{1}H NMR (500 MHz, $CDCl_3$) of crude product:

Assigned to *E*-isomer: δ 8.05 (d, J = 15.9, 1H, ArC*H*=C), 7.24-7.20 (m, 1H, aromatic), 6.39 (d, J = 16.0, 1H, C(O)C*H*=C), 3.82 (s, 3H, OMe) [51, 52].

Assigned to Z-isomer: δ 7.20-7.15 (m, 1H, aromatic), 7.09 (d, J = 12.2, 1H, ArC*H*=C), 6.06 (d, J = 12.2, 1H, C(O)C*H*=C), 3.65 (s, 3H, OMe) [51].

Integrations show a *Z/E* ratio of 83:17.

Ethyl 3-(2-bromophenyl)prop-2-enoate [53]

From (ethoxycarbonylmethyl)methyldiphenylphosphonium bromide (79 mg, 0.22 mmol), KHMDS (41 mg, 0.21 mmol) and 2-bromobenzaldehyde (0.025 ml, 40 mg, 0.21 mmol) by procedure B.

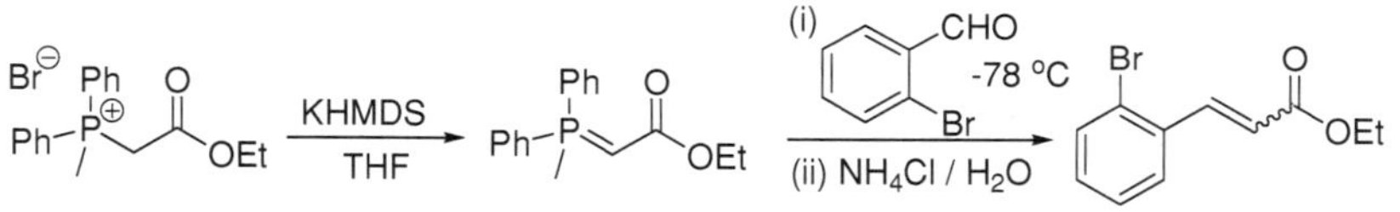

^{1}H NMR (500 MHz, $CDCl_3$) of crude product:

Assigned to *E*-isomer: δ 8.05 (d, J = 15.9, 1H, ArC*H*=C), 6.38 (d, J = 15.9, 1H, C(O)C*H*=C), 4.28 (q, J = 7.1, 2H, CH_2), 1.34 (t, J = 7.1, 3H, CH_3) [53].

Assigned to Z-isomer: δ 7.07 (d, J = 12.2, 1H, ArC*H*=C), 6.05 (d, J = 12.2, 1H, C(O)C*H*=C), 4.10 (q, J = 7.1, 2H, CH_2), 1.16 (t, J = 7.1, 3H, CH_3) [53].

Integrations show a *Z/E* ratio of 83:17.

***tert*-butyl 3-(2-bromophenyl)prop-2-enoate** [54]

From (*tert*-butoxycarbonylmethyl)methyldiphenylphosphonium bromide (108 mg, 0.273 mmol), NaHMDS (48 mg, 0.26 mmol) and 2-bromobenzaldehyde (0.031 ml, 49 mg, 0.26 mmol) by procedure B.

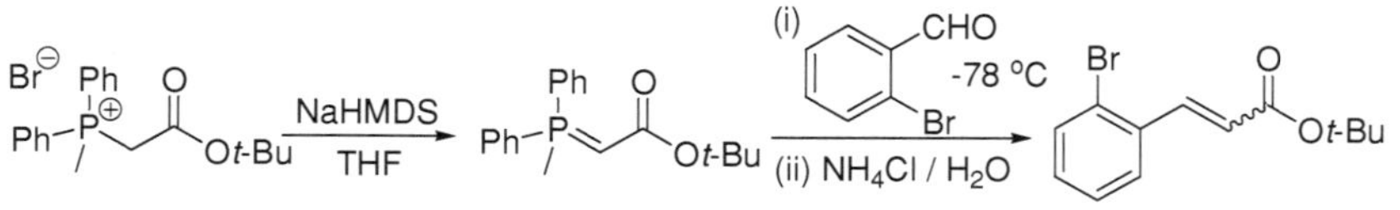

^{1}H NMR (500 MHz, $CDCl_3$):

Assigned to E-isomer: δ 7.96 (d, *J* = 15.9, 1H, ArC*H*), 6.32 (d, *J* = 15.9, 1H, COC*H*), 1.54 (s, 9H, C(CH_3)$_3$) [54].

Assigned to Z-isomer: δ 6.97 (d, *J* = 12.1, 1H, ArC*H*), 5.98 (d, *J* = 12.1, 1H, COC*H*), 1.33 (s, 9H, C(CH_3)$_3$).

Integrations show a *Z/E* ratio of 85:15.

Chromatographic conditions facilitating the separation of the *Z* and *E* isomers of this alkene could not be found. This is in keeping with earlier reports of the *Z* and *E* isomers of various alkyl cinnamates being inseparable by column chromatography [45, 46, 51]. The 85:15 mixture of *Z* and *E* isomers of the alkene was separated from the other material in the crude product by column chromatography on neutral alumina (Brockmann grade I) using 95:5 pentane/diethyl ether (combined yield 62 mg, 84 %). This product was characterised fully by ^{1}H, ^{13}C, gCOSY, zTOCSY, 1D and 2D NOESY gHSQC, gHMBC NMR. Stereochemical assignments of the alkenes were confirmed by the observation of NOE contact between the *Z*-alkene hydrogens, and by the absence of such NOE contact for the *E*-isomer, although H-3 of the *E*-isomer did show NOE contact with the *t*-butyl group at longer mixing times.

Assigned to *E*-isomer:

^{1}H NMR (600 MHz, $CDCl_3$) δ 7.96 (d, *J* = 15.9, 1H, ArC*H*=C), 7.61-7.58 (m, 2H, ArH-3 & ArH-6), 7.30 (t, *J* = 7.6, 1H, ArH-5), 7.20 (td, *J* = 7.8, 1.6, 1H, ArH-4), 6.32 (d, *J* = 15.9, 1H, C=C*H*C(O)), 1.54 (s, 9H, C(CH_3)$_3$).

^{13}C NMR (151 MHz, $CDCl_3$) δ 165.7 (C=O), 140.7 (Ar*C*H=C), 134.7 (ArC-1), 133.3 (ArC-3), 130.9 (Ar-C-4), 126.7 (ArC-6), 127.6 (ArC-5), 125.2 (ArC-2), 123.0 (C=*C*HC(O)), 80.8 (*C*Me$_3$), 28.2 (C(*C*H$_3$)$_3$).

Assigned to *Z*-isomer:
^{1}H NMR (600 MHz, $CDCl_3$) δ 7.57 (dd, *J* = 8.0, 1.0, 1H, ArH-3), 7.40 (dd, *J* = 7.7, 1.3, 1H, ArH-4), 7.28 (td, partially obscured by $CHCl_3$ peak, *J* = 7.5, 0.9, ArH-5), 7.16 (td, *J* = 7.6, 1.5, 1H, ArH-6), 6.97 (d, *J* = 12.1, 1H, ArC*H*=C), 5.98 (d, *J* = 12.1, 1H, (C=C*H*C(O)), 1.33 (s, 9H, C(CH_3)$_3$).

^{13}C NMR (151 MHz, $CDCl_3$) δ 165.0 (C=O), 141.9 (Ar*C*H=C), 136.6 (ArC-1), 132.1 (ArC-3), 130.6 (ArC-4), 129.4 (ArC-6), 126.6 (ArC-5), 124.1 (C=*C*HC(O)), 122.8 (ArC-2), 80.7 (*C*Me$_3$), 27.8 (C(*C*H$_3$)$_3$).

HMRS (m/z, TOF MS EI +): Calc. for $[M + Na]^+$ = $C_{13}H_{15}O_2{}^{79}Br\ Na$ 305.0153; found 305.0164 (3.6 ppm), Calc. for [M + Na] = $C_{16}H_{19}O_4{}^{81}Br\ Na$ 307.0133; found 307.0161.

4.6 Wittig Reactions of Keto-Stabilised Ylides with Benzaldehydes

4.6.1 Synthesis of 4-arylbut-3-en-2-ones from Acetonylidenetriphenylphosphoranes

General Procedure

The appropriately substituted acetonyltriphenylphosphorane (1 equivalent) was added to a nitrogen flushed Schlenk flask. Dry THF was added to give a clear solution of the ylide after stirring (approximately 0.15 mol L^{-1}; the ylide is not fully soluble at higher concentration). Once all of the ylide had dissolved, the solution was cooled to −78 °C in a dry ice/acetone bath. The appropriate benzaldehyde (1 equivalent) was then added slowly to the salt free ylide solution either neat or as a solution in THF. The solution was allowed to stir for 15 min after aldehyde addition at −78 °C, and then allowed to warm up to room temperature. The solution was then left to stir overnight.

Work-up: The THF reaction solvent was removed from the reaction mixture *in vacuo*, and pentane (ca. 20 ml) was added to dissolve alkenes. The resulting mixture was then filtered through a cotton wool plug into a round-bottom flask of known mass to remove most of the phosphine oxide, giving a clear or faintly yellow solution. The pentane was then removed from the filtrate *in vacuo*. An oil was typically obtained, which in some cases crystallised upon standing. A ^{1}H NMR spectrum was taken of the crude product.

For previously uncharacterised alkenes, a sample of the crude mixture of isomers (plus phosphine oxide and residual aldehyde) was eluted through an alumina plug with pentane to remove aldehyde and phosphine oxide. This gave a sample of the mixture of *Z* and *E* isomers (NMR spectrum of this obtained in each case and the *Z*/*E* ratio determined and compared with that of the crude product), which was then subjected to HPLC on a preparative AS-H column with an autosampler using 90:10 pentane/ethanol as the eluting solvent. A sample of each isomer (in low yield) was thus obtained for characterisation purposes.

The *Z*/*E* ratio for each 4-arylbutenone was determined by appraisal of the integrations of the signals assigned to each isomer in the ^{1}H NMR spectrum of the crude product from the reaction. The results are shown in Table 2.5.

4-phenylbut-3-en-2-one [55]

Ph$_3$P=CHC(O)CH$_3$ + PhCHO —(THF, -78 °C to 20 °C)→ PhCH=CHC(O)CH$_3$

From acetonylidenetriphenylphosphorane (0.751 g, 2.36 mmol) and benzaldehyde (0.24 ml, 0.25 g, 2.4 mmol). See Table 2.4 entry 1.

^{1}H NMR (400 MHz, $CDCl_3$) of crude mixture of enones:

Assigned to *E*-isomer δ 6.72 (d, ^3J = 16.3, 1H, H-3), 2.38 (s, 3H, CH_3) [55].

Assigned to *Z*-isomer δ 6.90 (d, ^3J = 12.7, 1H, H-4), 6.20 (d, ^3J = 12.7, 1H, H-3), 2.14 (s, 3H, CH_3) [55].

Integrations indicate *Z/E* ratio of 3:97.

4-(2-bromophenyl)but-3-en-2-one [56]

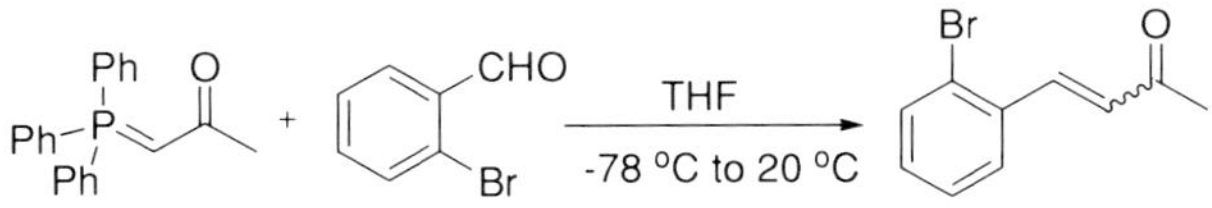

From acetonylidenetriphenylphosphorane (0.780 g, 2.45 mmol) and 2-bromobenzaldehyde (0.29 ml, 0.46 g, 2.48 mmol). Pure yield (combined mass of isolated, purified isomers after HPLC): 0.13 g (24 %). See Table 2.4 entry 2.

^{1}H NMR (400 MHz, $CDCl_3$) of crude mixture of enones:

Assigned to *E*-isomer: δ 7.91 (1H, d, ^3J = 16.3, H-4), 6.64 (d, ^3J = 16.3, 1H, H-3), 2.44 (s, 3H, CH_3) [56].

Assigned to *Z*-isomer: δ 7.04 (d, ^3J = 12.3, 1H, H-4), 6.26 (d, ^3J = 12.3, 1H, H-3), 2.05 (s, 3H, CH_3).

Integrations indicate *Z/E* ratio of 11: 89.

E-isomer:

Yield: 0.090 g.

^{1}H NMR (500 MHz, $CDCl_3$): δ 7.89 (d, ^3J = 16.3, 1H, H-3), 7.65-7.61 (overlapping doublets, 2H, ArH-3 and ArH-6), 7.37-7.31 (m, 1H, ArH-5), 7.28-7.22 (m, 1H, ArH-4), 6.62 (d, ^3J = 16.3, 1H, H-4), 2.42 (s, 3H, CH_3) [56].

Z-isomer:

Yield: 0.040 g.

^{1}H NMR (500 MHz, $CDCl_3$): δ 7.61 (dd, ^3J = 8.0, ^4J = 1.1, 1H, ArH-3), 7.36 (dd, ^3J = 7.6, ^4J = 1.6, 1H, ArH-6), 7.31-7.28 (qd, 1H, ArH-5), 7.23-7.19 (qd, 1H, ArH-4), 7.04 (d, ^3J = 12.3, 1H, H-3), 6.26 (d, ^3J = 12.3, 1H, H-4), 2.05 (s, 3H, CH_3).

Small signals due the presence of some *E*-isomer also observed: 7.89 (d, ^3J = 16.3, 1H, H-3), 6.62 (d, ^3J = 16.3, 1H, H-4), 2.42 (s, 3H, CH_3).

^{13}C NMR (126 MHz, $CDCl_3$): δ 199.2 (C = O), 138.4 (enone C-3), 135.2 (ArC-1), 131.6 (ArC-3), 129.9 (ArC-6), 129.8 (enone C-4),129.2 (ArC-4), 126.1 (ArC-5), 121.9 (ArC-2), 28.7 (enone C-1, CH_3).

IR: 2927 (C–H stretch), 1690 (C = O), 1,454, 1,422, 1,359, 1,304, 1,261, 1,107, 1,078, 1,026, 798, 745.

4-(*o*-methoxyphenyl)but-3-en-2-one [57]

Ph Ph–P Ph O + CHO OMe THF -78 °C to 20 °C OMe O

From acetonylidenetriphenylphosphorane (0.750 g, 2.36 mmol) and 2-methoxybenzaldehyde (0.29 ml, 0.34 g, 2.4 mmol). Pure yield (combined mass of isolated, purified isomers after HPLC): 0.110 g (26 %). See Table 2.4 entry 3.

^{1}H NMR (400 MHz, $CDCl_3$) of crude mixture of enones:

Assigned to *E*-isomer: δ 7.88 (d, ^3J = 16.6, 1H, H-4), 7.02 (d, ^3J = 16.6, 1H, H-3), 3.90 (s, 3H, OCH_3), 2.39 (s, 3H, CH_3) [57].

Assigned to *Z*-isomer: δ 7.14 (d, ^3J = 12.3, 1H, H-4), 6.20 (d, ^3J = 12.3, 1H, H-3), 3.85 (s, 3H, OCH_3), 2.10 (s, 3H, CH_3).

Integrations indicate *Z/E* ratio of 10: 90.

E-isomer: [57].

^{1}H NMR (400 MHz, $CDCl_3$): δ 7.89 (1H, d, ^3J = 12, enone H-3), 7.55 (1H, dd, ArH-6), 7.36 (1H, qd, ArH-4), 6.97 (1H, qd, ArH-5), 6.92 (1H, dd, ArH-3), 6.76 (d, ^3J = 12.0, 1H, enone H-4), 3.90 (s, 3H, OCH_3), 2.39 (s, 3H, CH_3).

Z-isomer:

A pure sample of this isomer was not obtained; the sample was contaminated with *E*-isomer, 2-methoxybenzaldehyde and other material. This made assignment of peaks in the ^{13}C spectrum impossible. A few distinctive signals in the ^{1}H spectrum could be observed.

^{1}H NMR (400 MHz, $CDCl_3$): δ 2.10 (s, 3H, CH_3), 3.85 (s, 3H, OCH_3), 6.19 (d, ^3J = 12, 1H, enone H-4), 7.14 (d, ^3J = 12, 1H, enone H-3). Doublets with ^3J = 16 Hz also visible at 6.76 and 7.89 ppm (*E*-isomer).

IR: 3,050-2,849 (C–H stretch), 1,690 (C = O stretch), 1,600 (C = C stretch), 1,487, 1,465, 1,438, 1,355, 1,289, 1,249, 1,177, 1,108, 1,049, 1,026, 756.

HRMS (ESI +): Calc. for $[M+H]^+$= $C_{11}H_{13}$ O_2 177.0916; found 177.0914 (1.1 ppm).

Synthesis and attempted characterisation of 4-(*o*-bromophenyl)-1-methoxy-but-3-en-2-one

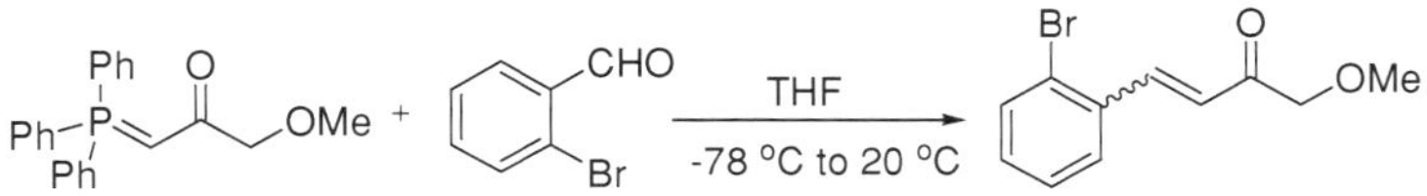

From 3-methoxyacetonylidenetriphenylphosphorane (0.465 g, 1.33 mmol) and 2-bromobenzaldehyde (0.16 ml, 0.25 g, 1.35 mmol). See Table 2.3 entry 4.
^{1}H NMR (400 MHz, $CDCl_3$) of crude mixture of enones:

Assigned to *E*-isomer: δ 8.05 (d, ^3J = 16.4, 1H, H-4), 6.87 (d, ^3J = 16.4, 1H, H-3), 4.28 (s, 2H, H-1), 3.48 (s, 3H, OCH_3).

Assigned to *Z*-isomer: δ 7.08 (d, ^3J = 12. 5, 1H, H-4), 6.37 (d, ^3J = 12.5, 1H, H-3), 3.98 (s, 2H, H-1), 3.34 (s, 3H, OCH_3).

Integrations indicate *Z/E* ratio of 17:83. The separation of the alkene isomers by HPLC was attempted, but the compound decomposed on the HPLC column.

4.6.2 Synthesis of 4-arylbut-3-en-2-ones from Acetonylidenemethyldiphenylphosphoranes

General Procedure C: Using solid NaHMDS or KHMDS at −78 °C

Dry phosphonium salt (1 equivalent) and NaHMDS or KHMDS (0.95 equivalents) were placed in a flame-dried Schlenk flask in a glove box under an atmosphere of argon. The flask was sealed, removed form the glove box and attached to a nitrogen supply via a Schlenk manifold using the pump & fill technique [1]. Dry THF was added, giving a solution of ylide (pale yellow) and a precipitate of NaBr and stirring for approximately 1 h. This was cooled to −78 °C in a dry ice/acetone bath, and then the aldehyde (0.95 equivalents, neat or as a solution in dry THF) was added dropwise. The solution was stirred at low temperature for 10 min and then a 5 % w/v aqueous solution of HCl was added to the reaction mixture to quench any remaining ylide or base.

General Procedure D: Using NaHMDS solution at −45 °C or −78 °C

The appropriately substituted acetonylmethyldiphenylphosphonium chloride (1 equivalent) was added to a Schlenk flask in a glove box under an argon atmosphere. The flask was sealed, removed form the glove box and attached to a nitrogen supply via a Schlenk manifold using the pump & fill technique [1]. Dry THF was added to give a suspension of the salt (approximately 0.15 mol l^{-1}). A 1 mol l^{-1} solution of NaHMDS in THF (containing 0.95 equivalents of base) was added, causing a coloured solution of the acetonylidenemethyldiphenylphosphorane to form, as well as a precipitate of NaCl. This was cooled either to −78 °C (dry ice-acetone bath) or to −45 °C (dry ice-acetonitrile bath). The appropriate benzaldehyde (1 equivalent) was then added slowly to the ylide solution by syringe. The solution was then left to stir for approximately 20 min at low temperature before the addition of a 5 %w/v aqueous solution of HCl (3 ml) to quench any remaining ylide and base.

General Procedure E: Using NaHMDS solution at 20 °C

According to Vedejs and coworkers [58, 59], the Wittig reactions of stabilised ylides with aliphatic aldehydes occur under kinetic control at room temperature, and so some reactions were repeated using a modification of procedure A, described below, which differs from it in that the reactions were carried out at 20 °C and the work-up is slightly changed. We wished to investigate if similar trends in the observed *Z/E* ratios would be obtained in the reactions carried out at room temperature and at low temperature.

The appropriately substituted acetonylmethyldiphenylphosphonium chloride (1 equivalent) was added to a Schlenk flask in a glove box under an argon atmosphere. The flask was sealed, removed form the glove box and attached to a nitrogen supply via a Schlenk manifold using the pump & fill technique [1]. Dry THF was added to give a suspension of the salt (approximately 0.15 mol l^{-1}). A 1 mol l^{-1} solution of NaHMDS in THF (containing 0.95 equivalents of base) was added, causing a coloured solution of the acetonylidenemethyldiphenylphosphorane to form, as well as a precipitate of NaCl. The appropriate benzaldehyde (1 equivalent) was then added drop-wise to the ylide solution by syringe. The solution was then left to stir for 12 h to 2 days. A solution of HCl (5 %w/v) was added and the mixture was worked up.

Work-up (for procedures C–E): Diethyl ether (10 ml) was added to the reaction mixture. The mixture was shaken in a separatory funnel and the phases separated. The aqueous phase was washed twice more with diethyl ether (10 ml) and the layers separated. The combined organic phases were dried over Na_2SO_4. The resulting mixture was then filtered through a cotton wool plug into a round-bottom flask of known mass to remove the drying agent, giving a clear or faintly yellow solution. The ether and THF was then removed from the filtrate using a rotary evaporator. An oil was typically obtained. A ^1H NMR spectrum was taken of the crude product. The *Z/E* ratio was determined by comparison of of the integrations of all the characteristic signals of the *E* and *Z* enones that showed sufficient baseline separation to permit accurate integration.

4-phenylbut-3-en-2-one [55]

Reaction at −45 °C.

Cl⊖ Ph O, Ph–P⊕ — NaHMDS / THF → Ph–P= ... — (i) PhCHO, -45 °C; (ii) HCl / H₂O →

The ylide was generated from acetonylmethyldiphenylphosphonium chloride (0.204 g, 0.69 mmol) & NaHMDS (0.63 ml of a 1 mol L^{-1} solution in THF, 0.63 mmol) and reacted at −45 °C[8] with benzaldehyde (0.07 ml, 0.07 g, 0.7 mmol) according to procedure D. Integration of signals assigned to each alkene isomer in the ^{1}H NMR of the crude product indicated a Z/E ratio of 20:80. See Table 2.4 entry 1.

Reaction at 20 °C.

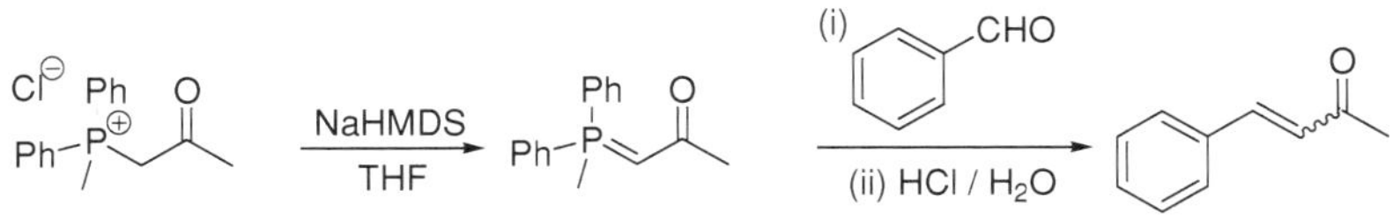

The ylide was generated from acetonylmethyldiphenylphosphonium chloride (0.63 g, 2.15 mmol) & NaHMDS (2.05 ml of a 1 mol L^{-1} solution in THF, 2.05 mmol) and reacted at 20°C[8] with benzaldehyde (0.20 ml, 0.21 g, 2.0 mmol) according to procedure E. Integration of signals assigned to each alkene isomer in the ^{1}H NMR of the crude product indicated a *Z/E* ratio of 19:81. See Table 2.4 entry 2. The same signals in each case were assigned to the alkene isomers in the ^{1}H NMR spectrum. ^{1}H NMR (300 MHz, $CDCl_3$) of crude mixture of enones:

Assigned to *E*-isomer: δ 6.72 (d, ^3J = 16.3, 1H, H-3), 2.38 (s, 3H, CH_3) [55]. H-4 obscured by phosphine oxide.

Assigned to *Z*-isomer): δ 6.90 (d, ^3J = 12.7, 1H, H-4), 6.20 (d, ^3J = 12.7, 1H, H-3), 2.14 (s, 3H, CH_3) [55].

4-(2-chlorophenyl)but-3-en-2-one [60, 61]

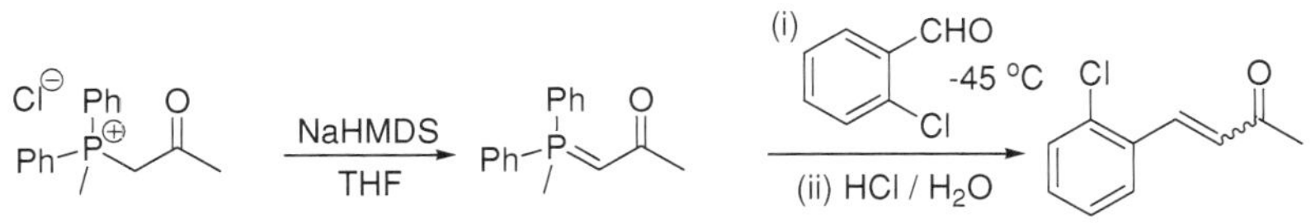

The ylide was generated from acetonylmethyldiphenylphosphonium chloride (0.218 g, 0.745 mmol) & NaHMDS (0.70 ml of a 1 mol L^{-1} solution in THF, 0.70 mmol) and reacted at −45°C[8] with 2-chlorobenzaldehyde (0.08 ml, 100 mg, 0.7 mmol) according to procedure D. See Table 2.4 entry 3.

[8] Acetonylmethyldiphenylphosphorane was found to be insoluble in THF below ca. −50 °C. Addition of 2-chlorobenzaldehyde to a biphasic salt-free mixture of ylide and THF at −78 °C followed by addition of HCl at low temperature resulted in no reaction.

^{1}H NMR (400 MHz, CDCl3) of crude mixture of enones:

Assigned to *E*-isomer: δ 6.61 (d, ^{3}J = 16.4, 1H, H-3), 2.35 (s, 3H, CH_3) [60, 61].

Assigned to *Z*-isomer: δ 7.00 (d, ^{3}J = 12.4, 1H, H-4), 6.22 (d, ^{3}J = 12.4, 1H, H-3), 2.26 (s, 3H, CH_3).

Integration of signals assigned to each alkene isomer in the ^{1}H NMR of the crude product indicated a *Z/E* ratio of 33:67. The *Z*-enone was found to be extremely prone to isomerisation to *E*-enone (presumably induced by light). The *Z*-isomer was synthesised selectively by a separate route, as shown below, although was again found to isomerise very readily.

Bis(2,2,2-trifluoroethyl)-2-oxopropylphosphonate [62]

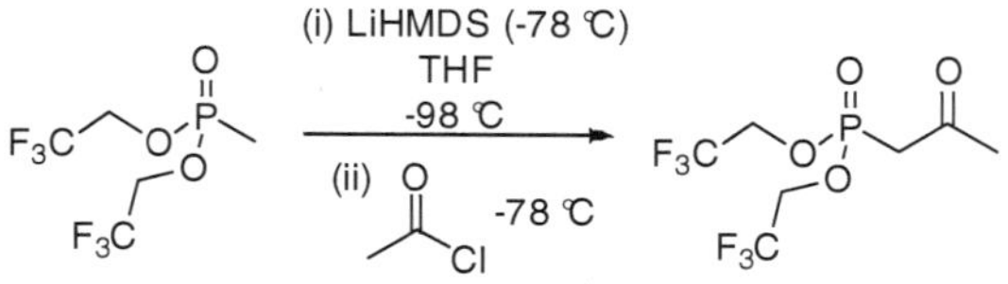

(i) LiHMDS (-78 °C)
THF
-98 °C
(ii) acetyl chloride, -78 °C

Bis(2,2,2-trifluoroethyl)methylphosphonate (2.647 g, 10.18 mmol, supplied by Aldrich) was placed in a 100 ml Schlenk tube under an atmosphere of nitrogen, and to this was added dry THF (16 ml). The resulting solution was cooled to − 98 °C in a liquid nitrogen/methanol bath.

To a second Schlenk flask in a glove box under an atmosphere of argon was added LiHMDS (3.747 g, 22.4 mmol). This flask was removed from the glove box, and charged with nitrogen using a nitrogen/vacuum manifold by standard pump and fill technique. To this flask was added dry THF (10 ml), and the resulting solution was cooled to −78 °C in a dry ice/acetone bath. This solution was then transferred over 20 min by cannula (under flow of nitrogen) into the solution of bis(2,2,2-trifluoroethyl)methylphosphonate at −98 °C to give a solution of phosphono ylide (only stable at very low temperature).

A solution of acetyl chloride (0.80 ml, 11 mmol) in dry THF (20 ml) in a third Schlenk flask was cooled to −78 °C, and transferred slowly by cannula to the solution of phosphono ylide at −98 °C, giving a yellow solution. This was maintained at −98 °C for 2 h, and then quenched by slowly adding saturated aqueous ammonium chloride (5 ml). Ethyl acetate (20 ml) was added to the mixture, which was then transferred to a separatory funnel. The phases were separated, and the aqueous layer was extracted a further two times with ethyl acetate (20 ml aliquots). The organic phases were combined, dried over Na_2SO_4, and filtered, and then concentrated on a rotary evaporator to give an oil containing some side-products. This was purified by column chromatography on silica using 3:1 cyclohexane/ethyl acetate to give the product as an oil (2.68 g, 89 %).

^{1}H NMR (400 MHz, $CDCl_3$): δ 4.45 (app quintet, $^3J_{PH} = {}^3J_{FH} = 8.2$, 4H, CH_2CF_3), 3.29 (d, $J = 21.8$, 2H, P-CH_2), 2.32 (s, 3H, CH_3) [63].

^{31}P NMR (162 MHz, $CDCl_3$): δ 23.5.

Still-Gennari reaction [64, 65]

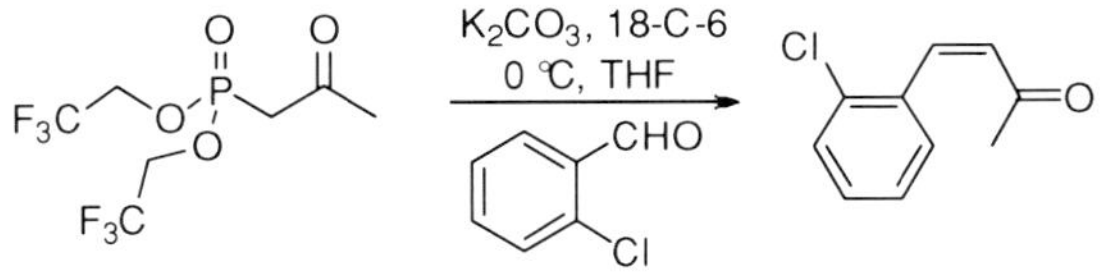

K_2CO_3 (0.103 g, 0.75 mmol), 18-crown-6 (0.090 g, 0.34 mmol), and dry THF (5 ml) were added to a Schlenk flask (flame dried and cooled under vacuum) under an atmosphere of nitrogen, and stirred at 0 °C for 3 h in an ice bath. To this was added slowly from a second Schlenk flask (under nitrogen) by cannula a solution of bis(2,2,2-trifluoroethyl)-2-oxopropylphosphonate (0.10 g, 0.33 mmol) in dry THF (2 ml), and then from a third Schlenk flask also by cannula a solution of 2-chlorobenzaldehyde (0.06 ml, 0.08 g, 0.5 mmol) in dry THF (1.5 ml). This mixture was stirred at 0 °C for 6 h, then quenched by addition of saturated aqueous ammonium chloride solution (4 ml). Diethyl ether (10 ml) was added, and the mixture was transferred to a separatory funnel. The phases were separated, and the aqueous phase was extracted twice with diethyl ether (10 ml aliquots). The combined ether phases were washed sequentially with saturated aqueous ammonium chloride solution and water until the ether was neutral. The ether phase was dried over Na_2SO_4, filtered, and then concentrated on a rotary evaporator to give an oil. ^{1}H NMR (500 MHz, $CDCl_3$) of this crude oil showed it to contain 2-chlorobenzaldehyde, and only the *Z*-isomer of 4-(2′-chlorophenyl)but-3-en-2-one:

^{1}H NMR (500 MHz, $CDCl_3$) δ 7.31-7.27 (td, 1H), 7.26-7.21 (m, 1H), 7.07 (d, $J = 12.4$, 1H, H-4), 6.28 (d, $J = 12.4$, 1H, H-3), 2.08 (s, 3H, CH_3). Other signals obscured by aldehyde.

The crude oil was chromatographed on silica, protected from light while on the column by aluminium foil, using 95:5 pentane/diethyl ether, and the enone was obtained as the second fraction, but had isomerised partially to the *E*-isomer (*Z/E* ratio 39:61). The sample was exposed to light for approximately 1 h, during which time it isomerised further to a mixture having a *Z/E* ratio of 20:80.

^{1}H NMR (500 MHz, $CDCl_3$), with integrations relative to the signal at δ 7.94 (*E*–isomer H-4) being 1H:

Assigned to *E*-isomer: δ 7.94 (d, $J = 16.4$, 1H, H-4), 7.64 (dd, $J = 7.5$, 1.9, 1H, ArH-6), 6.67 (d, $J = 16.4$, 1H, H-3), 2.43 (s, 3H, CH_3) [60].

Assigned to *Z*-isomer: δ 7.26-7.22 (m, 0.66H, partially obscured by $CHCl_3$ signal, ArH-5), 7.08 (d, $J = 12.4$, 0.65H, H-4), 6.29 (d, $J = 12.4$, 0.65, H-3), 2.08 (s, 1.97H, CH_3).

Other multiplets from overlapping signals were also present:

7.45-7.37 (m, 2.3H, contains *E*-ArH-3, Z-ArH-3 & ArH-4), 7.36-7.27 (m, 2.6H, contains *E*-ArH-4 & ArH-5, *Z* ArH-6).

^{13}C NMR (126 MHz, $CDCl_3$):

Assigned to *E*-isomer: δ 198.4 (C = O), 139.2 (C-4), 135.1 (ArC-2), 132.7 (ArC-1), 131.2 (ArC-4), 130.2 (ArC-3), 129.6 (C-3), 127.6 (ArC-6), 127.2 (ArC-5), 27.2 (CH_3).

Assigned to *Z*-isomer: δ 200.3 (C = O), 137.2 (C-4), 134.3 (ArC-1), 133.1 (ArC-2), 131.0 (ArC-6), 130.8 (C-3), 130.0 (ArC-4), 129.4 (ArC-3), 126.5 (ArC-5), 30.5 (CH_3).

4-(2-bromophenyl)but-3-en-2-one [66][9]

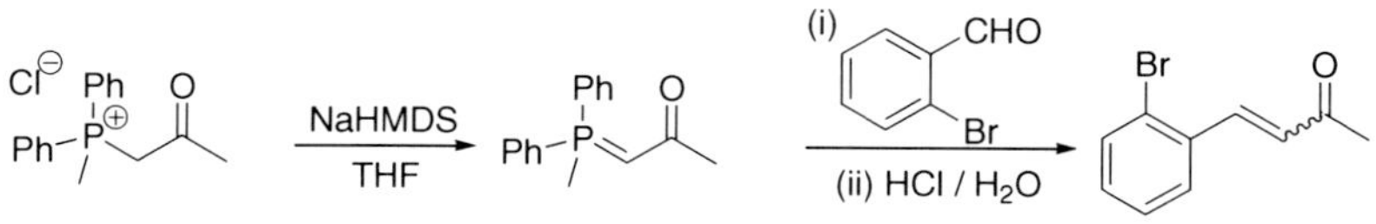

The ylide was generated from acetonylmethyldiphenylphosphonium chloride (0.90 g, 3.1 mmol) & NaHMDS (3.0 ml of a 1 mol L^{-1} solution in THF, 3.0 mmol) and reacted at 20°C[g] with 2-bromobenzaldehyde (0.36 ml, 0.57 g, 3.1 mmol) according to procedure E. See Table 2.4 entry 4. The *Z*-enone was found to be extremely prone to isomerisation to *E*-enone (presumably induced by light).

^{1}H NMR (400 MHz, $CDCl_3$) of crude mixture of enones:

Assigned to *E*-isomer (300 MHz, $CDCl_3$): δ 6.62 (d, ^{3}J = 16.3, 1H, H-3), 2.42 (s, 3H, CH_3) [66].

Assigned to *Z*-isomer (300 MHz, $CDCl_3$): δ 7.02 (d, ^{3}J = 12.3, 1H, H-4), 6.26 (d, ^{3}J = 12.3, 1H, H-3). Methyl peak at 2.05 ppm is obscured by methyldiphenylphosphine oxide doublet at 2.0 ppm.[8]

Integration of signals assigned to each alkene isomer in the ^{1}H NMR of the crude product indicated a *Z/E* ratio of 40:60.

1-chloro-4-phenylbut-3-en-2-one [67, 68]

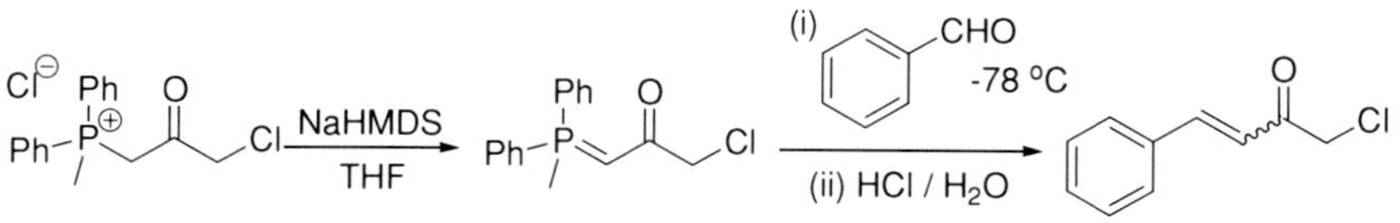

The ylide was generated from 3-chloroacetonylidenemethyldiphenylphosphonium chloride (0.70 g, 2.1 mmol) & NaHMDS (2.1 ml of a 1 mol L^{-1} solution in THF,

[9] For characterisation details for the *Z*-isomer of this compound, see Sect. 4.6.1.

2.1 mmol) and reacted at −78 °C with benzaldehyde (0.22 ml, 0.23 g, 2.16 mmol) according to procedure D. The reaction mixture became dark brown while stirring, and the product enone was found to be unstable over time if left to stand. See Table 2.4 entry 5.

^{1}H NMR (400 MHz, $CDCl_3$) of crude mixture of enones:

Assigned to *E*-isomer: δ 7.0 (d, ^3J = 16.1, 1H, H-3), 4.3 (s, 2H, CH_2Cl) [67].

Assigned to *Z*-isomer: δ 6.3 (d, ^3J = 12.7, 1H, H-3).

Integrations indicate a *Z/E* ratio of 12:88.

2,2-dimethyl-5-phenylpent-4-en-3-one [69, 70]

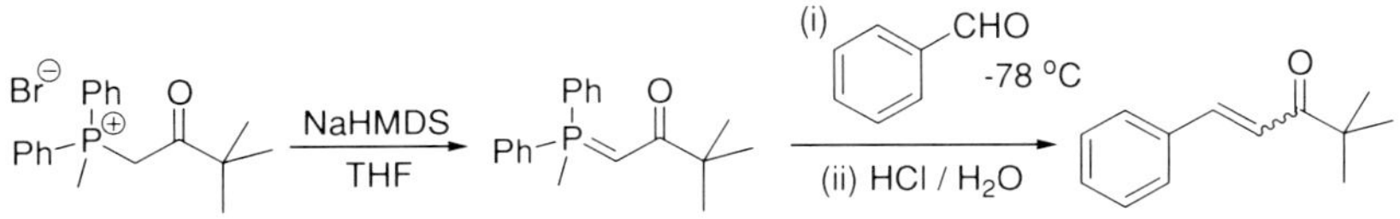

The ylide was generated from (3,3-dimethylbutan-2-onyl)methyldiphenylphosphonium chloride (0.899 g, 2.37 mmol) & NaHMDS (396 mg, 2.16 mmol) and reacted at −78 °C with benzaldehyde (0.22 ml, 0.23 g, 2.16 mmol) according to procedure C. See Table 2.4 entry 6.

^{1}H NMR (300 MHz, $CDCl_3$) of crude mixture of enones:

Assigned to *E*-isomer: δ 7.13 (d, ^3J = 15.6, 1H, H-3), 1.23 (s, 9H, $C(CH_3)_3$) [69].

Assigned to *Z*-isomer: δ 6.78 (d, ^3J = 12.7, 1H, H-4), 6.45 (d, ^3J = 12.7, 1H, H-3), 1.20 (s, 9H, $C(CH_3)_3$, peak not integrable due to overlap with another signal) [70].

Integrations indicate a *Z/E* ratio of 17:83.

4-(2-chlorophenyl)-1-chlorobut-3-en-2-one

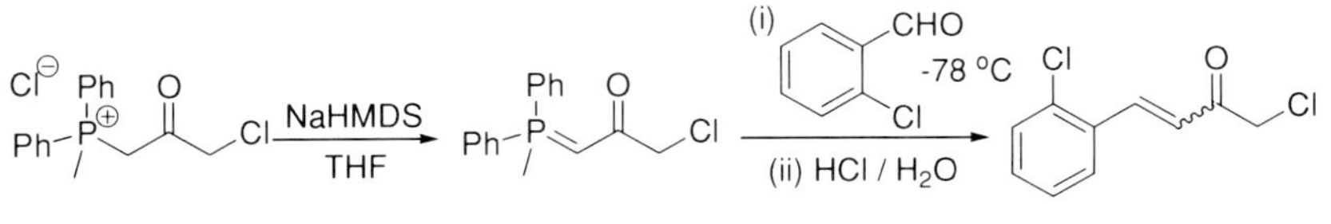

The ylide was generated from 3-chloroacetonylidenemethyldiphenylphosphonium chloride (0.497 g, 1.52 mmol) & NaHMDS (1.38 ml of a 1 mol L^{-1} solution, 1.38 mmol), and reacted at −78 °C with 2-chlorobenzaldehyde (0.17 ml, 0.21 g, 1.5 mmol) according to procedure D. See Table 2.4 entry 7.

^{1}H NMR (300 MHz, $CDCl_3$) of crude mixture of enones:

Assigned to *E*-isomer: δ 8.12 (d, *J* = 16.2, 1H, H-4), 6.95 (d, *J* = 16.1, 1H, H-3), 4.33 (s, 2H, CH_2Cl).

Assigned to *Z*-isomer: δ 7.23 (d, $J = 12.4$, baseline obscured—only identifiable by comparison with a sample purified by alumina plug, H-4), 6.45 (d, $J = 12.4$, 1H, H-3), 4.07 (s, 2H, CH_2Cl).

Integrations show a *Z/E* ratio of 51:49. Chromatographic conditions to facilitate the separation of the *E* and *Z* isomers could not be found. Column chromatography using 9:1 cyclohexane/acetone was employed in attempt to isolate each isomer, or at least obtain an uncontaminated mixture of the two, but all fractions retrieved contained small quantities of both isomers and large amounts of other material, assumed to be decomposition product.

4-(2-bromophenyl)-1-chlorobut-3-en-2-one

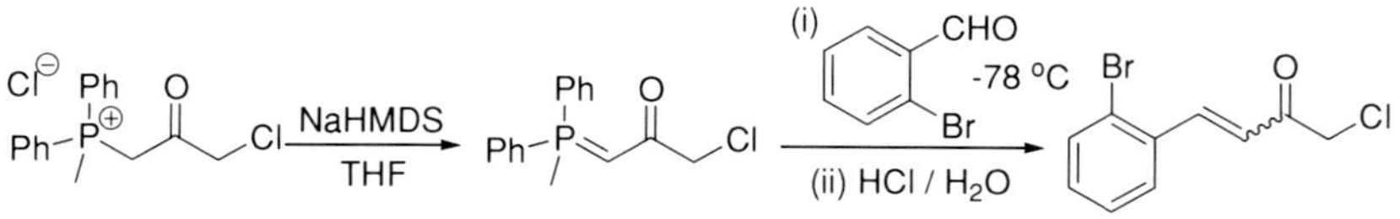

The ylide was generated from 3-chloroacetonylidenemethyldiphenylphosphonium chloride (57 mg, 0.17 mmol) & NaHMDS (0.15 ml of a 1 mol L^{-1} solution, 0.15 mmol), and reacted at −78 °C with 2-bromobenzaldehyde (0.02 ml, 0.03 g, 0.17 mmol) according to procedure D. See Table 2.4 entry 8.

^1H NMR (400 MHz, $CDCl_3$) of crude product:

Assigned to *E*-isomer: δ 6.81 (d, $J = 16.1$, 1H, H-3), 4.23 (s, 2H, CH_2Cl).

Assigned to *Z*-isomer: 6.33 (d, $J = 12.3$, 1H, H-3), 3.95 (s, 2H, CH_2Cl).

Integrations show a *Z/E* ratio of 50:50.

Chromatographic conditions to facilitate the separation of the *E* and *Z* isomers could not be found. Column chromatography of the crude product using 95:5 cyclohexane/ethyl acetate on neutral alumina gave a sample heavily enriched in the *E*-isomer and containing a large amount of impurity. The product appears to be prone to isomerisation & decomposition in contact with chromatographic stationary phases.

5-(2-bromophenyl)-2,2-pent-4-en-3-one [71]

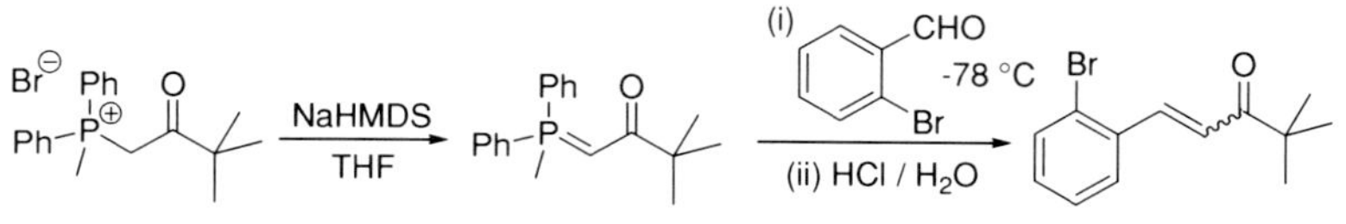

The ylide was generated from (3,3-dimethylbutan-2-onyl)methyldiphenylphosphonium chloride (0.898 mg, 2.37 mmol) & NaHMDS (400 mg, 2.18 mmol), and reacted at −78 °C with 2-bromobenzaldehyde (0.25 ml, 0.40 g, 0.22 mmol) according to procedure C. See Table 2.4 entry 9.

^{1}H NMR (300 MHz, $CDCl_3$) of crude mixture of enones:

Assigned to *E*-isomer: δ 8.02 (d, ^{3}J = 15.6, 1H, H-5), 7.06 (d, ^{3}J = 15.6, 1H, H-4), 1.24 (s, 9H, $C(CH_3)_3$) [71].

Assigned to *Z*-isomer: δ 6.93 (d, ^{3}J = 12.4, 1H, H-5), 6.60 (d, ^{3}J = 12.4, 1H, H-4), 1.16 (s, 9H, $C(CH_3)_3$).

Integrations indicate a *Z/E* ratio of 48:52.

The crude product was subjected to preparative TLC on silica using 90:10 pentane/diethyl ether. The enones did not fully separate under these conditions, but a sample heavily enriched in the *Z*-isomer could be obtained by scraping off the top ca. 30 % of the band that resulted from the elution. Several such samples were combined and subjected to a second round of prep TLC under the same conditions. Scraping off the top ca. 50 % of the band that resulted from this elution gave a sample of *Z*-isomer containing no more than 5 % of the *E*-isomer, which was characterised by ^{1}H, ^{13}C, COSY, HSQC and HMBC NMR, but was also found to be extremely sensitive to isomerisation to the *E*-isomer on exposure to light.

^{1}H NMR (600 MHz, $CDCl_3$): δ 7.55 (dd, J = 8.0, 0.9, 1H, ArH-3), 7.43 (dd, J = 7.7, 1.3, 1H, ArH-6), 7.25-7.22 (m, 1H, ArH-4), 7.14 (td, J = 7.5, 1.2, 1H, ArH-5), 6.93 (d, J = 12.5, 1H, H-5), 6.60 (d, J = 12.5, 1H, H-4), 1.16 (s, 9H, *t*-Bu). Spectrum also contains the following signals due the presence of *E*-isomer integrating for <5 % relative to 1H of the *Z*-enone: 8.02 (d, J = 15.6), 7.06 (d, J = 15.6), 1.24 (s) [71].

^{13}C NMR (151 MHz, $CDCl_3$): δ 206.3 (C = O), 139.5 (C-5), 136.3 (ArC-2), 132.2 (ArC-3), 130.6 (ArC-6), 129.6 (ArC-5), 126.8 (ArC-4), 125.7 (C-4), 123.1 (ArC-1), 44.0 (CMe$_3$), 26.2 (CH_3).

Spectrum also contains a small peak due to the *t*-Bu group of the *E*-isomer at δ 29.7.

4.7 Reactions of Non-stabilised Ylides

4.7.1 Synthesis of 1-Arylprop-1-enes

General Procedure and Work-Up

All reactions were carried out by using general procedure B, described in Sect. 4.3. Pentane or cyclohexane (5 ml) and distilled water (2 ml) were added, and the mixture was transferred to a separatory funnel. The phases were separated, and the aqueous phase was washed with two further aliquots of pentane (5 ml).

The organic phases were combined, dried over Na_2SO_4 and then filtered to remove the drying agent. The filtrate was concentrated *in vacuo* to give a yellow oil. A ^{1}H NMR of this oil was obtained and the *Z/E* ratio of the alkene product contained therein determined based on integration of characteristic baseline separated signals belonging to each of the isomers. The signals used to determine the *Z/E* ratio for each compound are listed below.

1-phenylprop-1-ene [72, 73]

(a) From ethyltriphenylphosphonium bromide (0.394 g, 1.06 mmol), NaHMDS (0.178 g, 0.99 mmol), and benzaldehyde (0.10 ml, 0.10 g, 0.99 mmol) by procedure B.

Br⁻ Ph$_3$P⁺Et —NaHMDS, THF→ Ph$_3$P=CHCH$_3$ —(i) PhCHO, -78 °C; (ii) -78 to 20 °C→ PhCH=CHCH$_3$

^{1}H NMR spectrum (300 MHz, $CDCl_3$) of the crude product:

Assigned to *E*-isomer: δ 6.23 (dq, $J = 15.6, 6.4$, 1H) [72].

Assigned to *Z*-isomer: δ 6.43 (dd, $J = 11.6, 1.5$, 1H), 5.79 (dq, $J = 11.6, 7.2$, 1H), 1.90 (dd, $J = 7.1$, baseline separation obscured by corresponding signal of *E*-isomer and water signal, 3H) [73].

Integrations indicate a *Z/E* ratio of 90:10.

(b) From diethyldiphenylphosphonium bromide, KHMDS, and benzaldehyde.

Br⁻ Ph$_3$P⁺Et —NaHMDS, THF→ Ph$_3$P=CHCH$_3$ —(i) PhCHO, -78 °C; (ii) -78 to 20 °C→ PhCH=CHCH$_3$

The OPA formed from these reactants was monitored by low temperature ^{31}P NMR. Details of that experiment and characterisation data for the 1-phenylprop-1-ene produced upon warming of the OPA are given in Sect. 4.7.4.

(c) From *P*-ethyl-*P*-phenyldibenzophospholium bromide (60 mg, 0.16 mmol), KHMDS (30 mg, 0.15 mmol), and benzaldehyde (0.016 ml, 16 mg, 0.15 mmol using 100 μl syringe) by procedure B.

Br⁻ *P*-ethyl-*P*-phenyldibenzophospholium —NaHMDS, THF→ ylide —(i) PhCHO, -78 °C; (ii) -78 to 20 °C→ PhCH=CHCH$_3$

^{1}H NMR (300 MHz, $CDCl_3$) of the crude product showed the same baseline separated signals as in part (a) above. Integration of these signals indicated a *Z/E* ratio of 53:47.

1-(2′-bromophenyl)prop-1-ene [74, 75]

(a) From ethyltriphenylphosphonium bromide (435 mg, 1.17 mmol), NaHMDS (198 mg, 1.07 mmol), and 2-bromobenzaldehyde (0.13 ml, 21 mg, 1.1 mmol using 100 μl syringe) by procedure B.

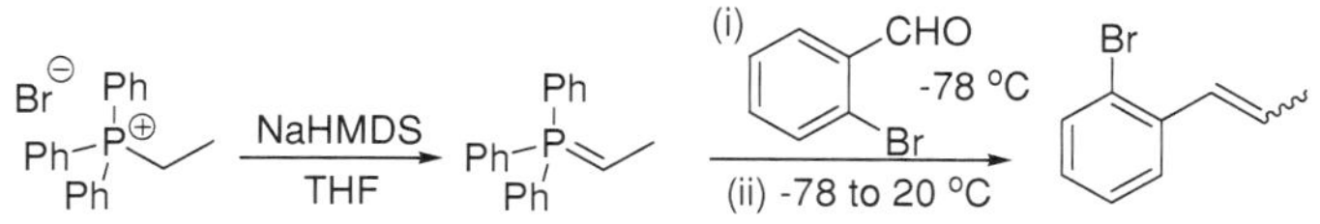

^{1}H NMR (300 MHz, $CDCl_3$) of the crude product:

Assigned to *E*-isomer: δ 6.72 (dd, J = 15.7, 1.3, 1H), 6.18 (dq, J = 15.5, 6.6, 1H), 1.92 (dd, J = 6.7, 1.6, 3H) [74].

Assigned to *Z*-isomer: δ 6.47 (dd, J = 11.5, 1.5, 1H), 5.88 (dq, J = 11.5, 7.1, 1H), 1.78 (dd, J = 7.1, 1.7, 3H) [75].

Integrations indicate a *Z/E* ratio of 79:21.

(b) From diethyldiphenylphosphonium bromide, KHMDS and 2-bromobenzaldehyde.

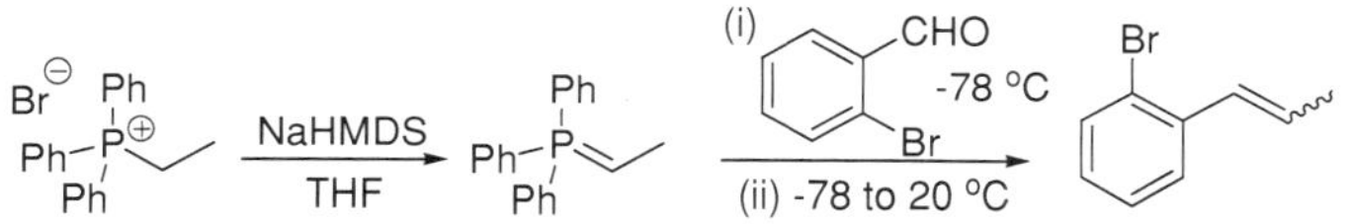

The OPA formed from these reactants was monitored by low temperature ^{31}P NMR. Details of that experiment and characterisation data for the 1-(2-bromophenyl)prop-1-ene produced upon warming of the OPA are given in Sect. 4.7.4.
(c) From *P*-ethyl-*P*-phenyldibenzophospholium bromide (56 mg, 0.15 mmol), KHMDS (28 mg, 0.14 mmol), and 2-bromobenzaldehyde (0.018 ml, 29 mg, 0.15 mmol using 100 μl syringe) by procedure B.

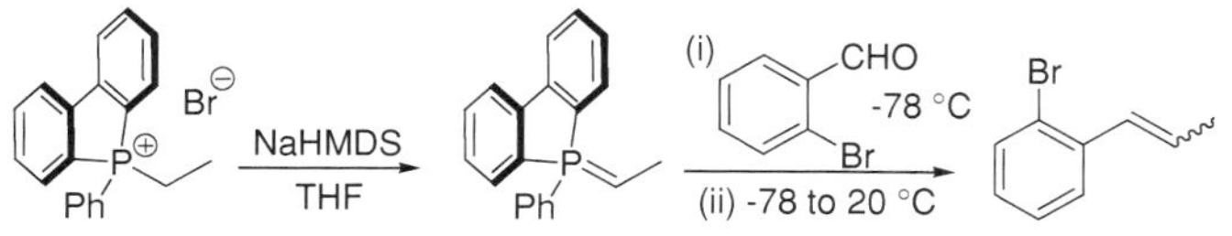

^{1}H NMR spectrum (600 MHz, $CDCl_3$) of the crude product:

Assigned to *E*-isomer: δ 6.17 (dq, J = 15.6, 6.7, 1H), 1.93-1.90 (dd, J = 6.7, 1.8, 1H 3H) [74].

Assigned to *Z*-isomer: δ 6.47 (dd, J = 11.5, 1.7, 1H), 5.92-5.84 (m, 1H), 1.77 (dd, J = 7.1, 1.8, 3H) [75].

Integrations indicate a *Z/E* ratio of 82:18.

4.7.2 Synthesis of (1-aryl-1-hydroxyprop-2-yl) Phosphonium Salts

General procedure for low temperature acid quenching of Wittig reaction

Dry phosphonium salt (1.1 equivalents) and NaHMDS (1.0 equivalent) were placed in an oven-dried Schlenk flask in a glove box under an atmosphere of argon, along with a magnetic stir bar [22, 76]. The flask was sealed and removed from the glove box, and then charged with nitrogen by attachment to a vacuum/nitrogen manifold using the standard Schlenk pump and fill technique [1].

For reactions of *P*-ethylidene-*P*-phenyldibenzophospholane, the mixture of phosphonium salt and base was cooled to −25 °C prior to addition of solvent. Dry THF was then added (to form a 0.1 mol L^{-1} solution of ylide) drop-wise to form ylide. The reaction mixture was then cooled to −45 °C, at which temperature it was stirred for 15 min, closed to the nitrogen supply. The maroon solution that resulted was then cooled to −78 °C, open to the nitrogen supply.

For all other ylides, dry THF (to form a 0.1 mol L^{-1} solution of ylide) was added at 20 °C. The flask was sealed from the nitrogen supply and the mixture was stirred, resulting in the formation of solution of phosphonium ylide (maroon for ylides derived from *P*-phenyl-5*H*-dibenzophosphole, orange for those derived from ethyldiphenylphosphine and triphenylphosphine). The ylide solution was stirred in the sealed flask for 30 min, then cooled to −78 °C by immersion of the flask in a dry ice/acetone bath while under a flow of nitrogen.

The appropriately substituted benzaldehyde (1.0 equivalent) was added drop-wise to the cooled solution of ylide. The ylide colour faded once one equivalent of aldehyde had been added. The reaction mixture in its entirity was then transferred via nitrogen-flushed, oven dried cannula into a second Schlenk flask cooled to − 78 °C by immersion in a separate dry ice/acetone bath containing a 1.25 mol L^{-1} solution of HCl in MeOH (2 equivalents) diluted with dry THF (5 ml), resulting in the formation of a colourless or slightly yellow liquid containing a white precipitate. This was stirred for 5 min at −78 °C, and then allowed to warm to room temperature. The solvent was removed from this mixture on a rotary evaporator. DCM and an aqueous solution of $NaHCO_3$ were added and the mixture transferred to a separatory funnel. The phases were separated and then the aqueous phase was washed a further two times with DCM. The DCM phases were combined, dried over Na_2SO_4 and then filtered to remove the drying agent. The filtrate was concentrated by rotary evaporation to give a yellow oil.

^{1}H, ^{13}C, ^{31}P, selectively decoupled $^{1}H\{^{31}P\}$, $^{13}C\{^{1}H,^{31}P\}$, COSY, TOCSY, HSQC, HMBC, ^{1}H-^{31}P HMBC and NOESY NMR spectra of this oil were obtained. Using these techniques, it was possible for each phosphorus-containing compound present in the crude product to determine all of the distinguishable signals belonging to that compound in both the ^{1}H and ^{31}P NMR spectra of the crude product. For this purpose, the selectively decoupled $^{1}H\{^{31}P\}$, COSY, TOCSY, and ^{1}H-^{31}P HMBC NMR techniques were the most useful. The identity of each compound was determined based on the connectivity established from the set of NMR data obtained for the crude product, and by reference to literature precedents. In particular, a set of signals was assigned to *erythro* or *threo* β-HPS by comparison with the spectra of similar compounds that have previously been reported [22, 76], or by analogy with *erythro*-(1-(2′-bromophenyl)-1-hydroxyprop-2-yl)triphenylphosphonium bromide, which was synthesised in this project (see below). In general, it was also possible to assign the major diastereomer of β-HPS to be the *erythro*-isomer since all of the Wittig reactions studied are predominantly *Z*-selective [22]. The *erythro/threo* ratio of the β-hydroxyphosphonium salt product contained in the oil was determined by integration of the characteristic signals belonging to each of the isomers in the ^{1}H and ^{31}P NMR spectra. A list of the signals assigned to each β-HPS isomer and used to determine the *erythro/threo* ratio is given below for each compound.

(1-hydroxy-1-phenylprop-2-yl)triphenylphosphonium bromide [76]

(i) NaHMDS Solvent (ii) PhCHO -78 °C (iii) HCl/MeOH Solvent -78 °C

From ethyltriphenylphosphonium bromide (180 mg, 0.485 mmol), NaHMDS (88 mg, 0.48 mmol), and benzaldehyde (0.050 ml, 52 mg, 0.49 mmol using 100 μl syringe). The crude product was observed by NMR to contain *erythro* and *threo* β-HPS isomers, PhCHO and Ph_3PO, and a significant amount of the starting phosphonium salt $[EtPh_3P]Br$.

^{1}H NMR (400 MHz, $CDCl_3$) of crude product:

Assigned to *threo*-isomer: δ 4.73 (t, J = 8.0, 1H, OC*H*), 1.52 (dd, J = 19.4, 7.2, 3H, CH_3) [76].

Assigned to *erythro*-isomer: δ 7.25 (t, $J = 7.5$, 2H), 7.17 (t, $J = 7.3$, 1H), 5.42 (dd, $J = 7.6$, 2.5, 1H, OC*H*), 3.86 −3.75 (m, 1H, PC*H*, overlaps slightly with signal of [$EtPh_3P$]Br) [76].

Other signals (integrations with respect to *erythro*-OC*H* being 1H): δ 3.71 (dq, $J = 12.7$, 7.4, 1.6H, [$EtPh_3P$]Br CH_2), 1.40-1.22 (m, 5.3 H, contains dt, $J = 20.0$, 7.4, [$EtPh_3P$]Br CH_3 and dd, $J = 18.8$, 7.2, *erythro*-β-HPS CH_3).

^{31}P NMR (162 MHz, $CDCl_3$) of crude product: δ 31.9 (0.05P), 30.8 (1P, *erythro*-β-HPS), 29.8 (0.08P), 29.2 (0.11P), 26.2 (0.78P, [$EtPh_3P$]Br).

Integrations of these signals in the ^{1}H NMR spectrum indicates an *erythro/threo* ratio of 90:10.

(1-hydroxy-1-phenylprop-2-yl)ethyldiphenylphosphonium bromide

From diethyldiphenylphosphonium bromide (120 mg, 0.36 mmol), NaHMDS (66 mg, 0.36 mmol), and benzaldehyde (0.040 ml, 42 mg, 0.39 mmol by 100 μl syringe) in THF (4 ml). Signals in the ^{1}H and ^{31}P NMR spectra were assigned using the two-dimensional NMR techniques mentioned in the general procedure, and by reference to selectively decoupled ^{1}H{^{31}P} spectra of the crude product from the reaction.

(i) NaHMDS THF (ii) PhCHO -78 °C (iii) HCl/MeOH THF -78 °C

NMR signals assigned to *threo* isomer:

^{1}H NMR (600 MHz, $CDCl_3$) δ 4.43-4.35 (m, 1H, PC*H*), 4.35-04.28 (app t, $J =$ ca. 9.5 Hz, 1H, OC*H*), 3.71-3.62 (m, obscured by side-product signal, one of diastereotopic PCH_2 hydrogens), 3.43-3.33 (m, 1H, one of diastereotopic PCH_2 hydrogens), 0.83 (dd, $J = 18.2$, 7.1, 3H, OCH-CHCH_3).

^{31}P NMR (243 MHz, $CDCl_3$) δ 38.7.

NMR signals assigned to *erythro* isomer:

^{1}H NMR (600 MHz, $CDCl_3$) δ 5.87 (dd, $J = 6.4$, 2.7, 1H, OC*H*), 3.91-3.80 (m, 2H, contains PC*H* and one of diastereotopic PCH_2 hydrogens), 3.59-3.49 (m, slightly obscured by side-product signal, one of diastereotopic PCH_2 hydrogens).

^{31}P NMR (243 MHz, $CDCl_3$) δ 36.5.

Integrations of these signals in the ^{1}H and ^{31}P NMR spectra indicate an *erythro/threo* ratio of 54:46. The identity of the β-HPS isomers was assigned by analogy with the ^{1}H NMR spectra of similar compounds (1-hydroxy-1-phenylprop-2-yl) triphenylphosphonium bromide [76] and (1-(2-bromophenyl)-1-hydroxyprop-2-

yl)triphenylphosphonium bromide (see below). The O*CH* signal of the major diastereomer shows $^{3}J_{PH}$ = 6.4 Hz, and has a higher chemical shift than the minor diastereomer's OC*H* signal, which shows $^{3}J_{PH}$ >9 Hz. The $PCHCH_3$ signal of the major diastereomer, although obscured by other signals, can be observed by COSY NMR to be higher in chemical shift than the corresponding signal of the minor diastereomer. On the basis of these observations, the major diastereomer is assigned to be of *erythro* configuration. This assignment was confirmed by deprotonation of a pure sample of the major diastereomer (see below), which resulted in the formation of *Z*-alkene.

Other signals were also present in the ^{1}H and ^{31}P NMR spectra of the crude product that could be assigned to specific species. A signal in the ^{1}H spectrum at δ 1.03-0.92 (m, 8.6H) was shown to contain the *erythro*-β-HPS OCH–CHCH_3 and PCH$_2$CH_3 hydrogens, as well as the *threo*-β-HPS PCH$_2$CH_3 hydrogens, which all together integrate for 8.6H with respect to the *erythro*-β-HPS OC*H* signal, which is consistent with the observed *erythro/threo* ratio. Diethyldiphenylphosphonium bromide was also shown to be present by the signals at δ_H 3.24 (dq, J = 15.1, 7.5, 0.73H vs. *erythro*-β-HPS OCH) and δ_P 32.3 (0.15P vs. *erythro*-β-HPS) [16]. The PCH$_2$CH_3 hydrogens of diethyldiphenylphosphonium bromide were obscured by other signals.

The crude product from the reaction was washed with ether, then dissolved in a minimum of hot ethanol in a pear shaped flask. To this was added hot ethyl acetate, and the flask was stoppered and the solution allowed to cool. The resulting solution was left open in a fume hood. Slow evaporation of the solvent gave white crystals in contact with a yellow oil. Addition of dichloromethane caused the oil to dissolve, while the crystals remained. The crystals were washed with six 5 ml aliquots of dichloromethane, with the washings being decanted after each wash. The white crystals were then dried *in vacuo* (50 mg, 32 % based on starting phosphonium salt). Characterisation by ^{1}H, ^{13}C, ^{31}P, ^{13}C{^{1}H,^{31}P}, COSY, TOCSY, HSQC, and HMBC NMR showed the crystals to be one isomer of β-HPS—the one that has been determined to be the *erythro* isomer.

^{1}H NMR (600 MHz, CD$_3$OD) δ 7.98-7.83 (m, 6H, *P*-PhH-2 & *P*-PhH-4), 7.78-7.70 (m, 4H, *P*-PhH-3), 7.37 (m, 4H, *C*(OH)-PhH-2 & PhH-3), 7.32-7.26 (m, 1H, *C*(OH)-PhH-4), 5.14 (dd, J = 6.9, 2.8, 1H), 3.60 (dqd, $^{2}J_{PH}$ = 14.5, J_{HH} = 7.2, 2.8, 1H), 3.30-3.23 (m, 1H, one of diastereotopic PCH_2, partially obscured by CH$_3$OH signal), 3.21-3.11 (m, 1H, one of diastereotopic PCH_2), 1.27 (dt, $^{3}J_{PH}$ = 19.1, $^{3}J_{HH}$ = 7.5, 3H, PCH$_2$CH_3), 1.17 (dd, $^{3}J_{PH}$ = 17.8, $^{3}J_{HH}$ = 7.2, 3H, PCHCH_3).

^{31}P{^{1}H} NMR (243 MHz, CD$_3$OD) δ 34.5.

^{13}C NMR (151 MHz, CD$_3$OD) δ 140.9 (d, J = 12.8, OCH-PhC-1), 134.4 (d, J = 2.9, one of diastereotopic *P*-PhC-4), 134.3 (d, J = 2.9, one of diastereotopic *P*-PhC-4), 133.7 (d, J = 8.7, one of diastereotopic *P*-PhC-2), 133.5 (d, J = 8.8, one of diastereotopic *P*-PhC-2), 129.8 (d, J = 11.9, one of diastereotopic *P*-PhC-3), 129.62 (d, J = 12.0, one of diastereotopic *P*-PhC-3), 128.0 (s, OCH-PhC-2 or

C-3), 127.5 (s, PhC-4), 125.5 (s, OCH-PhC-2 or C-3), 117.9 (d, $J = 26.3$, one of diastereotopic *P*-PhC-1), 117.4 (d, $J = 26.1$, one of diastereotopic *P*-PhC-1), 70.1 (d, $J = 3.4$, *C*HO), 35.6 (d, $^1J_{PC} = 45.6$, P*C*HMe), 14.0 (d, $^1J_{PC} = 49.5$, P*C*H_2), 6.9 (s, PCH*C*H_3), 5.8 (d, $^2J_{PC} = 5.6$, PCH_2CH_3).

HRMS (m/z): Calc. for $[M]^+ = C_{22}H_{25}PO$ 349.1721; found 349.1717 (1.2 ppm).

MP (crystallised from chloroform/ethyl acetate) 130–134 °C.

A sample of the *erythro*-β-HPS (22 mg, 0.06 mmol) was mixed with dry THF (0.6 ml) in a Schlenk flask, and cooled to −78 °C. A solution of NaHMDS in THF (1 mol L^{-1}, 0.06 ml, 0.06 mmol) was added dropwise to the cooled solution. The reaction mixture was stirred for 5 min at −78 °C, then allowed to warm to room temperature, and subjected to aqueous work-up using the procedure described at the start of Sect. 4.7.1. The crude product was analysed by ^1H NMR. Only signals due to *Z*-1-phenylprop-1-ene could be detected: ^1H NMR (300 MHz, $CDCl_3$) δ 6.37 (dd, $J = 11.6$, 1.8 Hz, 1H), 5.72 (dd, $J = 11.6$, 7.1 Hz, 1H) [73].

(1-(2′-bromophenyl)-1-hydroxyprop-2-yl)triphenylphosphonium bromide

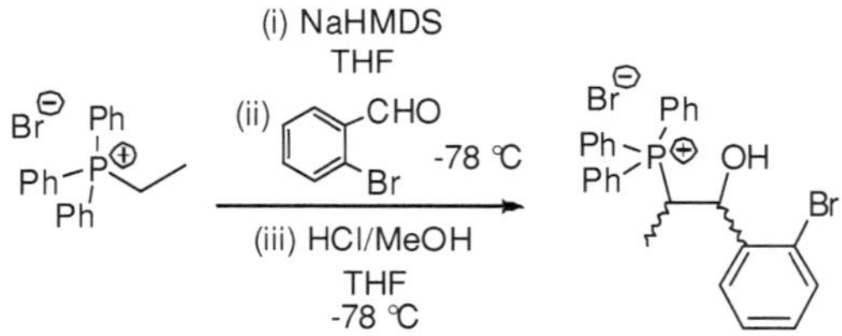

From ethyltriphenylphosphonium bromide (360 mg, 0.97 mmol), NaHMDS (173 mmol, 0.94 mmol), and 2-bromobenzaldehyde (0.11 ml, 0.17 g, 0.94 mmol) in THF (8 ml).

NMR signals assigned to *threo* isomer:

^1H NMR (300 MHz, $CDCl_3$) δ 5.15 (dd, $J = 14.5$, 7.4, 0.03H), 4.82-4.75 (m, 0.03H), 1.50 (dd, $J = 19.4$, 7.2, 0.03H).

^{31}P NMR (121 MHz, $CDCl_3$) δ 31.0.

NMR signals assigned to *erythro* isomer:

^1H NMR (300 MHz, $CDCl_3$) δ 7.28 (m, 1H), 7.08 (td, $J = 7.7$, 1.6, 1H), 5.41 (dd, $J = 7.5$, 2.3, 1H), 3.88 (dqd, $J = 14.0$, 7.2, 2.4, 1H), 1.22 (dd, $J = 18.8$, 7.2, 3H).

^{31}P NMR (121 MHz, $CDCl_3$) δ 31.4.

Integrations of these signals in the ^1H and ^{31}P NMR spectra indicate an *erythro/threo* ratio of 95:5. Also present in the reaction mixture were some small signals due to other species (integrations relative to *erythro*-β-HPS OC*H* and phosphorus in ^1H and ^{31}P spectra respectively).

1-(2′-bromophenyl)prop-1-ene: ^{1}H NMR (600 MHz, $CDCl_3$) δ 6.74-6.63 (m, 0.04H, *E*-alkene H), 6.45 (d, J = 11.5, 0.04H, *Z*-alkene H), 6.16 (ddd, J = 15.6, 13.4, 6.7, 0.04H, *E*-alkene H), 5.86 (m, 0.04H, *Z*-alkene H), 1.89 (dd, J = 6.7, 1.7, 0.13H, *E*-CH_3), 1.75 (dd, J = 7.1, 1.8, 0.12H, *Z*-CH_3) [74, 75].

Ethyltriphenylphosphonium bromide: ^{1}H NMR (600 MHz, $CDCl_3$) δ 3.78 (dq, J = 12.5, 7.5, 0.32H), 1.36 (ddd, J = 19.3, 15.0, 7.4, 0.48H), ^{31}P NMR (243 MHz, $CDCl_3$) δ 26.3 (0.16P) [15].

Triphenylphosphine oxide ^{31}P NMR (243 MHz, $CDCl_3$) δ 29.2 (0.14P) [77].

The crude product was washed with ether and recrystallised from hot acetonitrile to give (after standing in refrigerator) white crystals of *erythro*-(1-(2′-bromophenyl)-1-hydroxyprop-2-yl)triphenylphosphonium bromide (320 mg, 60 %), which was characterised by ^{1}H, ^{13}C, ^{31}P, ^{13}C{^{1}H,^{31}P}, COSY, TOCSY, HSQC, and HMBC NMR.

^{1}H NMR (600 MHz, $CDCl_3$) δ 7.90-7.84 (m, 7H, contains 2-BrPh H-6, *P*-PhH-2), 7.81 (m, 3H, *P*-PhH-4), 7.71 (td, J = 7.9, 3.4, 6H, *P*-PhH-3), 7.42 (dd, J = 7.9, 0.9, 1H, 2-BrPh H-5), 7.34 (t, J = 7.5, 1H, 2-BrPh H-3), 7.12 (td, J = 7.7, 1.6, 1H, 2-BrPh H-4), 6.82 (d, J = 4.8, 1H, O*H*), 5.47 (m, 1H, OC*H*), 3.93 (dqd, J = 14.3, 7.0, 2.0, 1H, PC*H*), 1.25 (dd, J = 18.9, 7.2, 3H, CH_3).

^{13}C NMR (151 MHz, $CDCl_3$) δ 139.4 (d, J = 13.7, 2-BrPh C-1), 134.9 (d, J = 3.0, *P*-PhH-4), 134.7 (d, J = 9.5, *P*-PhH-2), 132.3 (s, 2-BrPh C-5), 130.6 (s, 2-BrPh C-6), 130.3 (d, J = 12.4, *P*-PhH-3), 129.4 (s, 2-BrPh C-4), 127.8 (s, 2-BrPh C-3), 120.5 (s, 2-BrPh C-2), 118.1 (d, J = 83.8, *P*-PhH-1), 68.8 (d, J = 2.8, J = OC*H*), 35.7 (d, J = 46.3, PC*H*), 10.3 (d, J = 1.1, CH_3).

^{31}P NMR (243 MHz, $CDCl_3$) δ 31.5.

HRMS (m/z): Calc. 475.0826, found 475.0809 (3.7 ppm); calc. 477.0806, found 477.0799 (1.4 ppm).

MP (crystallised from chloroform/ethyl acetate) 227–231 °C.

The compound was confirmed as having the *erythro* configuration by X-ray crystallography, and the counter-ion was confirmed as being bromide.[10]

---

[10] CCDC-883627 contain the X-ray crystallographic data for this compound. This data can be obtained free of charge from The Cambridge Crystallographic Data Centre via http://www.ccdc.cam.ac.uk/data_request/cif.

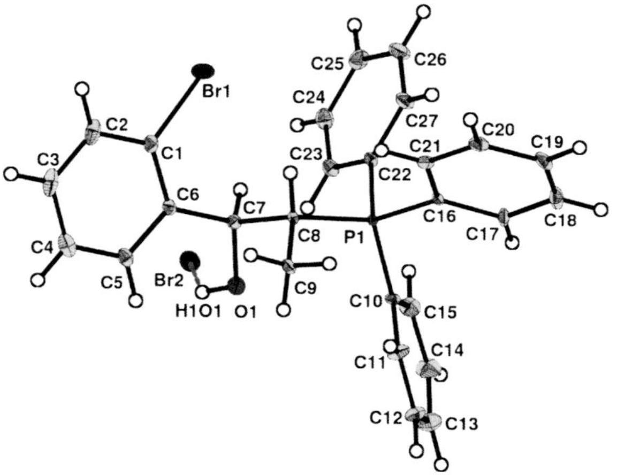

(1-(2′-bromophenyl)-1-hydroxyprop-2-yl) ethyldiphenylphosphonium bromide

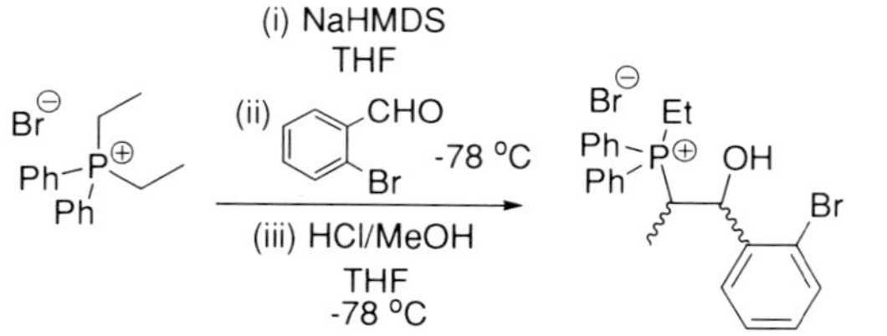

From diethyldiphenylphosphonium bromide (183 mg, 0.566 mmol), NaHMDS (104 mg, 0.566 mmol), and 2-bromobenzaldehyde (0.065 ml, 0.10 g, 5.6 mmol) in THF (6 ml).

NMR signals assigned to *threo* isomer:

^{1}H NMR (600 MHz, $CDCl_3$) δ 8.38 (d, $J = 6.7$, 1H, *O*H), 7.34 (dd, $J = 8.0$, 1.1, 1H), 7.07 (td, $J = 7.9$, 1.6, 1H), 5.02 (dd, $J = 12.0$, 10.0, 1H, OC*H*), 4.95-4.85 (m, 1H, PC*H*), 0.83 (dd, $J = 18.0$, 7.4, 3H).

^{31}P NMR (243 MHz, $CDCl_3$) δ 39.5.

NMR signals assigned to *erythro* isomer:

^{1}H NMR (600 MHz, $CDCl_3$) δ 7.28 (td, $J = 7.8$, 0.9, 1H), 7.12 (td, $J = 7.7$, 1.7, 1H), 5.79 (dd, $J = 7.0$, 2.3, 1H, OC*H*), 3.67 (dqd, $J = 14.4$, 7.2, 2.4, 1H, PC*H*), 1.31 (dt, $J = 19.6$, 7.5, 3H, PCH_2CH_3), 1.05 (dd, $J = 17.8$, 7.3, slightly obscured by *threo*-PCH_2CH_3 dt signal, OCH-$CHCH_3$).

^{31}P NMR (243 MHz, $CDCl_3$) δ 36.3

Integrations of these signals in the ^{1}H and ^{31}P NMR spectra indicate an *erythro/threo* ratio of 65:35. This ratio agrees exactly with the OPA *cis/trans* ratio determined for this reaction by low temperature ^{31}P NMR monitoring of the Wittig reaction.

Other overlapping signals of the *erythro* and *threo* isomers can also be seen in the ^{1}H NMR (integrations relative to *erythro*-β-HPS OCH in ^{1}H): δ 3.94-3.78 (m, 1.6H, one each of diastereotopic PCH_2 signals of *erythro* & *threo* β-HPS), 3.59-3.43 (m, 1.6H, each of diastereotopic PCH_2 signals of *erythro* & *threo* β-HPS). Also present in the reaction mixture were some signals due to other species (integrations relative to *erythro*-β-HPS OCH and phosphorus in ^{1}H and ^{31}P spectra respectively):

A phosphine oxide, probably diethylphenylphosphine oxide (ylide hydrolysis product): $\delta_H = 2.27$ (dq, $J = 11.4$, 7.6, 0.4H, CH_2); $\delta_P = 32.5$ (0.1P). Diethyldiphenylphosphonium bromide: $\delta_P = 32.5$ (0.5P) [16].

Crystallisation of the crude product from ethanol/ethyl acetate yielded a sample enriched in the *erythro*-β-HPS (*erythro/threo* ratio of 90:10) and containing a small amount ($\sim$10 %) of diethyldiphenylphosphonium bromide with significant loss of yield (mass of sample recovered was 80 mg).

Data for crude (1-hydroxy-3-methy-1-phenyllbut-2-yl) triphenylphosphonium bromide

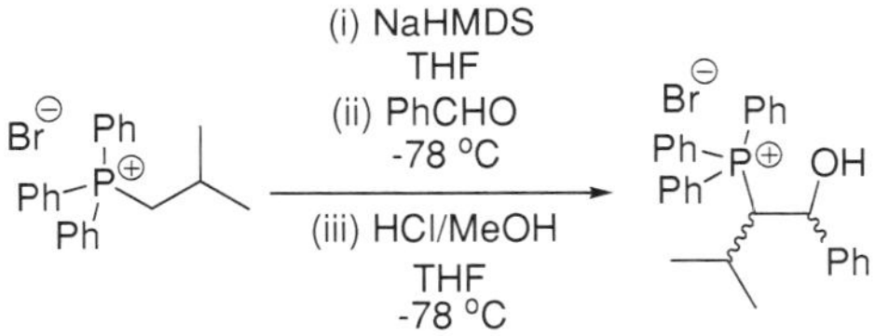

From *iso*-butyltriphenylphosphonium bromide (120 mg, 30.1 mmol), NaHMDS (54 mg, 0.29 mmol), and benzaldehyde (0.030 ml, 31 mg, 0.30 mmol by 100 μl syringe).

NMR signals assigned to *erythro* isomer:

^{1}H NMR (600 MHz, $CDCl_3$) δ 7.14 (m, 1H), 5.49 (dd, $J = 8.5$, 3.2, 1H, OCH), 4.17-4.06 (m, 1H, PCH), 2.46-2.34 (m, 1H, CHMe$_2$), 0.74 (d, $J = 7.0$, 3H, one of diastereotopic CH_3 groups), 0.67 (d, $J = 7.0$, 3H, one of diastereotopic CH_3 groups).

^{31}P NMR (243 MHz, $CDCl_3$) δ 30.3 (assigned using ^{1}H-^{31}P HMBC NMR experiments).

The major β-HPS was assigned to be *erythro* by analogy with similar compounds, as described in Sect. 2.4.2. No signals could be unambiguously assigned to *threo*-β-HPS in the ^{1}H NMR, and all four signals in the ^{31}P NMR were shown conclusively using ^{1}H-^{31}P HMBC not to be *threo*-β-HPS. The discernible spectral characteristics for the other compounds present are given below (integrations relative to *erythro*-β-HPS OC*H* ^{1}H signal or ^{31}P signal):

Z-1-phenyl-3-methylbut-1-ene: δ_H 6.28 (d, $J = 11.7$, 0.2H), 5.45 (dd, $J = 11.5$, 10.3, 0.2H), 2.93-2.85 (m, 0.2H), 1.02 (d, $J = 6.5$, obscured by signal of [(*i*-Bu)Ph_3P]Br) [78].

P-(*iso*-butyl)triphenylphosphonium bromide: δ_P 22.9 (0.4P); δ_H 3.67 (dd, $J = 12.9$, 6.3, 0.8H), 1.03 (d, $J = 6.7$, obscured by signal of alkene).[11]
Triphenylphosphine oxide: δ_P 29.8 (0.2P) [77].

(*iso*-Butyl)triphenylphosphine oxide: δ_P 31.1 (0.26P); δ_H 2.18 (dd, $J = 11.1$, 6.4, obscured by water signal), 0.98 (d, $J = 6.6$, 1.6H) [79].

C*H*Me_2 signals of [(*i*-Bu)Ph_3P]Br and *i*-Bu)Ph_2PO overlap (δ 2.2-2.0), and are also obscured by a signal due to water.

Data for crude (1-(2′-bromophenyl)-1-hydroxy-3-methylbut-2-yl)triphenylphosphonium bromide

(i) NaHMDS THF; (ii) 2-bromobenzaldehyde (CHO, Br), -78 °C; (iii) HCl/MeOH THF -78 °C

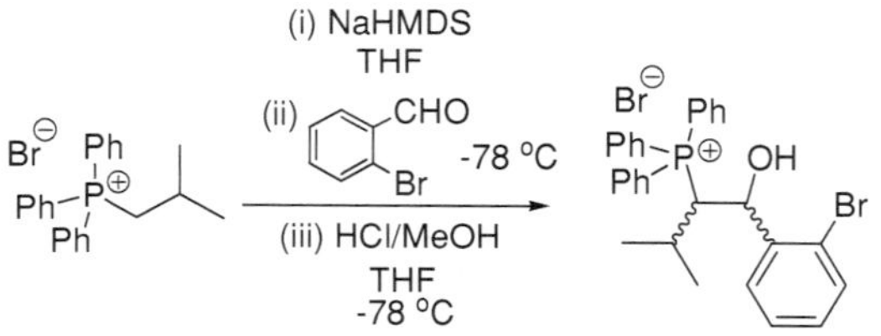

From *iso*-butyltriphenylphosphonium bromide (104 mg, 0.260 mmol), NaHMDS (46 mg, 0.25 mmol), and benzaldehyde (0.025 ml, 26 mg, 0.25 mmol by 100 μl syringe).

NMR signals assigned to *erythro* isomer:

^{1}H NMR (600 MHz, $CDCl_3$) δ 8.01 (dd, $J = 7.8$, 0.9, 1H), 7.35-7.30 (m, 1H), 7.11 (td, $J = 7.7$, 1.5, 1H), 5.40 (dd, $J = 7.3$, 3.2, 1H, C*H*OH), 4.52-4.42 (m, 1H, PC*H*), 2.48-2.38 (m, 1H, C*H*Me_2), 0.69 (d, $J = 7.0$, 3H, diastereotopic CH_3), 0.62 (d, $J = 6.9$, 3H, diastereotopic CH_3).

^{31}P NMR (243 MHz, $CDCl_3$) δ 30.6 (assigned using ^{1}H-^{31}P HMBC NMR experiments).

The major β-HPS was assigned to be *erythro* based on the high *Z*-selectivity in the corresponding unquenched Wittig reaction (see Sect. 4.9, reaction 18), and also by analogy with similar compounds, as described in Sect. 2.4.2. No signals could be

[11] See Sect. 4.2.5, for characterisation details of this compound.

unambiguously assigned to *threo*-β-HPS in the ^{1}H NMR, and all four signals in the ^{31}P NMR were shown conclusively using ^{1}H-^{31}P HMBC not to be *threo*-β-HPS. The discernible spectral characteristics for three other phosphorus containing compounds present are given below (integrations relative to *erythro*-β-HPS OC*H* ^{1}H signal or ^{31}P signal):

P-(*iso*-butyl)triphenylphosphonium bromide: δ_P 22.9 (0.44P); δ_H 3.80 (dd, J = 12.9, 6.2, 0.88H, PCH_2), 2.18-1.99 (m, 0.27H, C*H*Me$_2$), 1.08 (d, J = 6.7, 2.7H, CH(CH_3)$_2$).

Triphenylphosphine oxide: δ_P 29.8 (0.2P) [77].

(*iso*-Butyl)triphenylphosphine oxide: δ_P 31.6 (0.2P); δ_H 2.18 (dd, J = 10.7, 6.6 Hz, 0.44H, PCH_2), 2.21-2.11 (m, obscured by signal of [(*i*-Bu)Ph$_3$P]Br, C*H*Me$_2$) 1.01 (d, J = 6.6 Hz, 1.4H, CH(CH_3)$_2$) [79].

Data for crude *P*-(1-hydroxy-1-phenylprop-2-yl) *P*-phenyldibenzophospholium bromide

(i) NaHMDS
THF
(ii) CHO
-78 °C
(iii) HCl/MeOH
THF
-78 °C

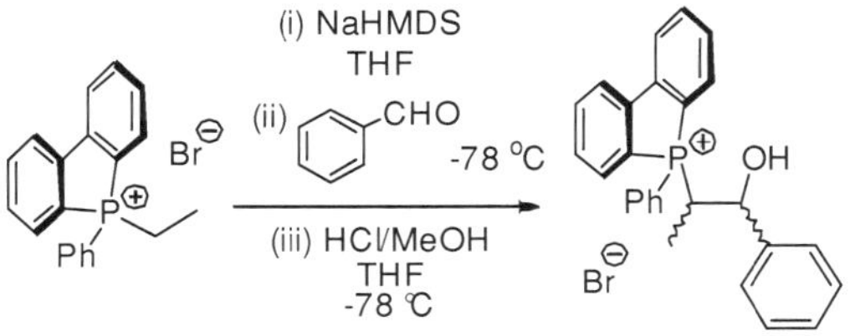

From *P*-ethyl-*P*-phenyldibenzophospholium bromide (30 mg, 0.081 mmol), KHMDS (16 mg, 0.080 mmol), and benzaldehyde (0.008 ml, 8 mg, 0.08 mmol by 25 μl syringe). The major β-HPS diasteromer was assigned to be the *erythro*-isomer based on the predominant selectivity for *Z*-alkene in the corresponding unquenched Wittig reaction (see Sect. 4.7.1).

NMR signals assigned to *threo* isomer:

^{1}H NMR (600 MHz, CDCl$_3$) δ 4.97 (t, J = 8.2, 1H, OC*H*), 4.39 (m (broad), 1H, PC*H*), 1.05-0.97 (m, 3H).

^{31}P NMR (243 MHz, CDCl$_3$) δ 33.8.

NMR signals assigned to *erythro* isomer:

^{1}H NMR (600 MHz, CDCl$_3$) δ 5.44 (d, J = 7.0, 1H, OC*H*), 0.72 (dd, J = 21.4, 5.4, 3H, CH_3).

^{31}P NMR (243 MHz, CDCl$_3$) δ 36.8.

Integrations of these signals in the ^{1}H and ^{31}P NMR spectra indicate an *erythro*/*threo* ratio of 72:28. Two overlapping signals for *erythro* & *threo* β-HPS DBP-ring hydrogens that together integrate for 1.4H (relative to *erythro* OC*H*) at δ 8.87-8.78 can also be distinguished. Other species can be observed in the ^{1}H and ^{31}P NMR spectra of the crude product (integrations relative to *erythro*-β-HPS OC*H* or phosphorus signal):

P-ethyl-*P*-phenyldibenzophospholium bromide: δ_H 3.82 (dd, $J = 13.0, 6.7$, 2.3H), 1.16-1.05 (m, 3.3H); $\delta_P = 32.4$ (1.15P).

Data for crude *P*-(1-(2′-bromophenyl)-1-hydroxyprop-2-yl)-*P*-phenyldibenzophospholium bromide

From *P*-ethyl-*P*-phenyldibenzophospholium bromide (126 mg, 0.341 mmol), KHMDS (65 mg, 0.33 mmol), and 2-bromobenzaldehyde (0.040 ml, 63 mg, 0.35 mmol by 100 μl syringe) in THF. The major β-HPS diasteromer was assigned to be the *erythro*-isomer based on the predominant selectivity for *Z*-alkene in the corresponding unquenched Wittig reaction (see Sect. 4.7.1).

NMR signals assigned to *erythro* isomer:

^{1}H NMR (600 MHz, $CDCl_3$) δ 8.93 (t, $J = 8.6$, 1H), 5.52 (dd, $J = 7.1, 2.5$, 1H, OC*H*), 0.74 (dd, $J = 21.3, 7.0$, 3H, CH_3).

^{31}P NMR (243 MHz, $CDCl_3$) δ 35.7

NMR signals assigned to *threo* isomer:

^{31}P NMR (243 MHz, $CDCl_3$) δ 34.0.

No signals in the ^{1}H NMR could be integrated satisfactorily, but a small cross-peak in the ^{1}H-^{31}P HMBC spectrum could be observed linking what appears to be the *threo*-β-HPS OC*H* signal to the ^{31}P signal at $\delta_P = 34.0$. Integration of the peaks in the ^{31}P NMR spectra indicate an *erythro/threo* ratio of 95:5. A large amount of *P*-ethyl-*P*-phenyldibenzophospholium bromide[8] was present in the crude product (as well as 2-bromobenzaldehyde, judging from the ^{1}H spectrum), showing signals at $\delta_H = 3.94–3.78$ (m, not baseline separated), 1.12 (dt, $J = 22.3, 7.1$, 4.0H) and $\delta_P = 32.0$ (1.3P), where the integrations are relative to the OC*H* and phosphorus signals of the *erythro*-β-HPS in the ^{1}H and ^{31}P NMR spectra, respectively.

4.7.3 *Low Temperature Acid Quenching of Wittig Reactions of* P-*Alyklidene*-P-*Phenyldibenzophospholanes to Give β-HPS and Subsequent Generation of OPA and Alkene*

General Procedure

The general procedure for low temperature acid quenching experiments described in Sect. 4.7.2 was used up to the point of carrying out the aqueous work-up. Rather than doing the aqueous work-up described there, the THF and methanol solvent was removed from the reaction mixture *in vacuo* to leave a crude white solid. The crude product was dissolved in $CDCl_3$, and a sample of the resulting solution was filtered of inorganic salts into an NMR tube, and was then characterised by ^{1}H, ^{13}C, ^{31}P, gCOSY, zTOCSY, 2D NOESY, gHSQC, gHMBC and ^{1}H-^{31}P HMBC NMR. These NMR techniques were used to assign signals to all compounds present in the crude product, and in particular to β-HPS. The major diastereomer was in all cases assigned to be *erythro* since the corresponding unquenched Wittig reactions are predominantly *Z*-selective. Assignments made in this manner were confirmed by low temperature deprotonation of the β-HPS, which gave *cis*-OPA as the major diastereomer of the intermediate (as judged by NOE experiments), and ultimately *Z*-alkene on heating of the OPA solution. The β-HPS *erythro/threo* ratio was determined by integration of signals assigned to each diastereomer in the ^{1}H and ^{31}P NMR spectra. In addition to β-HPS, the crude product also contained phosphonium salt starting material (quenched ylide), phosphine oxide (from reaction of ylide or phosphonium salt with water), and a small amount of aldehyde.

The crude product was dissolved in a mixture of chloroform and ethyl acetate, and β-HPS precipitated by gradual partial solvent removal using rotary evaporator. The white precipitate was collected by filtration of the mixture. The filtrate could be similarly treated to yield further β-HPS. ^{1}H and ^{31}P NMR analysis of the precipitate showed it to contain β-HPS diastereomers almost exclusively. The precipitate was placed in a Schlenk flask under a flow of nitrogen, dried under vacuum, and the flask was placed in a glove box under an argon atmosphere.

In the glove box, NaHMDS (1 equivalent) was added to the Schlenk flask, and the sealed flask was removed from the glove box and charged with nitrogen on a Schlenk manifold as before. The flask was cooled to −78 °C, and dry toluene-*d8* (0.8 ml) was added, resulting in the formation of a pale yellow/orange solution and a precipitate of sodium salt (bromide or chloride). After stirring for 5 min at low temperature, the flask was allowed to warm to room temperature and the OPA solution was filtered by cannula under nitrogen flow into an NMR tube contained in a long Schlenk flask under an atmosphere of nitrogen. The NMR tube was sealed with a rubber septum inside the Schlenk tube under a flow of nitrogen, and immediately placed in an NMR spectrometer at −20 °C. The OPA solution was characterised by ^{1}H, ^{13}C, ^{31}P, gCOSY, zTOCSY, 1D & 2D NOESY, gHSQC, gHMBC and ^{1}H-^{31}P gHMQC NMR. The OPA *cis/trans* ratio was also determined by integration of signals assigned to each diastereomer in the ^{1}H and ^{31}P NMR spectra.

The contents of the NMR tube were poured into a flame-dried Schlenk flask under an atmosphere of nitrogen, and the solution was heated to 80 °C for 2 h to effect the decomposition of the OPA to alkene and phosphine oxide. ^{1}H and ^{31}P NMR analysis of the solution after this time showed it to contain only alkene and phosphine oxide. The alkene *Z/E* ratio was determined by integration of characteristic signals in the ^{1}H NMR spectrum.

Reaction of *P*-(isobutylidene)-*P*-phenyldibenzophospholane and benzaldehyde

(i) NaHMDS THF (ii) CHO -78 °C (iii) HCl/MeOH THF -78 °C

NaHMDS THF

2 hrs. 80 °C

P-(isobutyl)-*P*-phenyldibenzophospholium bromide (33 mg, 0.083 mmol) and NaHMDS (16 mg, 0.087 mmol) were used to generate the ylide in THF (1.2 ml). The ylide was reacted at low temperature with benzaldehyde (0.008 ml, 8 mg, 0.08 mmol). The ^{1}H and ^{31}P spectra of the crude product showed the β-HPS *erythro/threo* ratio to be 89:11. The identity of the major isomer was assigned to be *erythro* based on the fact that its deprotonation results predominantly in *Z*-alkene (see below).

NMR signals assigned to *threo* isomer:

^{1}H NMR (600 MHz, $CDCl_3$) δ 9.07-9.01 (m, 1H), 5.49 (dd, $J = 17.1, 3.4$, 1H, OC*H*), 4.04 (d, $J = 13.6$, 1H, PC*H*).

^{31}P NMR (243 MHz, $CDCl_3$) δ 30.6.

NMR signals assigned to *erythro* isomer:

^{1}H NMR (600 MHz, $CDCl_3$) δ 9.15 (m, 1H), 8.98-8.90 (m, 1H), 5.43-5.35 (m, 1H, OC*H*), 5.15 (m, $^{2}J_{PH}$ = 15.4, 1H, PCH), 0.59-0.55 (2 heavily roofed doublets, J = 5.5, 6H, diastereotopic CH_3 groups).

^{31}P NMR (243 MHz, $CDCl_3$) δ 31.7.

Integration of the assigned signals indicates an *erythro/threo* ratio of 89:11. Also observable in ^{31}P NMR is *P*-(isobutyl)-*P*-phenyldibenzophospholium bromide at δ_P = 28.0 (s, 0.17P relative to *erythro*-β-HPS signal).[12] Signals for this compound are obscured by other signals in the ^{1}H NMR. The C*H*Me_2 signal of the *erythro*-β-HPS and that of *P*-(iso-butyl)-*P*-phenyldibenzophospholium bromide overlap to form a multiplet at δ_H = 1.97-1.85 (integrates fro 1.17 H relative to OC*H* of *erythro*-β-HPS).

The yield of precipitated β-HPS was 32 mg. The sample conatined *erythro* and *threo* β-HPS and *P*-(isobutyl)-*P*-phenyldibenzophospholium bromide.[12] The solid sample was treated with with NaHMDS (14 mg, 0.08 mmol)) in toluene-*d8* (0.8 ml) to yield a solution of OPA.

Assigned to *cis*-OPA:

^{1}H NMR (500 MHz, toluene-*d8*, 30 °C) δ 5.29 (t, J = 7.6, 1H), 4.80 (broad m, 1H), 0.62 (d, J = 7.0, 3H), 0.54 (d, J = 7.1, 3H).

^{31}P NMR (162 MHz, toluene-*d8*, 30 °C) δ -71.6.

No signals could be assigned to the *trans*-OPA; at room temperature the ^{31}P signal of this species is broadened very significantly, and so when present in small quantities (as is the case here) this signal cannot be accurately integrated. Partial decomposition of OPA to 3-methyl-1-phenylbut-1-ene in the NMR spectrometer at 30 °C was noted over the course of one day. The OPA solution was heated to effect formation of *P*-phenyldibenzophosphole oxide and 3-methyl-1-phenylbut-1-ene [78], whose *Z/E* ratio was determined to be 89:11.

^{1}H NMR (300 MHz, toluene-*d8*) of crude alkene product:

Assigned to *E*-isomer: δ 6.14 (dd, J = 16.0, 6.7, 1H) [78].

Assigned to *Z*-isomer: δ 6.35 (d, J = 11.7, 1H), 5.45 (t, J = 10.8, 1H), 3.00 (d, J = 6.7, 1H) [78].

[12] Characterisation details for *P*-(isobutyl)-*P*-phenyldibenzophospholium bromide are given in Sect. 4.2.5.

Reaction of *P*-(iso-butylidene)-*P*-phenyldibenzophospholane and 2-bromobenzaldehyde

(i) NaHMDS THF (ii) 2-bromobenzaldehyde (CHO, Br) -78 °C (iii) HCl/MeOH THF -78 °C; NaHMDS THF; 2 hrs. 80 °C

P-(iso-butyl)-*P*-phenyldibenzophospholium bromide (33 mg, 0.083 mmol) and NaHMDS (15 mg, 0.083 mmol) were used to generate the ylide in THF (1.2 ml). The ylide was reacted at low temperature with 2-bromobenzaldehyde (0.010 ml, 16 mg, 0.085 mmol). The spectra of the crude product showed the β-HPS diasteromeric ratio to be 91:9. The major diastereomer was assigned to be the *erythro*-isomer based on the fact that *cis*-OPA was the major diastereomer produced upon deprotonation of the β-HPS, and that *Z*-alkene was the major product after heating of the OPA (see below).

Signals from NMR of crude product assigned to *erythro*-β-HPS:

^{1}H NMR (600 MHz, $CDCl_3$) δ 9.19-9.11 (m, 1H), 7.13-7.06 (m, 1H), 5.35-5.30 (m, 1H), 4.33 (ddd, $J = 14.2, 3.8, 2.9$, 1H), 1.87-1.74 (m, 1H), 0.56-0.52 (overlapping doublets—appears as triplet, $J = 7.0$, 6H, diastereotopic CH_3 groups). ^{31}P NMR (243 MHz, $CDCl_3$) δ 30.7.

Signals from NMR of crude product assigned to *threo*-β-HPS:

^{1}H NMR (600 MHz, $CDCl_3$) δ 9.10-9.05 (m, 1H), 8.60-8.54 (m, 1H), 6.94 (t, $J = 7.2$, 1H), 6.86 (t, $J = 7.3$, 1H), 5.64 (dd, $J = 17.2, 4.2$, 1H), 4.47 (d, $^{2}J_{PH} = 12.6$, 1H).

^{31}P NMR (243 MHz, $CDCl_3$) δ 31.0

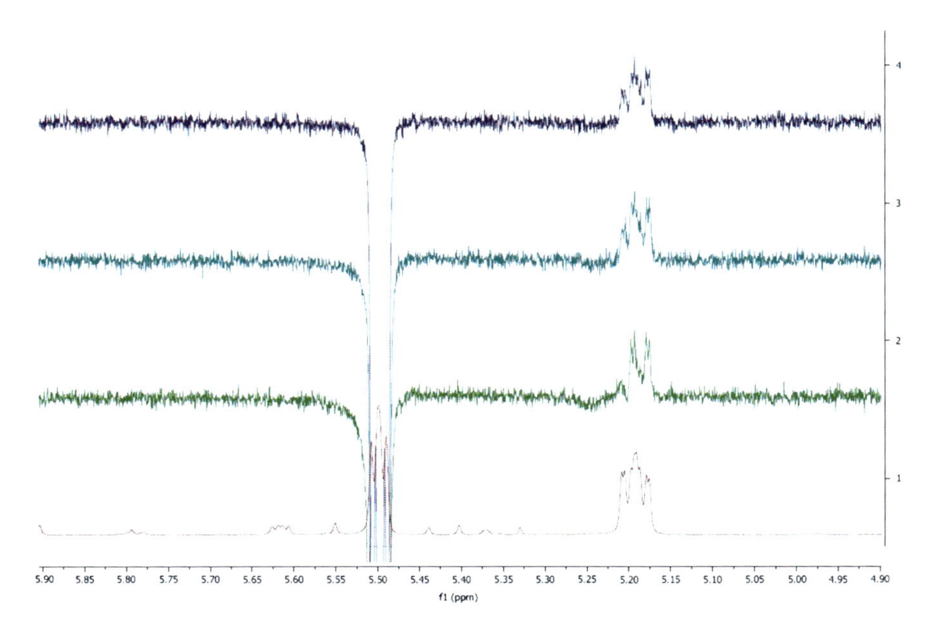

Fig. 4.2 The ^{1}H (*bottom*) and 1D NOESY spectra (with mixing times of 0.5, 1.0 and 1.5 s from the bottom up) involving irradiation at the resonance frequency of the OC*H* signal. NOE contact with the PC*H* signal is observed, showing that the two hydrogens are *cis* to each other

P-(iso-butyl)-*P*-phenyldibenzophospholium bromide could be observed in the ^{31}P NMR at $\delta_P = 28.0$. The yield of precipitated β-HPS was 40 mg. It was reacted with NaHMDS (15 mg, 0.083 mmol) in toluene-*d8* (0.8 ml) to yield a solution of OPA, whose *cis/trans* ratio was found to be 93:7 based on integrations of signals assigned to each isomer in the ^{1}H and ^{31}P NMR.

Signals from NMR of OPA solution assigned to *cis*-isomer:

^{1}H NMR (600 MHz, toluene-*d8*, −20 °C) δ 9.01-8.93 (m, 1H), 8.34 (d, $J = 7.6$, 1H), 8.01-7.92 (m, 2H), 7.57-7.48 (m, 2H), 7.37 (d, $J = 7.9$, 1H), 7.33-7.25 (m, 2H), 6.79 (t, $J = 7.1$, 1H), 5.53-5.47 (m, 1H, OC*H*), 5.23-5.16 (m, 1H, PC*H*), 2.04-1.89 (m, 1H, C*H*Me$_2$), 0.68 (d, $J = 7.2$, 3H, one of diastereotopic C*H*$_3$ groups), 0.63 (d, $J = 6.9$, 3H, one of diastereotopic C*H*$_3$ groups).

^{13}C NMR (151 MHz, toluene-*d8*, −20 °C) δ 147.1 (d, $J = 12.1$), 140.6 (d, $J = 20.0$), 139.1 (d, $J = 12.9$), 136.2 (d, $J = 15.5$), 135.7 (d, $J = 10.4$), 133.8 (d, $J = 48.2$), 118.5 (d, $J = 12.5$), 118.0 (d, $J = 4.0$), 73.3 (d, $J = 85.9$, PC*H*), 64.8 (d, $J = 16.0$, OCH), 25.5 (d, $J = 7.4$, CHMe$_2$), 21.1 (d, $J = 2.1$, one of diastereotopic *C*H$_3$ groups), 18.8 (d, $J = 16.1$, one of diastereotopic CH$_3$ groups).

^{31}P NMR (243 MHz, toluene-*d8*, −20 °C) δ −72.7.

The relative *cis* stereochemistry of the *cis*-OPA ring hydrogens was proven by 1D NOESY, as shown by Fig. 4.2.

Signals from NMR of OPA solution assigned to *trans*-isomer:

^{1}H NMR (600 MHz, toluene-*d8*, −20 °C) δ 8.86 (dd, J = 11.6, 7.3, 1H), 6.75 (t, J = 7.7, 1H), 5.62 (dd, J = 7.3, 4.4, 1H), 4.79 (dd, J = 19.8, 4.9, 1H), 0.88 (d, J = 5.7, 3H), 0.78 (d, J = 6.8, 3H).

^{13}C NMR (151 MHz, toluene-*d8*, −20 °C) δ 76.1 (d, J = 82.0, P*C*H), 62.6 (d, J = ca. 15).

^{31}P NMR (243 MHz, toluene-*d8*, −20 °C) δ −69.8.

The OPA solution was heated to effect formation of phenyldibenzophosphole oxide and 1-(2-bromophenyl)-3-methylbut-1-ene [80], whose *Z/E* ratio was determined as 91:9.

^{1}H NMR (300 MHz, toluene-*d8*) of crude alkene product:

Assigned to *Z*-isomer: δ 6.81-6.70 (m, 1H), 6.43 (d, J = 11.5, 1H), 5.54-5.42 (m, 1H), 2.65 (ddd, J = 17.2, 13.4, 7.0, 1H) [80].

Assigned to *E*-isomer: δ 6.06-5.94 (m, 1H) [80].

Based on the establishment that the *cis*-OPA and *Z*-alkene are the major diastereomers produced in this sequence of reactions, the major β-HPS diastereomer can be assigned to be the *erythro* isomer, and so the *erythro/threo* ratio of the β-HPS produced in the acid quenching of the Wittig reaction was assigned to be 91:9.

Reaction of *P*-(isobutylidene)-*P*-phenyldibenzophospholane and 1,2-O-isopropylidene-3-O-methyl-α-D-xylopentodialdofuranose-(1,4)

(i) NaHMDS THF (ii) Aldehyde -78 °C (iii) HCl/MeOH THF -78 °C

NaHMDS | THF

2 hrs. 80 °C

P-(isobutyl)-*P*-phenyldibenzophospholium bromide (22 mg, 0.055 mmol) and NaHMDS (10 mg, 0.055 mmol) were used to generate the ylide in THF (1 ml). The ylide was reacted at low temperature with 1,2-O-isopropylidene-3-O-methyl-α-D-xylopentodialdofuranose-(1,4) (0.11 ml, 0.055 mmol, as a 0.5 mol l^{-1} solution in THF). A definitive β-HPS ratio could not be determined from the ^{1}H and ^{31}P spectra as no no signals could be assigned to the *threo* isomer with certainty; suffice to say that the *eryhtro*-β-HPS is strongly predominant.

NMR signals assigned to *erythro*-β-HPS: ^{1}H NMR (400 MHz, $CDCl_3$): δ 9.02-8.93 (m, 1H), 5.73 (d, J = 3.6, 1H), 4.44 (d, J = 3.7, 2H), 4.44 (d, J = 3.7, 1H), 4.21 (dd, J = 9.2, 2.2, 1H), 4.08 (d, J = 2.4, 1H), 3.38 (d, J = 10.6, 3H), 3.29 (s, 1H), 2.44-2.26 (m, 1H), 1.41 (s, 1H), 1.31 (s, 3H), 1.20 (s, 3H), 0.85 (d, J = 7.1, 2H), 0.75 (d, J = 7.0, 3H).

^{31}P NMR (162 MHz, $CDCl_3$): δ 32.0.

The β-HPS was treated with KHMDS (16 mg, 0.08 mmol) in toluene-*d8* to give OPA. As with the β-HPS, only signals due to *cis*-OPA could be unambiguously be assigned; however, there do exist some minor signals that may be attributable to *trans*-OPA.

NMR signals assigned to *cis*-OPA: ^{1}H NMR (500 MHz, toluene-*d8*, 30 °C): δ 5.93 (d, J = 3.7, 1H, furan ring H-1), 4.96 (app d, J = 18.4, 1H, PC*H*), 4.71 (ddd, J = 23.7, 9.6, 7.9, 1H, OC*H*), 4.56 (app d, J = 9.3, 1H, furan ring H-4), 4.43 (d, J = 3.8, 1H, furan ring H-2), 4.15 (d, J = 2.5, 1H, furan ring H-3), 3.45 (s, 3H, OCH_3), 2.34-2.20 (m, 1H, C*H*Me_2), 1.17 (s, 3H, one of acetonide CH_3). ^{31}P NMR (121 MHz, toluene-*d8*, 30 °C): δ −66.3.

The OPA solution was heated to 80 °C for 2 h to bring about alkene formation. ^{1}H NMR (500 MHz, $CDCl_3$) of crude alkene product (*Z/E* ratio of 90:10):

Assigned to *E*-isomer: δ 5.84 (dd, J = 15.6, 6.5, 0.11H, *E*–H-6).

Assigned to Z-isomer: δ 5.45 (dd, J = 11.1, 8.2, 1H, Z-H-5), 4.94 (dd, J = 8.2, 3.0, 1H, Z-H-4), 3.41 (s, 3H, Z-OCH_3), 2.64 (m, 1H, Z-C*H*Me_2).[13]

Other signals were also discernable in the spectrum, but not baseline separated (integrations are relative to 1H of Z-alkene): δ 5.93-5.89 (1.05H contains d, J = 3.9, Z-H-.

1 and d, J = 3.9, *E*-H-1), 5.59-5.49 (m, 1.08H, contains app ddd, J = 15.6, 8.0, 1.1., ca. 0.1H *E*-H-5, and app t, ca. 1H, Z-H-6), 4.59 (m, 1.19H, contains d, J = 3.9, ca. 1H, Z-H-2; obscured signal of *E*-H-2; dd, J = 8.0, 2.8, *E*-H-4), 3.59 (1.1H, contains d, J = 3.1, *ca.* 1H, Z-H-3 and also signal for *E*-H-3), 1.53 (s, acetonide CH_3), 1.33 (s, acetonide CH_3), 1.01 (m, contains two doublets, J = 6.5, 2 × Z-CH_3).

[13] See Sect. 4.9 of this thesis for details of the characterisation of this compound.

Characterisation data has been obtained for the *Z*-isomer of this alkene, which was produced by Wittig reaction of 1,2-O-isopropylidene-3-O-methyl-α-D-xylopento-dialdofuranose-(1,4) with *P*-(*iso*-butylidene)triphenylphosphorane.[13]

4.7.4 Observation of Kinetic OPA cis/trans Ratio for Wittig Reactions of Non-stabilised Ylides

General Procedure

Non-stabilised ylides (especially *P*-(ethylidene)phenyldibenzophospholane) were found to be very sensitive to water and/or oxygen, and so manipulations in these reactions were carried out very carefully under minimal nitrogen flow. It was important to ensure that water was rigourously excluded from the reaction mixture. Thus THF dried in a Grubbs press (found by Karl Fischer titration to contain <10 ppm water) was employed, the phosphonium salts (in principle hygroscopic) were dried over P_2O_5 in a vacuum dessicator, and reaction flasks were flame dried and cooled under vacuum before being charged with inert gas. The rubber tubing used to conduct nitrogen to the reaction flask from the Schenk manifold was dried by fitting a syringe and needle to the open end of the tube and inserting the needle through a septum into a sealed flask of oven-dried KOH when the tube was not in use. If this procedure is continuously employed over time, the tubing becomes free of moisture.

Phosphonium bromide salt (1 equivalent) and sodium or potassium hexamethyldisilazide (0.95 equivalents, base used in each case specified below) was added to a flame dried Schlenk flask under argon in a glove box. This flask was removed from the glove box, attached to a Schlenk manifold and charged with nitrogen using standard Schlenk pump and fill technique. Dry THF (to give a 0.1 mol l^{-1} solution of ylide) was added at the temperature specified below, giving a coloured solution of ylide after stirring. The ylide solution was stirred for 15 min at the temperature specified below for each example, and then cooled to –78 °C. The appropriate aldehyde (1 equivalent) was then added dropwise, causing the ylidic colour to dissipate once one equivalent of aldehyde had been added. A white, pale yellow or light brown solution of OPA resulted, which was stirred for 5 min at −78 °C.

Approximately 0.6 ml of the THF solution was filtered via nitrogen-flushed cannula into an NMR tube contained in a long Schlenk flask under an atmosphere of nitrogen. The best results were obtained if both the reaction flask and the NMR tube were maintained at temperatures below −20 °C. For reactions of *P*-(ethylidene)ethyldiphenylphosphorane, the reaction flask and NMR tube were maintained at −78 °C. The solution of OPA was diluted with approximately 0.2 ml toluene-*d8*, giving a yellow solution. The OPA was then characterised by ^{31}P NMR at –20 °C (unless specified otherwise), in general with inverse gated decoupling from ^{1}H. The diastereomeric OPA signals were observed between δ −60 and −80 ppm. Conversion to OPA was invariably high (at least 75 % by ^{31}P NMR), with the bulk of the remaining material being accounted for by ylide. A small amount of ylide and phosphine oxides were invariably present in the ^{31}P spectrum. The phosphine oxides were shown to be derived from the ylide (probably by hydrolysis) by control experiments in which no aldehyde was added.

The OPA solution was then transferred to a Schlenk flask under nitrogen. Ethyldiphenylphosphine-derived OPA was allowed to warm to room tempreature, while *P*-phenyl-5*H*-dibenzophosphole-derived OPA was heated to 80 °C for 2 h to effect alkene formation. After this time, the solvent was removed *in vacuo*, and the remaining yellow oil was dissolved in $CDCl_3$. The crude mixture of alkene isomers was characterised by ^{1}H and gCOSY NMR.

Reaction of *P*-(ethylidene)ethyldiphenylphosphorane and benzaldehyde

(i) KHMDS THF (ii) -78 °C -78 to 20 °C

Diethyldiphenylphosphonium bromide (190 mg, 0.59 mmol) and KHMDS (115 mg, 0.58 mmol) were used to generate the ylide in THF (4.5 ml) at 20 °C. The ylide was reacted at −78 °C with benzaldehyde (0.059 ml, 62 mg, 0.58 mmol added by 100 μl syringe). ^{31}P NMR spectra were obtained of the resulting OPA at −40 °C and −20 °C.

^{31}P NMR of OPA solution (202 MHz, THF/toluene-*d8*, −40 °C): δ 39.0 (0.07P), 28.5 (0.10P), −61.6 (1P).

The OPA diastereomers could not be resolved by ^{31}P NMR. After the NMR sample had been allowed to warm to 20 °C, the solvent was removed from the sample *in vacuo*. The residue was dissolved in $CDCl_3$ and a ^{1}H NMR spectrum of the crude alkene product was obtained.

^{1}H NMR spectrum of the crude 1-phenylprop-1-ene product:

Assigned to *E*-isomer (MHz, $CDCl_3$): δ 6.23 (roofed dq, J = 15.8, 6.4, 1H) [72].

Assigned to Z-isomer (MHz, $CDCl_3$): δ 5.79 (dq, J = 11.6, 7.2, 1H) [73].
Integrations indicate a *Z/E* ratio of 32:68. Other overlapping signals due to the *E* and Z-isomer could also be observed (integrations relative to 1H of Z-isomer): δ 6.48-6.35 (m, 3H, contains dd, J = 11.6, 1.8, Z-PhC*H*=C and J = 15.7, 1.3, *E*-PhC*H*=C), 1.89 (app td, J = 6.7, 1.6, 9H, *Z* and *E*-CH_3).

Reaction of *P*-(ethylidene)ethyldiphenylphosphorane and 2-bromobenzaldehyde

(i) KHMDS THF (ii) 2-bromobenzaldehyde, -78 °C; -78 to 20 °C

Diethyldiphenylphosphonium bromide (145 mg, 0.45 mmol) and KHMDS (86 mg, 0.43 mmol) were used to generate the ylide in THF (4.5 ml) at 20 °C. The ylide was reacted at −78 °C with 2-bromobenzaldehyde (0.051 ml, 81 mg, 0.44 mmol, added by 100 μl syringe). A ^{31}P NMR spectrum was obtained at each of −70, −40, −20, −10, 0, 10 and 20 °C.

^{31}P NMR of OPA solution (202 MHz, THF/toluene-*d8*, −40 °C) δ 39.0 (0.06P, Et_2PhPO), 28.6 (0.09P, $EtPh_2PO$), −60.3 (0.56P *trans*-OPA), −61.9 (1.0P, *cis*-OPA).

Integrations indicate an OPA *cis/trans* ratio of 64:36. The variable temperature NMR experiments show that OPA decomposition to alkene and $EtPh_2PO$ begins to occur at an observable rate between −20 °C and −10 °C. As the temperature is increased above −10 °C, the peak at 28.6 ppm continues to grow at the expense of the OPA. The OPA *cis/trans* ratio remains invariant throughout at 64:36. OPA decomposition was complete at 20 °C. The solvent was removed from the NMR sample *in vacuo*. The residue was dissolved in $CDCl_3$ and a ^{1}H NMR spectrum of the crude alkene product was obtained. The major OPA diastereomer was assigned to be the *cis*-isomer based on the fact that the major diasteromer of the alkene produced was the Z-isomer.

^{1}H NMR (300 MHz, $CDCl_3$) of the crude 1-(2-bromophenyl)prop-1-ene product: Assigned to *E*-isomer: δ 6.72 (dd, J = 15.6, 1.4, 1H), 6.18 (dq, J = 15.6, 6.6, 1H), 1.92 (dd, J = 6.7, 1.6, 4H) [74].

Assigned to Z-isomer: δ 6.48 (dd, J = 11.4, 1.6, 1H), 5.89 (dq, J = 11.5, 7.1, 1H), 1.78 (dd, J = 7.1, 1.8, 3H) [75].

Integrations indicate a *Z/E* ratio of 56:44.

Reaction of *P*-(*iso*-butylidene)-*P*-phenyldibenzophospholane and benzaldehyde

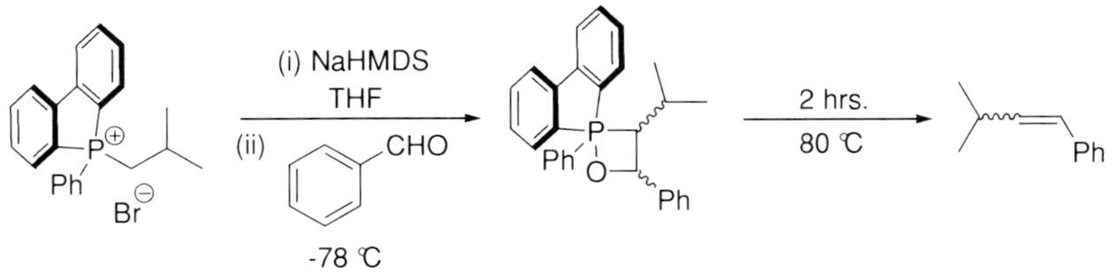

P-(*iso*-butyl)-*P*-phenyldibenzophospholium bromide (33 mg, 0.083 mmol) and NaHMDS (14 mg, 0.076 mmol) were used to generate the ylide in THF (0.8 ml) at 20 °C. The ylide was reacted at −78 °C with benzaldehyde (0.008 ml, 8 mg, 0.08 mmol, added by 25 μl syringe).

^{31}P NMR of OPA solution (121 MHz, THF/toluene-*d8*, 30 °C) δ 37.8 (0.10P), 37.7 (0.01P), 36.7 (0.01P), 29.4 (0.17P), 29.1 (0.05P), −10.0 (0.01P, ylide), −70.1 (0.12P, *trans*-OPA), −71.7 (1.0P, *cis*-OPA). OPA *cis/trans* ratio was found by integration to be 89:11. Based on integrations of ^{31}P NMR signals (which are relative to *cis*-OPA), the OPA isomers form 78 % of the phosphorus-containing material in the reaction mixture.

^{1}H NMR of crude 3-methyl-1-phenylbut-1-ene product (300 MHz, toluene-*d8*): δ 6.35 (d, J = 11.7, 1H, *Z*-Ph-C*H*, overlaps slightly with small signal for corresponding *E*-Ph-C*H*), 6.14 (dd, J = 16.0, 6.7, 0.12H, *E*- ^{i}Pr -C*H*), 5.45 (app t, $J \sim$ 11 Hz, 1H, *Z*-^{i}Pr-C*H*), 3.00 (m, 1H, *Z*-C*H*Me$_2$) [78].

Integration of these signals gave an alkene *Z/E* ratio of 88:12. Other signals due to the *iso*-propyl protons were also discernable in the spectrum, but not baseline separated: δ 1.08 (d, J = 6.7, *E*-CH(C*H*$_3$)$_2$), 1.02 (d, J = 6.6, *Z*-CH(C*H*$_3$)$_2$).

Reaction of *P*-(*iso*-butylidene)-*P*-phenyldibenzophospholane and 2-bromobenzaldehyde

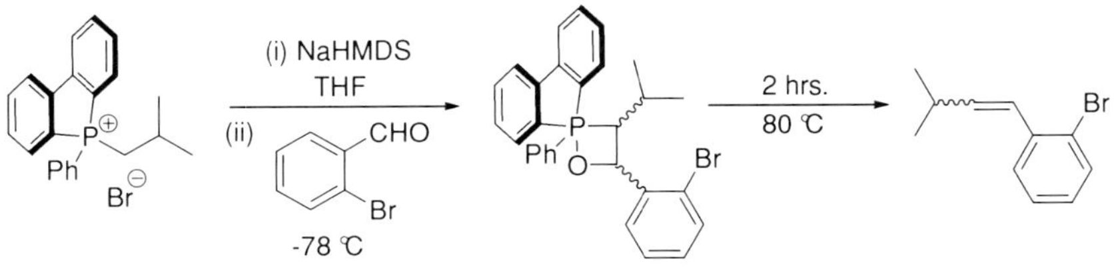

P-(*iso*-butyl)-*P*-phenyldibenzophospholium bromide (33 mg, 0.083 mmol) and NaHMDS (14 mg, 0.076 mmol) were used to generate the ylide in THF (0.8 ml) at 20 °C. The ylide was reacted at −78 °C with 2-bromobenzaldehyde (0.010 ml, 16 mg, 0.080 mmol, added by 25 μl syringe).

^{31}P NMR of OPA solution (121 MHz, THF/toluene-*d8,* 30 °C): δ 38.2 (0.09P), 29.5 (0.06P), 29.3 (0.01P), −69.7 (0.06P, broad, *trans*-OPA), −72.38 (1P, *cis*-OPA). OPA *cis/trans* ratio was found by integration to be 94:6. Integrations are relative to *cis*-OPA, and based on these OPA forms 87 % of the phosphorus-containing material in the reaction mixture.

^{1}H NMR of crude 1-(2-bromophenyl)-3-methyl-but-1-ene product (300 MHz, toluene-*d8*): δ 6.43 (d, J = 11.5, 1H, *Z*-Ar-C*H*), 6.06-5.94 (m, 0.09H, *E*-Ar-C*H*), 5.48 (dd, app t, J ~ 11 Hz, 1H, *Z*-^{i}Pr-C*H*), 2.64 (m, 1H, *Z*-C*H*Me$_2$). [80].

Integration of these signals gave an alkene *Z/E* ratio of 92:8. Signals at δ 1.06 (d, J = 6.5, *E*-CH(CH_3)$_2$) and 0.94 (d, J = 6.5, *Z*-CH(CH_3)$_2$) were assigned to the *iso*-propyl CH_3 protons of the *E* and *Z*-isomers of the alkene respectively. These were not baseline separated from impurity signals:

Reaction of *P*-(*iso*-butylidene)-*P*-phenyldibenzophospholane and 1,2-O-iso-propylidene-3-O-methyl-α-D-xylopentodialdofuranose-(1,4)

(i) NaHMDS
THF
(ii) OHC
MeO
-78 °C
Ph
Br
Ph

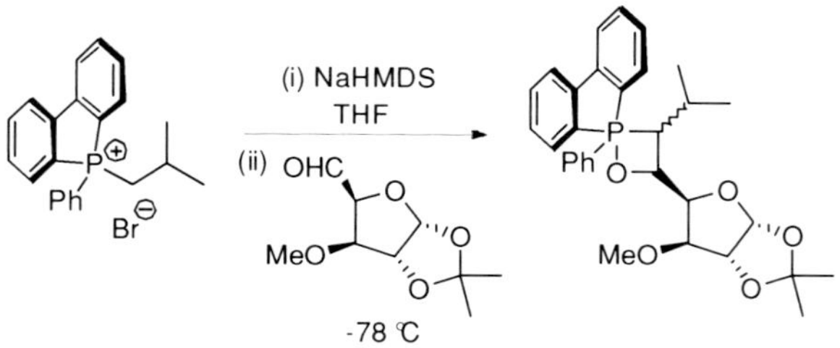

P-(*iso*-butyl)-*P*-phenyldibenzophospholium bromide (46 mg, 0.12 mmol) and KHMDS (22 mg, 0.11 mmol) were used to generate the ylide in THF (1.2 ml) at 20 °C. The ylide was reacted at −78 °C with 1,2-O-isopropylidene-3-O-methyl-α-D-xylopentodialdofuranose-(1,4), which was added as a 0.11 mol l^{-1} solution in THF (0.24 ml, 0.12 mmol).

^{31}P NMR of OPA solution (243 MHz, THF/toluene-*d8,* 30 °C): δ 38.1 (0.12P), 29.6 (0.18P), 29.0 (0.03P), −66.4 (1P, *cis*-OPA), −68.9 (0.06P, *trans*-OPA). OPA *cis/trans* ratio was found by integration to be 94:6. Integrations (relative to *cis*-OPA) show that OPA forms 77 % of phopsphorus-containing material in the reaction mixture.

Reaction of *P*-(*iso*-butylidene)-*P*-phenyldibenzophospholane and cyclopentanecarboxaldehyde

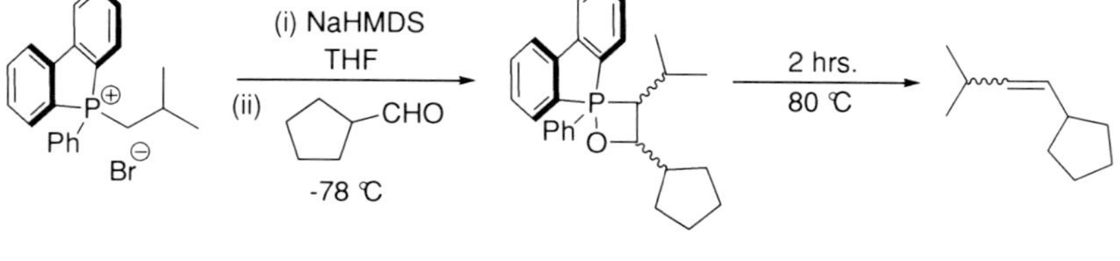

(i) NaHMDS
THF
(ii) CHO
-78 °C
2 hrs.
80 °C
Ph
Br
Ph

(*iso*-Butyl)phenyldibenzophospholium bromide (28 mg, 0.071 mmol) and NaHMDS (13 mg, 0.070 mmol) were used to generate the ylide in THF (0.7 ml) at 20 °C. The ylide was reacted at −78 °C with cyclopentanecarboxaldehyde (0.008 ml, 7 mg, 0.07 mmol by 25 μl syringe). After warming to room temperature, the THF solvent was removed under vacuum, and dry-toluene-*d8* was added. The resulting mixture was cannula filtered into an NMR tube under a nitrogen atmosphere as described in the general procedure.

^{31}P NMR of OPA solution (121 MHz, toluene-*d8*, 30 °C): δ 39.0 (0.13P), 37.1 (0.01P), 30.0 (0.58P), −71.2 (1.2P, broad, *trans*-OPA), −72.5 (1.0P, *cis*-OPA). The OPA *cis/trans* ratio was found by integration to be 45:55. Inegrations show that OPA forms 76 % of the phosphorus-containing material in the reaction mixture. The OPA diastereomers were assigned to be *cis* or *trans* based on the observed *Z/E* ratio of the alkene produced upon heating of the OPA, which correpsonds closely with the OPA *cis/trans* ratio.

^{1}H NMR of crude 1-cyclopentyl-3-methylbut-1-ene (500 MHz, toluene-*d8*): δ 5.48-5.44 (m, 2.89H, *E* HC=CH), 5.25 (quintet, J = 10.8, 2H, *Z HC*=*CH*), 2.79-2.64 (m, 2H, *Z HC*-C = C–*CH*), 2.46-2.38 (m, 1.15H, *E c*-C_5H_9 *CH*), 2.32-2.23 (m, 1.15H, *E CH*Me_2). The alkene *Z/E* ratio was found by integration of these signals to be 43:57.[13]

Other signals due to the *iso*-propyl protons were also discernable in the spectrum, but not baseline separated: δ 1.06 (d, J = 6.7, *Z*-CH(CH_3)$_2$), 1.03 (d, J = 6.7, *E*-CH(CH_3)$_2$). This compound was found to be unstable, and so could not be fully characterised. A crude sample of the *Z*-isomer was obtained by Wittig reaction of cyclopentanecarboxaldehyde with *P*-(*iso*-butylidene)triphenylphosphorane and subsequent purification of the crude product using the oxalyl chloride method (see Sect. 4.9 reaction 17).

Reaction of ethylidene-*P*-phenyldibenzophospholane and benzaldehyde

(i) KHMDS
THF
-25 °C to -45 °C
(ii) CHO
-78 °C
2 hrs.
80 °C
Ph

P-ethyl-*P*-phenyldibenzophospholium bromide (71 mg, 0.19 mmol) and KHMDS (36 mg, 0.18 mmol) were used to generate the ylide in THF (1.9 ml) at −25 °C. The reaction mixture was stirred at −25 °C for 2 min before being cooled to −45 °C, at which temperature it was stirred for 20 min, before being cooled to −78 °C. The ylide was reacted at −78 °C with benzaldehyde (0.019 ml, 20 mg, 0.19 mmol by 25 μl syringe).

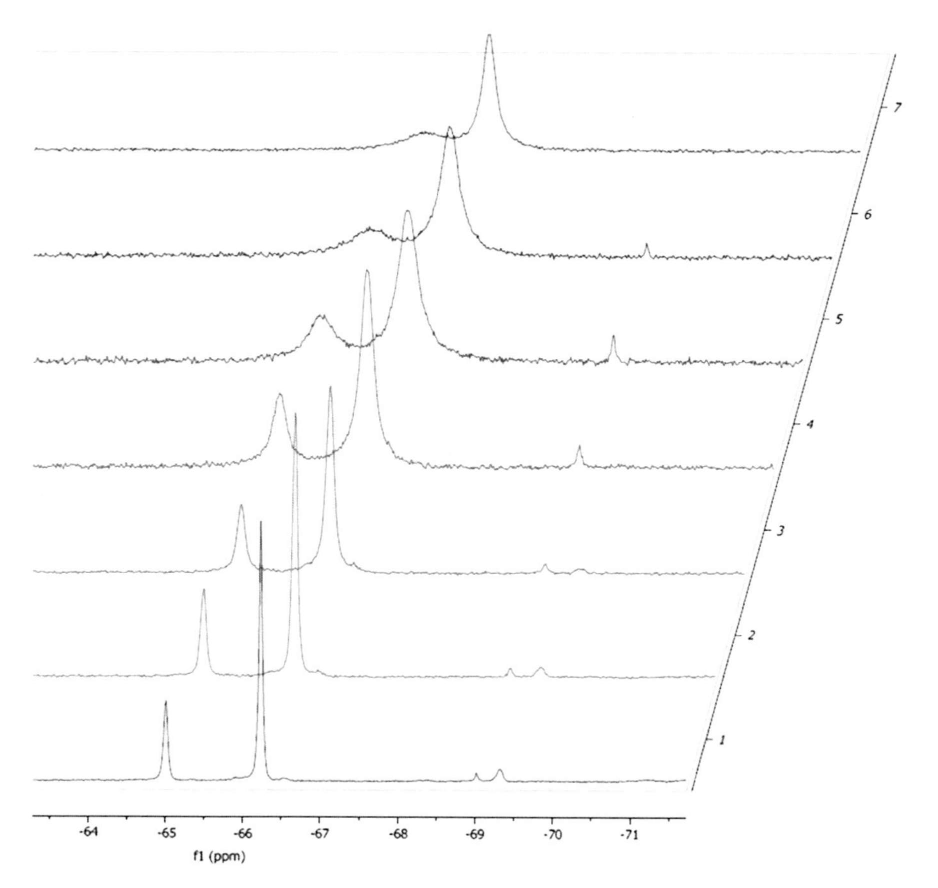

Fig. 4.3 OPA region of ^{31}P NMR of the OPA solution at −20 , −10, 0 , 10 , 20 , 30 and 40 °C (lowest spectrum obtained at −20 °C, highest at 40 °C)

^{31}P NMR (202 MHz, toluene-*d8*, −20 °C) δ 41.4 (0.01P), 35.0 (0.04P), 30.2 (0.07P), −10.0 (0.24P, ylide), −65.0 (0.36P, *trans*-OPA), −66.2 (1P, *cis*-OPA), −69.3 (0.06P), −76.9 (0.03P), −79.5 (0.02P). OPA *cis/trans* ratio was found by integration to be 72:28. Integrations indicate that OPA and ylide form 75 % and 13 % respectively of the phosphorus-containing material in the reactions mixture. The ^{31}P NMR of the OPA solution was obtained at −20, −10, 0, 10, 20, 30 and 40 °C. The spectra obtained are stacked for comparison in Fig. 4.3. The two OPA peaks retained sufficient baseline separation to allow integration of the individual signals up to 10 °C. The *cis/trans* ratio was invariant at least up to this temperature. The total amount of OPA present (*cis* + *trans*) relative to ylide and phosphine oxide was invariant over the whole range of temperatures at which NMR spectra were obtained.

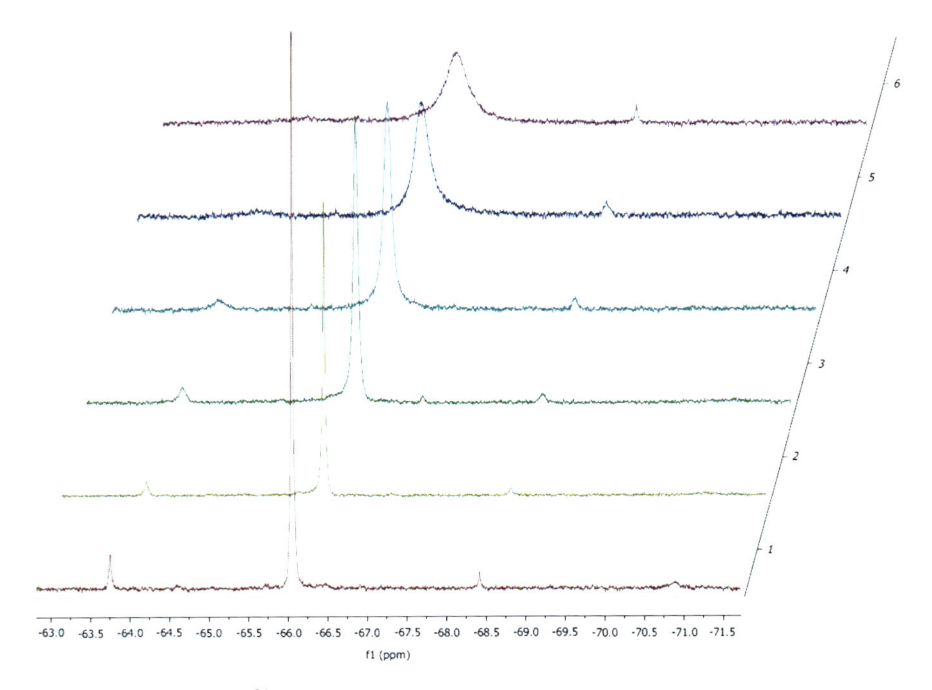

Fig. 4.4 OPA region of ^{31}P NMR of the OPA solution at −20, −10, 0, 10, 20, and 30 °C (lowest spectrum obtained at −20 °C, highest at 30 °C)

Reaction of ethylidene-*P*-phenyldibenzophospholane and 2-bromobenzaldehyde

(i) KHMDS THF -25 °C to -45 °C
(ii) CHO Br -78 °C
2 hrs. 80 °C
Ph P⊕ Br⊖ — Ph P O Br — Br

P-ethyl-*P*-phenyldibenzophospholium bromide (56 mg, 0.15 mmol) and KHMDS (29 mg, 0.15 mmol) were used to generate the ylide in THF (1.5 ml) at −25 °C. The reaction mixture was stirred at −25 °C for 2 min before being cooled to −45 °C, at which temperature it was stirred for 25 min, before being cooled to −78 °C. The ylide was reacted at −78 °C with 2-bromobenzaldehyde (0.018 ml, 29 mg, 0.15 mmol by 25 μl syringe).

^{31}P NMR (202 MHz, THF/toluene-*d8*, −20 °C) δ 41.4 (0.06P), 30.2 (0.08P), −10.0 (0.15P), −63.8 (0.06P, *trans*-OPA), −66.1 (1P, *cis*-OPA), −68.4 (0.02P).

OPA *cis/trans* ratio was found by integration to be 94:6. Integrations show that OPA and ylide form 77 and 11 % respectively of the phosphorus-containing material in the reactions mixture.

The ^{31}P NMR of the OPA solution was obtained at −20, −10, 0, 10, 20, and 30 °C. The spectra obtained are stacked for comparison in Fig. 4.4. The relative amounts (reported above for the spectrum at −20 °C) of OPA, ylide and phosphine oxide did not change over this temperature range.

4.8 Reactions of Semi-stabilised and Stabilised Ylides with 1,2-O-isopropylidene-3-O-methyl-α-D-xylopentodialdofuranose-(1,4)

4.8.1 General Procedure

Dry phosphonium salt (1 equivalent) and NaHMDS (0.95 equivalents) were placed in a flame-dried Schlenk flask in a glove box under an atmosphere of argon. The flask was sealed, removed from the glove box and charged with an atmosphere of nitrogen using a Schlenk manifold. Dry THF was added, giving a solution of ylide (coloured for semi-stabilised ylides, not coloured for stabilised ylides) and a precipitate of NaBr and stirring for approximately 1 h. This was cooled to −78 °C in a dry ice/acetone bath, and then a 0.5 mol L^{-1} solution of 1,2-O-isopropylidene-3-O-methyl-α-D-xylopentodialdofuranose-(1,4) in THF (containing 0.95 equivalents) was added dropwise.

For reactions with stabilised ylides, the reaction was quenched at −78 °C by addition of saturated aqueous ammonium chloride solution. For reactions with semi-stabilised ylides, the reaction was allowed to warm to room temperature before addition of water to quench any remaining ylide or base. The remaining work-up procedure was identical for all reactions; the quenched reaction mixture was poured into a separatory funnel and diethyl ether was added. The biphasic mixture was shaken and the phases separated. The aqueous phase was extracted twice more with ether and the ether phases were combined, dried over Na_2SO_4, filtered to remove the drying agent, and the solvent removed using a rotary evaporator. This gave an oil that was analysed by ^1H and gCOSY NMR to establish the *Z/E* ratio of the reaction.

Conversion was observed to be high in all cases by the large (relative) amount of alkene present in the crude product. In all but one case, this high conversion was confirmed by isolation of the alkene product. The yields (sum of yields of individual isomers) were greater than 80 % in the cases where the alkene was isolated. Isolation of the alkene product was done by column chromatography on neutral alumina using mixtures of cyclohexane/ethyl acetate as the eluting solvent, or by making use of the oxalyl chloride treatment and filtration technique that is described in Sect. 4.3 above, and discussed in full in Chap. 3.

(a) With bezylidenemethyldiphenylphosphorane

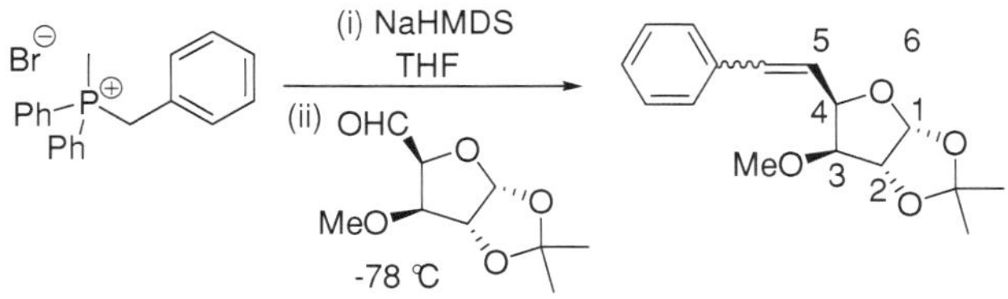

From benzylmethyldiphenylphosphonium bromide (0.077 g, 0.21 mmol), NaH-MDS (37 mg, 0.020 mmol) and 1,2-O-isopropylidene-3-O-methyl-α-D-xylopentodialdofuranose-(1,4) (0.40 ml of a 0.5 mol L^{-1} solution in THF, 0.20 mmol). The *Z/E* ratio was established by integration of the alkene signals in the ^{1}H NMR spectrum of the crude product. Signals in the ^{1}H NMR spectrum were assigned with the aid of ^{13}C, gCOSY, zTOCSY, 2D NOESY, gHSQC and gHMBC NMR spectra of the crude product.

^{1}H NMR (400 MHz, $CDCl_3$) of reaction mixture:

Assigned to *Z*-isomer: δ 6.73 (d, J = 11.6, 1H, H-6), 5.91-5.80 (m, 2H, H-1 & H-5), 4.94 (dd, J = 9.2, 2.9, 1H, H-4), 4.56 (d, J = 3.8, 1H, H-2), 3.59 (d, J = 3.0, 1H, H-3), 3.38 (s, 3H, OCH_3), 1.34 (s, 3H, acetonide CH_3), 1.25 (s, 3H, not baseline separated, acetonide CH_3).

Assigned to *E*-isomer: δ 6.67 (d, J = 16.1, 1H, H-6), 6.24 (dd, J = 16.1, 7.6, 1H, H-5), 4.71 (dd, J = 7.7, 3.0, 1H, H-4), 3.34 (s, 3H, OCH_3) [81].

Assigned to aldehyde: δ 9.64 (d, J = 5.0, 0H), 6.09 (t, J = 4.8, 1H), 4.55 (d, J = 3.7, 1H), 4.12 (d, J = 3.7, 1H), 3.36 (d, J = 5.6, 3H), 1.48 (s, 3H), 1.35 (s, 3H) [82].

Integration of the alkene signals indicates a *Z/E* ratio of 95:5.

The *cis* relative stereochemistry of H-3 and H-4, H-1 and H-2 and the *Z*-stereochemistry of the alkene was confrimed by the observation of NOE contact between these pais of hydrogens by 2D NOESY NMR. See Fig. 2.12a and b in Sect. 2.5. The *Z*-isomer was isolated as a white solid (41 mg, 71 %) by column chromatography on neutral alumina using 96:4 cyclohexane/ethyl acetate as eluting solvent.

^{1}H NMR (500 MHz, $CDCl_3$) δ 7.37-7.33 (m, 4H, PhH-2 & PhH-3), 7.28 (m, partially obscured by $CHCl_3$, PhH-4), 6.81 (d, J = 11.7, 1H, H-6), 5.96 (d, J = 3.9, 1H, H-1), 5.92 (dd, J = 11.6, 9.2, 1H, H-5), 5.01 (dd, J = 9.2, 2.9, 1H, H-4), 4.63 (d, J = 3.9, 1H, H-2), 3.66 (d, J = 3.0, 1H, H-3), 3.44 (s, 3H, OMe), 1.41 (s, 3H, acetonide CH_3), 1.32 (s, 3H, acetonide CH_3).

^{13}C NMR (101 MHz, $CDCl_3$) δ 136.4 (PhC-1), 134.9 (C-6), 128.7 &128.2 (PhC-2 & PhC-3), 127.5 (PhC-4), 124.4 (C-5), 111.5 (CMe$_2$), 104.7 (C-1), 86.0 (C-3), 82.3 (C-2), 75.9 (C-4), 58.2 (OCH$_3$), 26.7 (acetonide CH$_3$), 26.4 (acetonide CH$_3$).

HRMS: Calc. for $[M+Na]^+ = C_{16}H_{20}O_4Na$ 299.1259; found 299.1246 (4.4 ppm). IR: 3,056 (m, alkene & aromatic C–H stretch), 2,989-2,833 (s, aliphatic C–H stretch), 1,495 (m), 1,450 (m), 1,422 (m), 1,385-1,375 (s, CMe_2 bend), 1,262 (s), 1,218 (s), 1,194 (m), 1,165 (s), 1,119 (s), 1,096 (s), 1,018 (s), 953 (w), 893 (m), 858 (s), 798 (m), 737 (m), 701 (s).

MP (crystallised from cyclohexane) 83–84 °C.

(b) With 2-bromobezylidenemethyldiphenylphosphorane

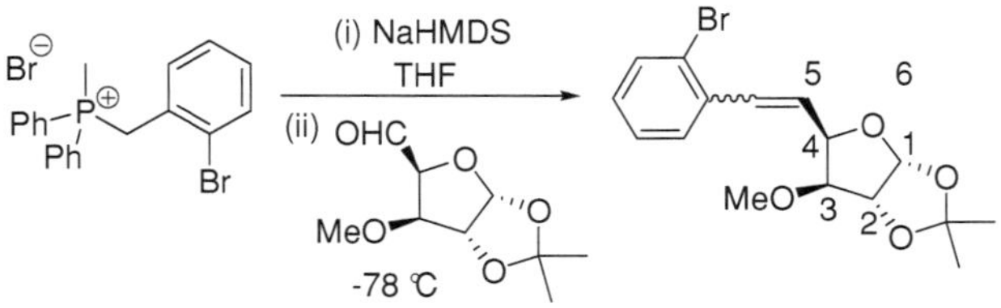

From 2-bromobenzylmethyldiphenylphosphonium bromide (90 mg, 0.20 mmol), NaHMDS (35 mg, 0.19 mmol) and 1,2-O-isopropylidene-3-O-methyl-α-D-xylopentodialdofuranose-(1,4) (0.40 ml of a 0.5 mol L^{-1} solution in THF, 0.20 mmol).

^{1}H NMR (500 MHz, $CDCl_3$) of reaction mixture:

Assigned to Z-isomer: δ 7.58 (dd, J = 7.6, 1.1, 1H), 7.29 (td, J = 7.6, 1.0, 1H), 7.14 (td, J = 7.6, 1.3, 1H), 6.77 (d, J = 11.5, 1H, H-6), 6.02 (dd, J = 11.5, 9.3, 1H, H-5), 5.86 (d, J = 3.8, 1H), 4.73 (dd, J = 9.3, 3.0, 1H, H-4), 4.53 (d, J = 3.8, 1H, H-2), 3.53 (d, J = 3.0, 1H, H-3), 3.36 (s, 3H, OCH_3), 1.26 (s, 3H acetonide CH_3), 1.22 (s, 3H, acetonide CH_3).

Assigned to *E*-isomer: δ 6.74, (d, J = 16.0, 1H, H-6), 6.20 (dd, J = 16.0, 7.7, 1H, H-5), 4.78 (dd, J = 7.3, 3.2, H-4), 3.34 (s, 3H, OCH_3).

Assigned to aldehyde: δ 9.57 (d, J = 1.6, 0.06H relative to 1H of *Z*-isomer) [82]. Integrations indicate an alkene *Z/E* ratio of 95:5.

The relative stereochemistry of the substituents at C-3 and C-4 was confirmed as *cis* by 1D and 2D NOESY. The 2D NOESY spectrum ($CDCl_3$, 500 HMz, 1,000 ms mixing time) in Fig. 4.5 shows NOE contact between H-1 and H-2, between H-5 and H-6 (therefore alkene confirmed as *Z*), between H-3 and H-4 (hydrogens *cis* relative to each other, therefore no epimerisation occurred at C-4 of the aldehyde under the reaction conditions), and between H-2 and each of H-3 and

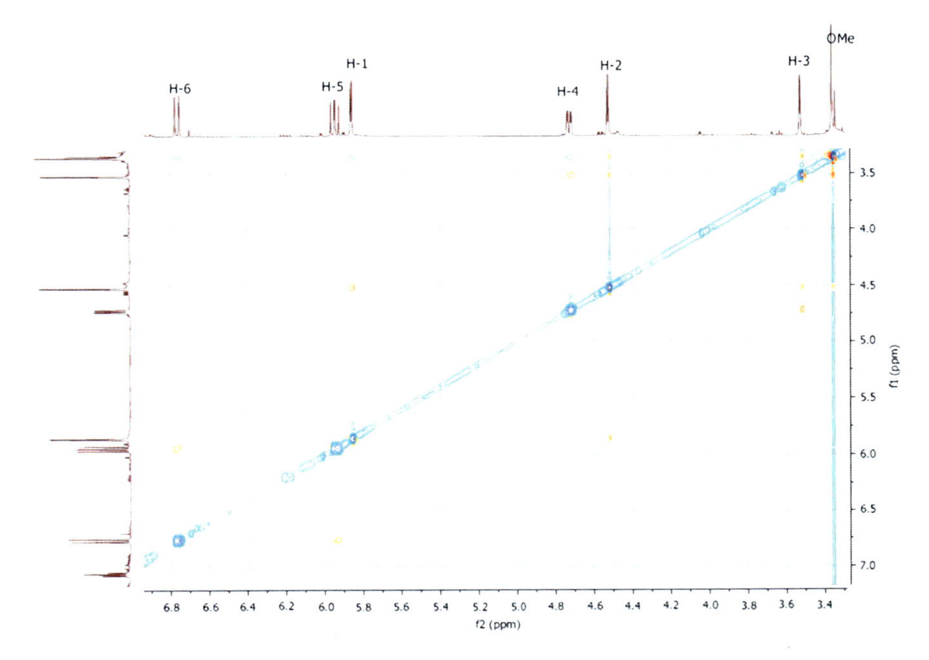

Fig. 4.5 2D NOESY spectrum of crude product. NOE contact can be seen between Z-alkene hydrogens H-5 and H-6, H-1 and H-2, and H-3 and H-4 respectively, showing that there is a *cis* relationship between each of these pairs of hydrogens. H-2 shows NOE contact with *both* H-3 and OCH_3

the OMe group. This shows that H-2 points between H-3 and OMe, which would give it a dihedral angle to H-3 close to 90°. This is consistent with the fact that no coupling is observed between these two protons.

The *Z*-isomer was isolated as a white solid by column chromatography on neutral alumina using 96:4 cyclohexane/ethyl acetate as eluent (yield 59 mg, 83 %).

Z-isomer

^{1}H NMR (600 MHz, $CDCl_3$): δ 7.58 (dd, J = 8.0, 1.0, 1H, aryl H-3), 7.47 (dd, J = 7.6, 1.4, 1H, aryl H-6), 7.29 (t, J = 7.5, 1H, aryl H-5), 7.14 (td, J = 7.7, 1.6, 1H, aryl H-4), 6.84 (d, J = 11.5, 1H, H-6), 6.02 (dd, J = 11.5, 9.3, 1H, H-5), 5.94 (d, J = 3.8, 1H, H-1), 4.80 (dd, J = 9.3, 2.9, 1H, H-4), 4.60 (d, J = 3.8, 1H, H-2), 3.60 (d, J = 3.0, 1H, H-3), 3.44 (s, 3H, OMe), 1.34 (s, 3H, acetonide CH_3), 1.30 (s, 3H, acetonide CH_3).

^{13}C NMR (151 MHz,$CDCl_3$): δ 136.4 (aryl C-1), 134.4 (C-6), 132.6 (aryl C-3), 130.5 (aryl C-6), 129.1 (aryl C-4), 127.0 (aryl C-5), 125.6 (C-5), 123.9 (aryl C-2), 111.6 (CMe_2), 104.8 (C-1), 85.8 (C-3), 82.3 (C-2), 75.9 (C-4), 58.3 (OMe), 26.7 (acetonide CH_3), 26.4 (acetonide CH_3).

HMRS (m/z): Calc. for [M+Na] = $C_{16}H_{19}O_4^{79}Br$ Na 377.0364; found 377.0378 (100 %) (3.6 ppm), Calc. for [M+Na] = $C_{16}H_{19}O_4^{81}Br$ Na 377.0364; found 377.0327 (97 %) (4.4 ppm).

IR: 3,069-3,000 (w, alkene & aromatic C–H stretch), 2,987-2,830 (m, aliphatic C–H stretch), 1,729 (w), 1,590 (w), 1,560 (w), 1,469 (m), 1,435 (m), 1,382-1,372 (2 bands, s, CMe_2 bend), 1,293 (m), 1,259 (s), 1,220 (s), 1,194 (s), 1,165 (s), 1,200 (s), 1,997 (s), 1,081 (s), 1,025 (s), 948 (m), 894 (m), 856 (s), 831 (m), 796-779 (s, 1,2-disubstituted arene) 751 (m), 717 (w).

MP 79–81 °C.

(c) With (ethoxycarbonylmethylidene)methyldiphenylphosphorane

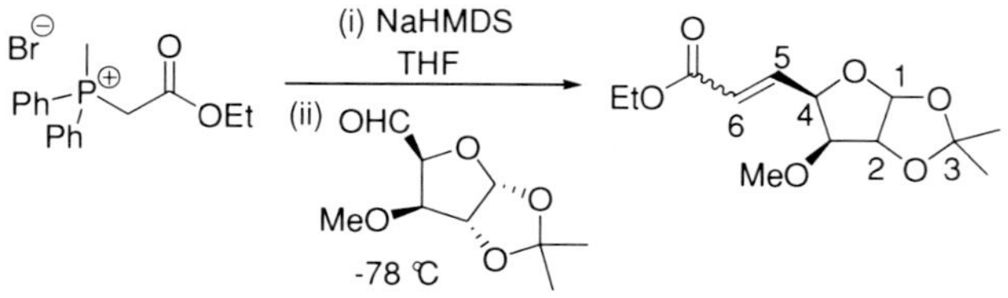

From (ethoxycarbonylmethyl)methyldiphenylphosphonium bromide (30 mg, 0.082 mmol), NaHMDS (15 mg, 0.082 mmol) and 1,2-O-isopropylidene-3-O-methyl-α-D-xylopentodialdofuranose-(1,4) (0.16 ml of a 0.5 mol L^{-1} solution in THF, 0.080 mmol) by procedure B. The reaction was quenched at −78 °C by addition of a saturated aqueous solution of NH_4Cl to ensure that alkene formation had actually occurred at low temperature. NMR coupling was confirmed by analysis of a gCOSY spectrum of the reaction mixture.

^1H NMR (500 MHz, $CDCl_3$) of reaction mixture.

Assigned to *Z*-isomer: δ 6.30 (dd, *J* = 11.8, 6.9, 1H. H-5), 5.96 (d, *J* = 3.9, 1H, H-1), 5.92 (dd, *J* = 11.8, 1.6, 1H, H-6), 5.63 (ddd, *J* = 6.9, 3.2, 1.6, 1H, H-4), 4.60 (d, *J* = 3.9, 1H, H-2), 4.04 (d, *J* = 3.3, 1H, H-3), 3.35 (s, 3H, OCH_3), 1.51 (s, 3H, one of diasterotopic CH_3 groups) [83].

Assigned to *E*-isomer: δ 6.94 (dd, *J* = 15.7, 5.0, H-5), 6.16 (dd, *J* = 15.7, 1.7, H-6), 4.79 (ddd, *J* = 4.9, 3.3, 1.7, H-4), 3.77 (d, *J* = 3.3, H-3), 3.37 (s, 3H, OCH_3), 1.50 (s, 3H, one of diasterotopic CH_3 groups) [83].

Integrations indicate a *Z/E* ratio 79:21.

Assigned to 1,2-O-isopropylidene-3-O-methyl-α-D-xylopentodialdofuranose-(1,4) (integrations relative to 1H of *Z*-alkene):

δ 9.62 (t, *J* = 1.6, 0.3H, CHO), 6.08 (d, *J* = 3.5, 0.3H, H-1), 4.54 (dd, *J* = 3.7, 1.5, 0.3H, H-4), 4.11 (d, *J* = 3.8, 0.3H, H-2), 3.35 (s, 0.9H, OCH_3), 1.47 (s, not baseline separated, one of diasterotopic CH_3 groups) [82].

Further signals were present due to the *Z* and *E* isomers of the alkene but could not be assigned due to the signals overlapping:

4.22-4.15 (m, 2.5H, *E* & *Z* CH_2CH_3), 1.25-1.34 (m, contains *E* & *Z* alkene CH_2CH_3 and other diastereotopic CH_3, and also aldehyde diasterotopic CH_3).

(d) With (methoxycarbonylmethylidene)methyldiphenylphosphorane

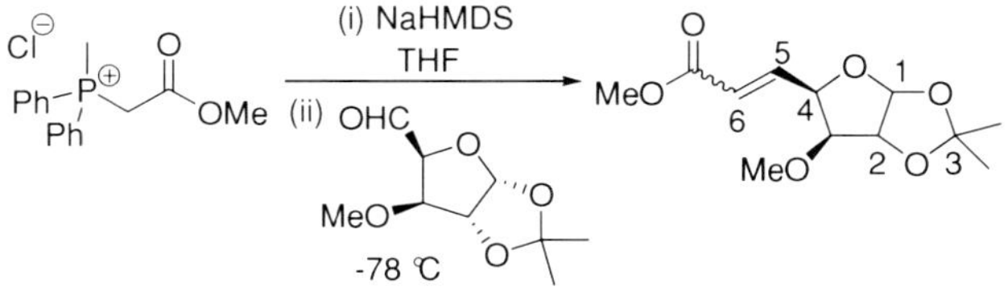

(i) NaHMDS THF (ii) OHC ... MeO ... -78 °C

From (methoxycarbonylmethyl)methyldiphenylphosphonium bromide (92 mg, 0.30 mmol), KHMDS (58 mg, 0.29 mmol) and 1,2-O-isopropylidene-3-O-methyl-α-D-xylopentodialdofuranose-(1,4) (0.60 ml of a 0.5 mol L^{-1} solution in THF, 0.30 mmol). The reaction was quenched at −78 °C by addition of a saturated aqueous solution of NH_4Cl to ensure that alkene formation had actually occurred at low temperature. NMR coupling was confirmed by analysis of a gCOSY spectrum of the reaction mixture.

^{1}H NMR (300 MHz, $CDCl_3$) of reaction mixture:

Assigned to *Z*-isomer: δ 6.33 (dd, *J* = 11.8, 6.9, 1H, H5), 6.09 (d, *J* = 3.5, 1H, H1), 5.64 (ddd, *J* = 6.8, 3.3, 1.6, 1H, H4), 4.05 (d, *J* = 3.3, 1H, H3).

Assigned to *E*-isomer: δ 6.18 (dd, *J* = 15.8, 1.7, 1H, H6), 4.84-4.76 (m, 1H, H4).

Integrations indicate a *Z/E* ratio 80:20. The alkene isomers were separated, and by-products and remaining starting materials removed, by column chromatography on slilica using 80:20 cyclohexane/ethyl acetate as eluting solvent. The *Z*-isomer eluted first, then the *E*-isomer.

Z-isomer (Yield 52 mg, 69 %)

^{1}H NMR (500 MHz, $CDCl_3$) δ 6.33 (dd, *J* = 11.8, 6.9, 1H, H-5), 5.99-5.91 (m, 2H, contains overlapping signals for H-6 (dd, small coupling constant visible, *J* = 1.6) and H-1), 5.64 (ddd, *J* = 6.9, 3.2, 1.6, 1H, H-4), 4.61 (d, *J* = 3.8, 1H, H-2), 4.05 (d, *J* = 3.3, 1H, H-3), 3.74 (s, 3H, CO_2CH_3), 3.35 (s, 3H, ring OCH_3), 1.53 (s, 3H, acetonide CH_3), 1.34 (s, 3H, acetonide CH_3).

^{13}C NMR (126 MHz, $CDCl_3$) δ 166.0 (C = O), 145.2 (C-6), 120.8 (C-5), 111.8 (acetonide CMe_2), 105.1 (C-1), 86.3 (C-3), 82.4 (C-2), 77.8 (C-4), 58.2 (ring OCH_3), 51.4 (CO_2CH_3), 26.9 & 26.4 (2 × acetonide CH_3).

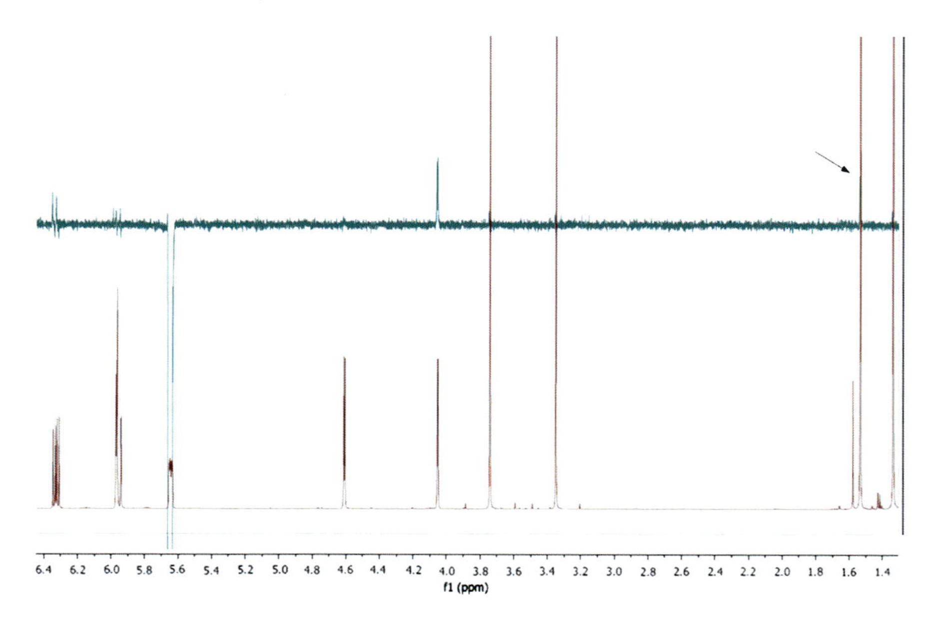

Fig. 4.6 1D-NOESY spectrum (500 ms mixing time) for irradiation at the frequency of the *Z*-isomer H-4 signal at δ 5.64. NOE contact observed with H-3 (so these hydrogens are mutually *cis*), and with one of the acetonide CH_3 groups

HRMS: Calc. for $[M{+}Na]^+ = C_{12}H_{18}O_6Na$ 281.1001; found 281.0996 (1.8 ppm). IR: 3,050-2,833 (s, aliphatic C–H stretch), 1,721 (s, C = O stretch), 1,655 (m, C = C stretch), 1,439 (m), 1,408 (m), 1,384-1,374 (m, CMe_2 stretch), 1,203 (s), 1,165 (s), 1,118 (s), 1,079 (s), 1,023 (s), 852 (m), 826 (m).

The non-epimerisation at C-4 of the *Z*-isomer during the reaction was shown by NOE contact between H-3 and H-4 in 1D NOESY experiments (NOE contact also seen between H-4 and one of acetonide CH_3 groups), as shown in Fig. 4.6.

E-isomer
(Yield 10 mg, 13 %)
^{1}H NMR (500 MHz, $CDCl_3$) δ 6.96 (dd, $J = 15.7, 4.9$, 1H, H-5), 6.18 (dd, $J = 15.8, 1.7$, 1H, H-6), 5.96 (d, $J = 3.8$, 1H, H-1), 4.80 (ddd, $J = 4.9, 3.3, 1.8$, 1H, H-4), 4.61 (d, $J = 3.8$, 1H, H-2), 3.78 (d, $J = 3.3$, 1H, H-3), 3.75 (s, 3H, CO_2CH_3), 3.38 (s, 3H, ring OCH_3), 1.51 (s, 3H, acetonide CH_3), 1.34 (s, 3H, acetonide CH_3).

^{13}C NMR (126 MHz, $CDCl_3$) δ 166.5 (C = O), 141.4 (C-5), 122.8 (C-6), 111.9 (acetonide CMe_2), 104.9 (C-1), 85.7 (C-4), 82.0 (C-2), 79.4 (C-3), 58.3 (ring OCH_3), 51.6 (CO_2CH_3), 26.8& 26.2 (2 × acetonide CH_3).

HRMS (TOF CI+): Calc. for $[M + H]^+ = C_{12}H_{19}O_6$ 259.1182; found 259.1194 (4.6 ppm).

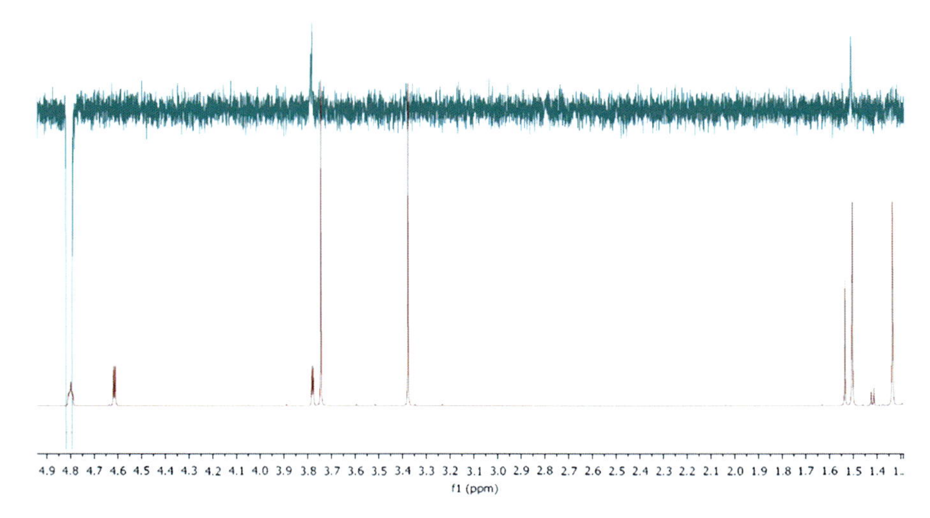

Fig. 4.7 1D-NOESY spectrum (500 ms mixing time) for irradiation at the frequency of the *E*-isomer H-4 signal at δ 4.80, showing NOE contact of H-4 with H-3 (these are thus mutually *cis*) and an acetonide CH_3 group

IR: 3,050-2,835 (m, alkene & aliphatic C–H stretch), 1,723 (s, C = O stretch), 1,667 (m, C = C stetch), 1,456 (m), 1,438 (m), 1,384-1,377 (m, CMe_2 bend), 1,307 (m), 1,267 (s), 1,218 (m), 1,196 (m), 1,166 (m), 1,134 (s), 1,082 (s), 1,020 (s), 918 (w), 889 (w), 859 (m), 738 (s).

The non-epimerisation at C-4 of the *E*-isomer during the reaction was shown by NOE contact between H-3 and H-4 in 1D NOESY experiments (NOE contact also seen between H-4 and one of acetonide CH_3 groups), for example the one shown in Fig. 4.7.

(e) With (*tert*-butoxycarbonylmethylidene)methyldiphenylphosphorane

(i) NaHMDS
THF
(ii) OHC
-78 °C

From (*tert*-butoxycarbonylmethyl)methyldiphenylphosphonium bromide (205 mg, 0.519 mmol), KHMDS (102 mg, 0.51 mmol) and 1,2-O-isopropylidene-3-O-methyl-α-D-xylopentodialdofuranose-(1,4) (1.0 ml of a 0.5 mol L^{-1} solution in THF, 0.50 mmol). The reaction was quenched at −78 °C by addition of a saturated aqueous solution of NH_4Cl to ensure that alkene formation had actually occurred at low temperature. NMR coupling confirmed by analysis of a gCOSY spectrum of the reaction mixture.

^{1}H NMR (500 MHz, $CDCl_3$) of reaction mixture:

Assigned to *Z*-isomer: δ 6.19 (dd, J = 11.9, 6.9, 1H, H5), 5.84 (dd, J = 11.9, 1.6, 1H, H6), 5.57 (ddd, J = 6.8, 3.2, 1.6, 1H, H4), 4.03 (t, J = 5.0, 1H, H3).

Assigned to *E*-isomer: δ 6.85 (dd, J = 15.7, 5.2, 1H, H6), 4.77 (ddd, J = 5.0, 3.2, 1.7, 1H, H4), 3.76 (d, J = 3.2, 1H, H3).

Integrations indicate a *Z/E* ratio 80:20.

The crude product of the reaction was purified by column chromatography on silica using 4:1 cyclohexane/ethyl acetate as the eluting solvent. The *Z*-isomer was first to elute, followed by a mixture of *Z* and *E* alkene, then *E*-alkene.

Z-isomer (105 mg, 70 %)

^{1}H NMR (500 MHz, $CDCl_3$) δ 6.19 (dd, J = 11.9, 6.9, 1H, H-5), 5.96 (d, J = 3.8, 1H, H-1), 5.84 (dd, J = 11.9, 1.6, 1H, H-6), 5.57 (ddd, J = 6.8, 3.2, 1.6, 1H, H-4), 4.60 (d, J = 3.8, 1H, H-2), 4.04 (d, J = 3.3, 1H, H-3), 3.36 (s, 3H, OMe), 1.53 (s, 3H, acetonide CH_3), 1.49 (d, J = 7.8, 9H, $C(CH_3)_3$), 1.33 (s, 3H, acetonide CH_3).

^{13}C NMR (126 MHz, $CDCl_3$) δ 165.0 (C = O), 142.6 (C-5), 123.4 (C-6), 111.6 (acetonide *C*Me_2), 105.0 (C-1), 86.3 (C-3), 82.5 (C-2), 80.9 (C-4), 77.9 (O*C*Me_3), 58.3 (OMe), 28.2 ($C(CH_3)_3$), 26.9 & 26.3 (2 × acetonide CH_3).

HRMS: Calc. for $[M+Na]^+$ = $C_{15}H_{24}O_6Na$ 323.1471; found 323.1461 (3.0 ppm). IR: 3,050-2,831 (m, alkene & aliphatic C–H stretch) 1,713 (s, C = O stretch), 1,648 (w, C = C stretch), 1,458 (w), 1,412 (w), 1,370 (m, CMe_2 bend), 1,293 (w), 1,238 (m), 1,216 (m), 1,160 (s), 1,118 (s), 1,079 (s), 1,024 (s), 853 (m), 827 (m).

E-isomer (23 mg, 15 %)

^{1}H NMR (500 MHz, $CDCl_3$) δ 6.84 (dd, J = 15.7, 5.2, 1H, H-5), 6.09 (dd, J = 15.7, 1.6, 1H, H-6), 5.95 (d, J = 3.8, 1H, H-1), 4.77 (ddd, J = 4.9, 3.2, 1.7, 1H, H-4), 4.61 (d, J = 3.8, 1H, H-2), 3.76 (d, J = 3.2, 1H, H-3), 3.38 (s, 3H, OMe), 1.50 (s, 3H, acetonide CH_3), 1.48 (s, 9H, $C(CH_3)_3$), 1.33 (s, 3H, acetonide CH_3).

^{13}C NMR (126 MHz, $CDCl_3$) δ 165.3 (C = O), 139.7 (C-5), 125.2 (C-6), 110.0 (acetonide *C*Me_2), 104.9 (C-1), 85.6 (C-3), 82.0 (C-2), 80.4 (*C*Me_3), 79.4 (C-4), 58.2 (OMe), 28.1 ($C(CH_3)_3$), 26.9 & 26.2 (2 × acetonide CH_3).

HRMS: Calc. for $[M + Na]^+$ = $C_{15}H_{24}O_6Na$ 323.1471; found 323.1479 (2.2 ppm).

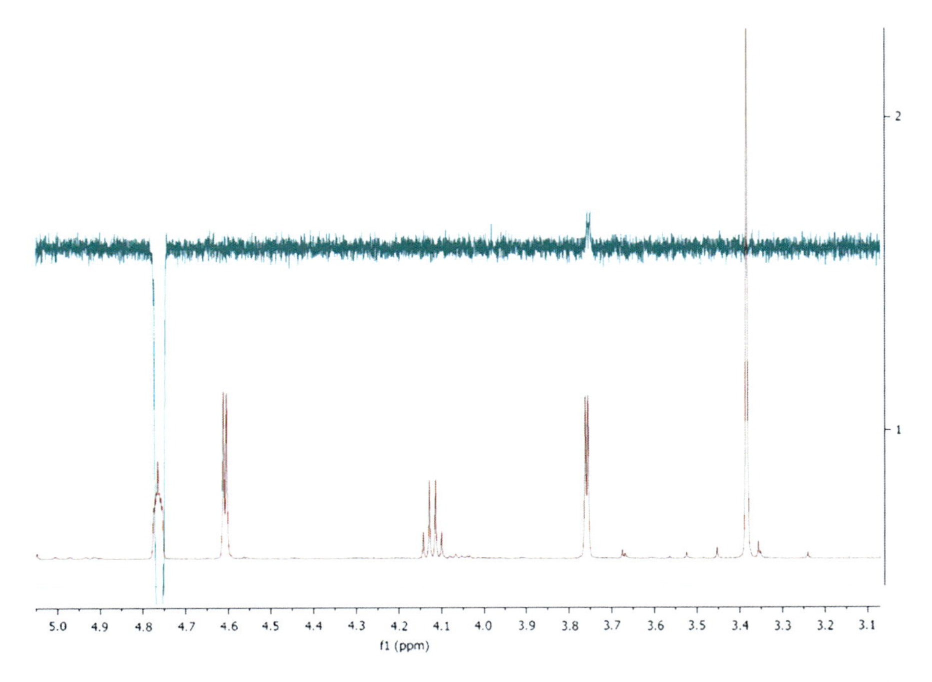

Fig. 4.8 1D NOESY spectrum for sample irradiated at the resonance frequency of H-4. NOE contact is observed between H-4 and H-3, indicating that they are mutually *cis*

IR: 3,050-2,800 (s, alkene & aliphatic C–H stretch), 1,715 (s, C = O stretch), 1,636 (m, C = C stretch), 1,458 (m), 1,370 (m, CMe_2 bend), 1,307 (m), 1,258 (s), 1,217 (m), 1,163 (s), 1,114 (s), 1,081 (s), 1,024 (s), 978 (m), 887 (m), 860 (m), 800 (w).

The relative stereochemistry of H-3 and H-4 was proven to be *cis* (as in the starting aldehyde) for the *E*-isomer by 1D NOESY NMR experiments (500 ms mixing time), in which NOE contact between these hydrogens was shown. The 1D-NOESY experiment for irradiation at the resonace frequency of H-4 is shown in Fig. 4.8.

4.9 Phosphine Oxide Removal from Crude Product of Wittig Reaction

A number of the Wittig reactions described here are the same reactions as were reported in Sects. 4.4–4.8. For each reaction where this is the case, a note has been explicitly made below in the appropriate place. These reactions are reported again here for the sake of completeness. Also, the NMR data quoted here is of the alkene after purification by oxalyl chloride treatment; in Chap. 3, wherever possible, the NMR data reported is that of the crude product, unpurified except for a minimal aqueous work-up.

4.9.1 Procedures for Wittig Reactions

Either procedure A or B, described in Sect. 4.3, was employed for all Wittig reactions carried out, with the exception of the reactions giving 4-phenylbut-3-en-2-one (for which a pre-formed stabilised ylide was used) and 1-cyclopentyl-3-methylbut-1-ene (where diethyl ether was used in place of THF).

Work up: In reactions of non-stabilised and semi-stabilised ylides, the reaction mixture was removed from the cold bath and allowed to warm to room temperature, and stirred for 12 h, after which time the THF solvent was removed *in vacuo* to give the crude product. For reactions of stabilised ylides, the reaction was quenched at low temperature by the addition of saturated aqueous NH_4Cl solution to ensure that alkene formation had actually occurred at low temperature. Diethyl ether (5 ml) was added to the mixture, which was transferred to a separatory funnel. The phases were separated and the aqueous layer was washed twice more with diethyl ether (5 ml). The combined ether phases were dried over Na_2SO_4, filtered, and concentrated to give the crude alkene, which was characterised by NMR.

4.9.2 Removal of Phosphine Oxide and Aldehyde from Crude Product by Oxalyl Chloride Treatment

This procedure has already been described in part in Sect. 4.3. Oxalyl chloride (1.1-1.5 equivalents) was added to the crude product by nitrogen-flushed syringe. The reaction mixture became vibrantly yellow and gas was evolved very vigourously. Cyclohexane (5 ml) was added immediately, causing the formation of a two-phase mixture; the upper phase consisted of a yellow cyclohexane solution, the lower an orange oil containing aldehyde mixed with chlorophosphonium salt. The cyclohexane phase was carefully decanted into another flask. The residue was washed a further five times with cyclohexane (5 ml aliquots), and the washings were transferred to the second flask. This (clear or yellow) solution was filtered via cannula (using a Whatman GFD grade glass microfibre filter). The cyclohexane filtrate was washed twice with saturated aqueous $NaHCO_3$ to quench and remove the remaining oxalyl chloride, and then twice with a 1 mol L^{-1} aqueous solution of HCl. The resulting solution was dried over $MgSO_4$, filtered, and concentrated *in vacuo* to give the product alkene. ^{1}H and ^{31}P NMR analysis of this product showed there to be no aldehyde or phosphorus-containing material present, only *Z* and *E* isomers of alkene. Comparison of the NMR spectra from before and after oxalyl chloride treatment in all cases showed the alkene *Z/E* ratio to be unchanged.

4.9.3 Characterisation of Purified Alkenes

The characterisation data for alkenes purified by oxalyl chloride treatment is listed below, and is presented in the order in which the compounds appear in Table 3.1.

Reactions of 1,2-O-isopropylidene-3-O-methyl-α-D-xylopentodialdofuranose-(1,4).

1. Reaction with *P*-(*iso*-butylidene)triphenylphosphorane. Table 3.1 entry 1.

From *iso*-butyltriphenylphosphonium bromide (61 mg, 0.15 mmol), KHMDS (28 mg, 0.14 mmol) and 1,2-O-isopropylidene-3-O-methyl-α-D-xylopentodialdofuranose-(1,4) (0.31 ml of a 0.5 mol L^{-1} solution in THF, 0.15 mmol) by procedure B. The crude product was treated with oxalyl chloride (0.014 ml, 0.17 mmol) to give the *Z*-alkene as a clear oil (34 mg, 92 %).

^{1}H NMR (500 MHz, $CDCl_3$) δ 5.91 (d, J = 3.9, 1H, H-1), 5.55-5.49 (m, 1H, H-6), 5.48-5.41 (m, 1H, H-5), 4.93 (dd, J = 8.2, 3.0, 1H, H-4), 4.59 (d, J = 3.9, 1H, H-2), 3.59 (d, J = 3.1, 1H, H-3), 3.40 (s, 3H, OMe), 2.70-2.58 (m, 1H, H-7), 1.52 (s, 3H, acetonide CH_3), 1.34 (s, 3H, acetonide CH_3), 1.01 & 0.99 (2 × d, J = 6.6, 6H, $CH(CH_3)_2$).

^{13}C NMR (101 MHz, $CDCl_3$) δ 142.4 (C-6), 120.7 (CMe_2), 111.3 (C-5), 104.6 (C-1), 86.2 (C-3), 82.3 (C-2), 75.8 (C-4), 58.2 (OMe), 27.5 ($CH(CH_3)_2$), 26.8 & 26.2 (diastereotopic acetonide CH_3), 23.3 & 23.0 (diastereotopic $CH(CH_3)_2$).

HRMS (m/z): Calc. for $[M]^+ = C_{13}H_{22}O_4$; found 242.1518 (0 ppm).

IR: 2,970-2,850 (m, C–H stretch), 1,735 (w), 1,658 (w), 1,466 (w), 1,380 (m, CMe_2 bend), 1,261 (m), 1,216 (m), 1,165 (m), 1,118)s), 1,081 (s), 1,022 (s), 863 (m), 796 (m).

Using 1D and 2D NOESY NMR spectroscopy (500 MHz, $CDCl_3$, 800 ms mixing time), the stereochemistry of the alkene double bond was demonstrated to be *Z* and the relative stereochemistry of H-3 and H-4 was shown to be *cis,* as it had been in the starting aldehyde. Thus no epimerisation can have occurred at C-4 under the reaction conditions. NOE contacts of H-4 with H-7 (showing that the ring and the *iso*-propyl group are *cis* relative to each other about the double bond), and of H-4 with H-3 (showing that H-3 and H-4 remain mutually *cis*) can be seen in the 2D NOESY spectrum in Fig. 4.9a, and in the 1D NOESY spectrum in Fig. 4.9b.

2. Reaction with (*tert*-butoxycarbonylmethylidene)methyldiphenylphosphorane. Table 3.1 entry 2.

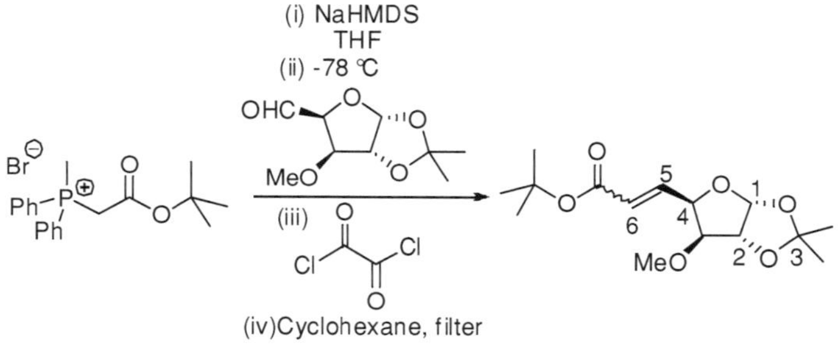

(i) NaHMDS THF
(ii) -78 °C
(iii)
(iv)Cyclohexane, filter

From (*tert*-butoxycarbonylmethyl)methyldiphenylphosphonium bromide (20 mg, 0.051 mmol), NaHMDS (9 mg, 0.05 mmol) and 1,2-O-isopropylidene-3-O-methyl-α-D-xylopentodialdofuranose-(1,4) (0.12 ml of a 0.5 mol L^{-1} solution in THF, 0.060 mmol) by procedure B. The reaction was quenched at −78 °C by addition of a saturated aqueous solution of NH_4Cl. The crude product was treated with oxalyl chloride (0.01 ml, 0.12 mmol) to give alkene[14] as a clear oil (mixture of isomers, 14 mg, 93 %).

^1H NMR (300 MHz, $CDCl_3$):

E-isomer: δ 6.85 (dd, J = 15.7, 5.2, 0.27H, H-5), 6.09 (dd, J = 15.7, 1.7, 0.27H, H-6), 4.77 (ddd, J = 5.0, 3.3, 1.7, 0.27H, H-4), 3.76 (d, J = 3.2, 0.27H, H-3), 3.38 (s, 0.76H, OMe), 1.48 (s, $C(CH_3)_3$). Acetonide signals obscured by other signals around 1.3–1.5 ppm.[14] Integrations are relative to 1H of *Z*-isomer.

[14] Characterisation data for this compound is given in Sect. 4.8.

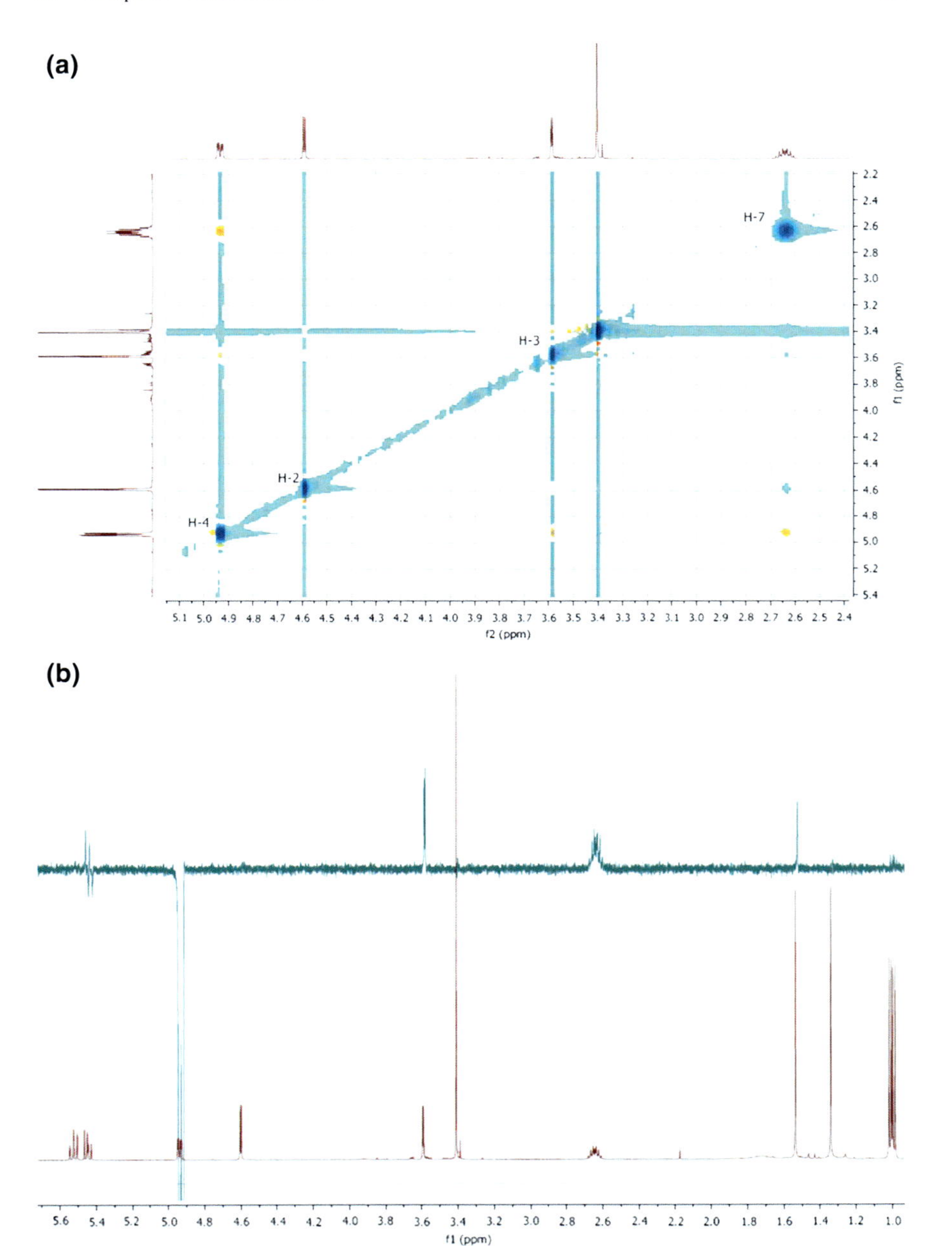

Fig. 4.9 **a** 2D NOESY of Z-alkene product showing NOE contact of H-4 with H-3 (showing that these are *cis* relative to each other) and with H-7 (confirming that the alkene is of Z-configuration). **b** 1D NOESY of purified Z-alkene. The sample was irradiated at the resonance frequency of H-4 (signal at δ 4.93 ppm). NOE contact can be seen with H-3 (δ 3.59), H-7 (δ 2.70-2.58), one of the acetonide methyl groups (δ 1.52) and with the *iso*-propyl methyl groups (δ 1.01-0.99, small contact)

Z-isomer: δ 6.19 (dd, J = 11.9, 6.8, 1H, H-5), 5.84 (dd, J = 11.9, 1.6, 1H, H-6), 5.57 (ddd, J = 6.8, 3.2, 1.6, 1H, H-4), 4.04 (d, J = 3.2, 1H, H-3), 3.36 (s, 3H, OMe), 1.53 (s, 3H, acetonide CH_3), 1.50 (s, $C(CH_3)_3$, 9H), 1.33 (s, 3H, acetonide CH_3).[14]

Relative integrations in the ^{1}H NMR indicate a *Z/E* ratio of 79:21. Other signals of the two isomers overlap with each other integrations relative to 1H of *Z*-alkene): ^{1}H NMR (300 MHz, $CDCl_3$): δ 5.95 (*E* & *Z* H-1, 1.27H), 4.60 (*E* & *Z* H-2, 1.27H).

3. Reaction with (methoxycarbonylmethyl)methyldiphenylphosphorane. Table 3.1 entry 3.

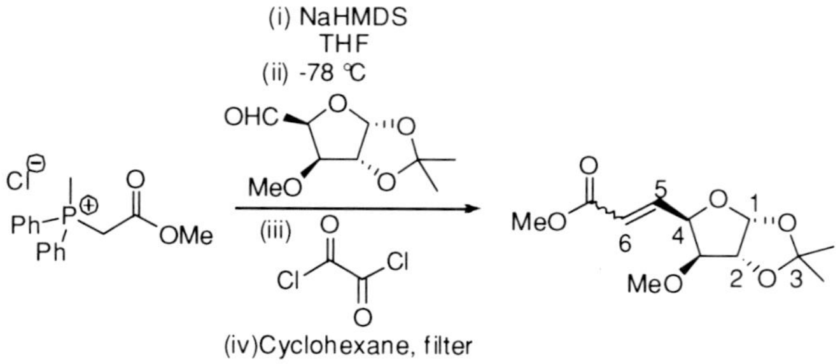

(i) NaHMDS
THF
(ii) -78 °C
OHC
MeO
(iii)
Cl
Cl
(iv)Cyclohexane, filter
Cl
Ph
Ph
OMe
MeO
MeO
5
6
4
1
2
3

From (methoxycarbonylmethyl)methyldiphenylphosphonium chloride (17 mg, 0.055 mmol), NaHMDS (10 mg, 0.054 mmol) and 1,2-O-isopropylidene-3-O-methyl-α-D-xylopentodialdofuranose-(1,4) (0.11 ml of a 0.5 mol L^{-1} solution in THF, 0.055 mmol) by procedure B. The reaction was quenched at −78 °C by addition of a saturated aqueous solution of NH_4Cl. The crude product was treated with oxalyl chloride (0.01 ml, 0.12 mmol) to give alkene[14] as a clear oil (mixture of isomers, 14 mg, 93 %).

^{1}H NMR (500 MHz, $CDCl_3$) of alkene mixture:

Assigned to *E*-isomer: δ 6.96 (dd, J = 15.7, 4.9, 1H, H-5), 6.18 (dd, J = 15.7, 1.7, 1H, H-6), 4.83-4.78 (m, 1H, H-4), 3.38 (s, 3H, ring OCH_3), 1.51 (s, 3H, acetonide CH_3).[14]

Assigned to Z-isomer: δ 6.33 (dd, J = 11.8, 6.9, 1H, H-5), 5.64 (ddd, J = 6.9, 3.2, 1.6, 1H, H-4), 4.05 (d, J = 3.3, 1H, H-3), 3.35 (s, 3H, ring OCH_3), 1.53 (s, 3H, acetonide CH_3).[14]

Integration of these signals indicates a *Z/E* ratio of 79:21. Overlapping signals (integrations relative to 1H of Z-alkene): δ 5.95 (dt, *J* = 8.6, 5.6, 2.3H, Z-H-1 & H-6, *E*-H-1), 4.61 (t, *J* = 3.9, 1.3H, H-2), 3.78-3.74 (2 doublets, *J* = 3.3 & *J* = 3.0, 4H, *Z* & *E* CO_2CH_3), 1.34 (s, 4H, *Z* & *E* acetonide CH_3).

4. Reaction with benzylidenemethyldiphenylphosphorane. Table 3.1 entry 4.

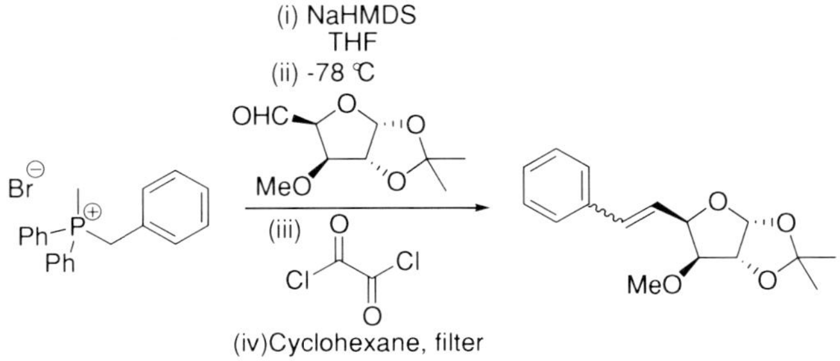

(i) NaHMDS THF
(ii) -78 °C
OHC
MeO
(iii)
Cl Cl
O O
(iv)Cyclohexane, filter
Br⊖
Ph–P⊕
Ph
MeO

From benzylmethyldiphenylphosphonium bromide (0.070 g, 0.19 mmol), NaH-MDS (35 mg, 0.019 mmol) and 1,2-O-isopropylidene-3-O-methyl-α-D-xylopen-todialdofuranose-(1,4) (0.38 ml of a 0.5 mol L^{-1} solution in THF, 0.19 mmol) by procedure B. The crude product was treated with oxalyl chloride (0.018 ml, 0.21 mmol) to give the product as a clear oil that crystallised on standing (46 mg, 87 %).[14]

^{1}H NMR (400 MHz, $CDCl_3$) of alkene mixture:

Assigned to *E*-isomer: δ 6.74 (d, *J* = 16.0, 1H, H-6), 6.31 (dd, *J* = 16.0, 7.6, 1H, H-5), 4.79 (dd, *J* = 7.7, 3.0, 1H, H-4), 3.71 (d, *J* = 3.0, 1H, H-3), 3.34 (s, 3H, OCH_3). Aromatic signals obscured by the corresponding signals of the *Z*-isomer [81].
Assigned to Z-isomer: δ 7.37-7.33 (app d, *J* = 4.2, 4H, PhH-2 and PhH-3), 6.81 (d, *J* = 11.6, 1H, H-6), 5.01 (dd, *J* = 9.1, 2.7, 1H, H-4), 4.63 (d, *J* = 3.8, 1H, H-2), 3.66 (d, *J* = 2.9, 1H, H-3), 3.44 (s, 3H), 1.41 (s, 3H, acetonide CH_3), 1.32 (s, 3H, acetonide CH_3). Acetonide hydrogens may overlap with the corresponding signals of *E*-isomer.[14]

Integration of these signals indicates a *Z/E* ratio of 95:5. Overlapping signals (integrations relative to 1H of Z-alkene): δ 5.91-5.80 (m, 2.05H, Z-H-1 & H-6, *E*-H-1).

5. Reaction with (2-bromobenzylidene)methyldiphenylphopsphorane.
Table 3.1 entry 5.

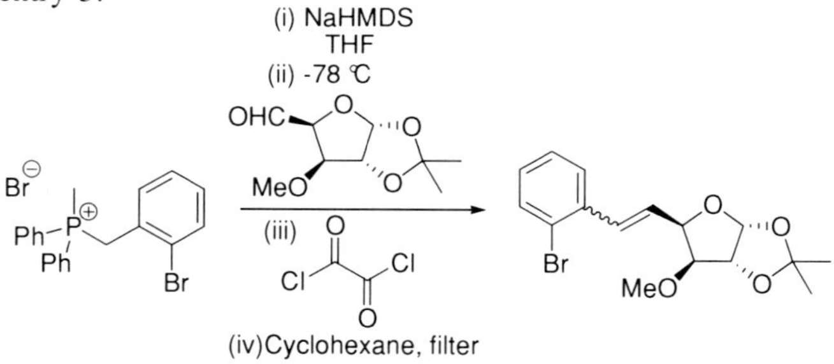

(i) NaHMDS THF
(ii) -78 °C
(iii)
(iv)Cyclohexane, filter

From 2-bromobenzylmethyldiphenylphosphonium bromide (32 mg, 0.071 mmol), NaHMDS (13 mg, 0.071 mmol) and 1,2-O-isopropylidene-3-O-methyl-α-D-xylopentodialdofuranose-(1,4) (0.14 ml of a 0.5 mol L^{-1} solution in THF, 0.07 mmol) by procedure B. The crude product was treated with oxalyl chloride (0.07 ml, 0.08 mmol) to give the product as a clear oil that crystallised on standing (24 mg, 96 %).[14]

^{1}H NMR (500 MHz, $CDCl_3$) of alkene mixture:

Assigned to *Z*-isomer: δ 7.47 (app dd, J = 7.6, 1.3, 1H, ArH-6), 7.29 (app td, J = 7.5, 0.8 Hz, 1H, ArH-5), 7.14 (app td, J = 7.6, 1.4, 1H, ArH-4), 6.77 (d, J = 11.5, 1H, H-6), 6.02 (dd, J = 11.5, 9.3, 1H, H-5), 5.94 (d, J = 3.8, 1H, H-1), 4.53 (d, J = 3.8, 1H, H-2), 3.52 (d, J = 3.0, 1H, H-3), 3.37 (s, 3H, OCH_3), 1.27 (s, 3H, acetonide CH_3), 1.23 (s, 3H, acetonide CH_3). Acetonide hydrogens may overlap with corresponding signals of *E*-isomer.[14]

Assigned *E*-isomer: δ 6.20 (dd, J = 16.0, 7.7, 1H, H-5), 5.90 (d, J = 3.8, 1H, H-1), 4.57 (d, J = 3.8, 1H, H-2), 3.67 (d, J = 3.1, 1H, H-3), 3.35 (s, 3H, OCH_3).[14]

Integration of these signals indicates a *Z/E* ratio of 95:5. Other overlapping signals were also observed (integrations relative to 1H of *Z*-isomer): δ 7.58 (app dd, J = 8.0, 1.0, 1H, *Z*-isomer ArH-3, partially overlaps with aromatic signal of *E*-isomer), 7.08 (d, J = 16.0, *E*-isomer H-6, partially obscured by aromatic signal of *E*-isomer), 4.83 (obscured by signal at δ 4.80, *E*-isomer H-4), 4.80 (m, dd, J = 9.3, 3.0, 1H, *Z*-isomer H-4).

Alkenes derived from other aldehydes.

6. Stilbene [24, 25]. Table 3.1 entry 6. This synthesis is also included in Sect. 4.4.

(i) NaHMDS THF (ii) -78 °C CHO (iii) Cl Cl O O (iv)Cyclohexane, filter

From benzylmethyldiphenylphosphonium bromide (986 mg, 2.66 mmol), NaHMDS (488 mg, 2.66 mmol), and benzaldehyde (0.27 ml, 0.28 g, 2.6 mmol) in dry THF (10 ml) by procedure B. The crude product was treated with oxalyl chloride (0.29 ml, 3.4 mmol) to give the product as a clear oil that crystallised on standing (0.46 g, 98 %). *Z/E* ratio = 15:85.

^{1}H NMR (300 MHz, $CDCl_3$) of product after treatment with oxalyl chloride.

Assigned to *E*-isomer: δ 7.56-7.46 (m, 4H), 7.41-7.31 (m, 4H), 7.11 (s, 2H, alkene H) [24].

Assigned to *Z*-isomer: δ 6.60 (s, 2H, alkene H) [25].

7. 2-fluoro-2′methylstilbene.[15] Table 3.1 entry 7. This synthesis is also included in Sect. 4.4.

(i) NaHMDS THF (ii) -78 °C CHO (iii) Cl Cl O O (iv)Cyclohexane, filter

From 2-fluorobenzylmethyldiphenylphosphonium bromide (1.069 g, 2.75 mmol), NaHMDS (2.6 ml of a 1 mol L^{-1} solution in THF, 2.6 mmol) and 2-methylbenzaldehyde (0.32 ml, 0.33 g, 2.8 mmol) by procedure A. The crude product was treated with oxalyl chloride (0.23 ml, 0.35 g, 2.8 mmol) to give the product as a clear oil that crystallised on standing (0.51 g, 90 %). *Z/E* ratio = 51:49.

[15] Characterisation data for this compound is given in Sect. 4.4.

^{1}H NMR (400 MHz, $CDCl_3$) of oxalyl chloride -purified product:

Assigned to *E*-isomer: δ 7.40 (d, $J = 16.4$, 1H), 2.43 (s, 3H).[15]

Assigned to *Z*-isomer: δ 2.28 (s, 3H).[15]

^{19}F NMR (376 MHz, $CDCl_3$) of crude product:

Assigned to *E*-isomer: δ −117.8-−118.0(m).[15]

Assigned to *Z*-isomer: δ −116.1-−116.3 (m).[15]

8. 2-bromo-2′methylstilbene [30][15]. See Table 3.1 entry 8. This synthesis is also included in Sect. 4.4.

(i) NaHMDS THF (ii) -78 °C CHO (iii) O Cl Cl O (iv)Cyclohexane, filter

From 2-bromobenzylmethyldiphenylphosphonium bromide (1.115 g, 2.48 mmol), NaHMDS (2.3 ml of a 1 mol L^{-1} solution in THF, 2.3 mmol) and 2-methylbenzaldehyde (0.30 ml,0.31 g, 2.6 mmol) by procedure A. The crude product was treated with oxalyl chloride (0.22 ml, 0.33 g, 2.6 mmol) to give the product as a clear oil that crystallised on standing (0.59 g, 94 %). ^{1}H NMR (300 MHz, $CDCl_3$) of oxalyl chloride-treated product (*Z/E* ratio = 75:25):

Assigned to *E*-isomer: δ 2.42 (s, 3H) [30].

Assigned to *Z*-isomer: δ 6.81-6.69 (AB system, 2H), 2.26 (s, 3H).[15]

9. 2, 2′dibromostilbene [36, 37]. Table 3.1 Entry 9. This synthesis is also included in Sect. 4.4.

(i) NaHMDS THF (ii) -78 °C CHO Br (iii) O Cl Cl O (iv)Cyclohexane, filter

From 2-bromobenzylmethyldiphenylphosphonium bromide (2.47 g, 5.49 mmol), NaHMDS (5.0 ml of a 1 mol L^{-1} solution in THF, 5.0 mmol) and 2-bromobenzaldehyde (0.64 ml, 1.01 g, 5.49 mmol) by procedure A. The crude product was treated with neat oxalyl chloride (0.50 ml, 0.75 g, 6.0 mmol), and the alkene was obtained as a clear oil (1.52 g, 90 %).

^{1}H NMR (500 MHz, $CDCl_3$) of purified stilbene mixture (*Z/E* ratio = 98:2):

Assigned to *E*-isomer: δ 7.71 (dd, J = 7.9, 1.5, 2H), 7.59 (d, J = 1.1, patially obscured by signal at 7.55), 7.39 (s, 2H, *HC*=*CH*), 7.35-7.30 (m, 2H), 7.15-7.11 (m, 2H) [36, 37].

Assigned to *Z*-isomer: δ 7.55 (dd, J = 7.8, 1.2, 2H), 7.06-6.94 (m, 6H), 6.77 (s, 2H, HC=CH) [36, 37].

10. 2-chloro-2′-methylstilbene [40]. Table 3.1 entry 10. This synthesis is also included in Sect. 4.4.

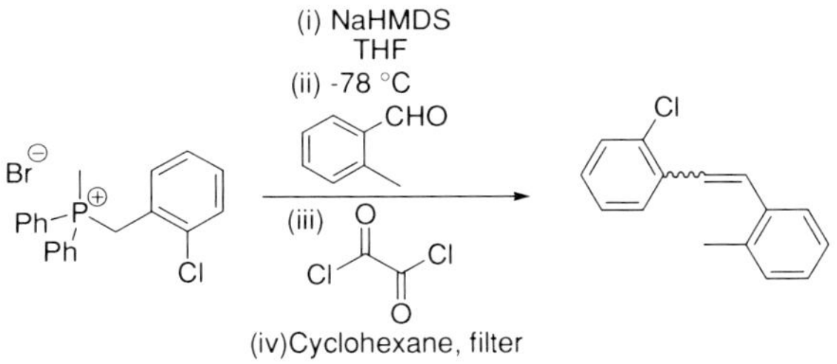

(i) NaHMDS THF
(ii) -78 °C
CHO
(iii)
(iv)Cyclohexane, filter

From 2-chlorobenzylmethyldiphenylphosphonium chloride (1.93 g, 4.76 mmol), NaHMDS (4.6 ml of a 1 mol L^{-1} solution in THF, 4.6 mmol) and 2-methylbenzaldehyde (0.55 ml, 0.57 g, 4.8 mmol) by procedure A. The crude product was purified by treatment with oxalyl chloride (0.44 ml, 5.2 mmol), giving a pure sample of the alkene as a clear oil (0.85 g, 78 %).

^{1}H NMR (300 MHz, $CDCl_3$) of purified product (*Z/E* ratio of 77:23): PBW292.

Assigned to *E*-isomer: δ 2.34 (s, 3H, Ar-CH_3) [40].

Assigned to *Z*-isomer: δ 6.83-6.73 (2 heavily roofed doublets, 2H, *HC*=*CH*), 2.24 (s, 3H, Ar–CH_3) [40].

11. 2-chlorostilbene [28, 29]. Table 3.1 entry 11. This synthesis is also included in Sect. 4.4

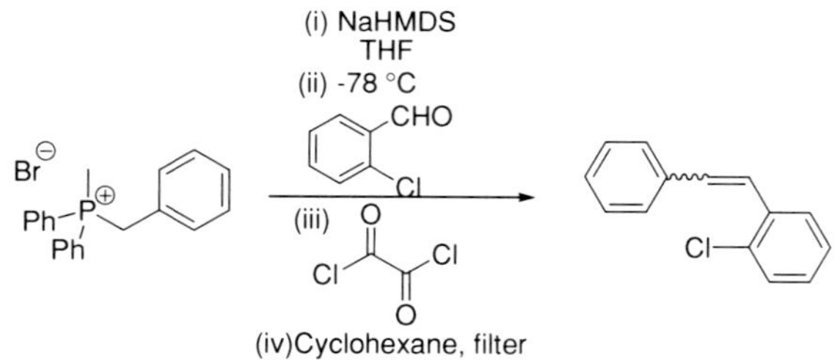

From benzylmethyldiphenylphosphonium bromide (160 mg, 0.431 mmol), KHMDS (82 mg, 0.41 mmol) and 2-chlorobenzaldehyde (0.046 ml, 57 mg, 0.41 mmol using 100 μl syringe) by procedure B. The crude product was purified by treatment with oxalyl chloride (0.04 ml, 0.06 g, 0.5 mmol) giving the alkene as a clear oil (77 mg, 85 %).

^{1}H NMR (300 MHz, $CDCl_3$) of oxalyl chloride-treated product (*Z/E* ratio of 92:8):

Assigned to *E*-isomer: δ 7.08 (d, $^{3}J = 16.4$, *trans* alkene H) [28].

Assigned to *Z*-isomer: δ 7.03 (app td, $J = 7.5, 0.7$, 1H), 6.73 (d, $^{3}J = 12.2$, *cis* alkene H, 1H), 6.69 (d, $^{3}J = 12.2$, *cis* alkene H, 1H) [29].

12. Table 3.1 entry 12.

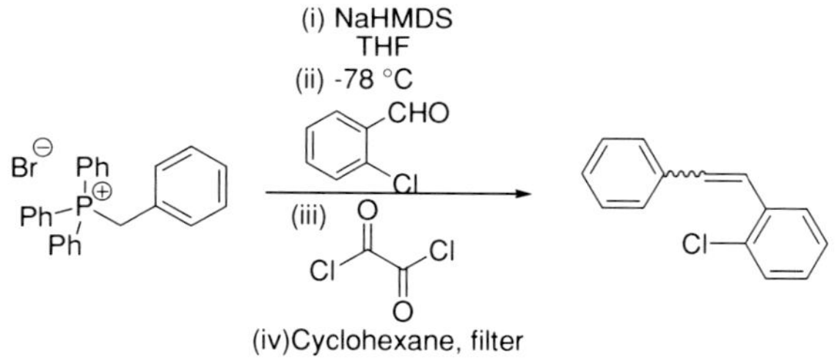

From benzyltriphenylphosphonium bromide (0.800 g, 1.85 mmol), NaHMDS (0.323 g, 1.76 mmol) and 2-chlorobenzaldehyde (0.20 ml, 0.25 g, 1.8 mmol) by procedure B. The crude product was purified by treatment with oxalyl chloride (0.16 ml, 0.24 g, 1.9 mmol) giving the alkene as a clear oil (0.29 g, 76 %).

^{1}H NMR (500 MHz, $CDCl_3$) signals assigned to *E* and *Z*-isomers are the same as for entry 11 above. Integrations give *Z/E* ratio of 90:10.

13. Table 3.1 entry 13.

(i) NaHMDS THF (ii) -78 °C (iii) (iv)Cyclohexane, filter

From 2-chlorobenzyltriphenylphosphonium bromide (0.910 g, 2.14 mmol), NaHMDS (0.374 g, 2.04 mmol) and benzaldehyde (0.21 ml, 0.22 g, 2.1 mmol) by procedure B. The crude product was purified by treatment with oxalyl chloride (0.20 ml, 0.30 g, 2.4 mmol) giving the alkene as a clear oil (0.40 g, 91 %).

^{1}H NMR (500 MHz, $CDCl_3$) signals assigned to *E* and *Z*-isomers are the same as for entry 11 above. Integrations give *Z/E* ratio of 51:49.

14. 2,2′-dichlorostilbene [34, 35]. Table 3.1 entry 14.

(i) NaHMDS THF (ii) -78 °C (iii) (iv)Cyclohexane, filter

From 2-chlorobenzyltriphenylphosphonium chloride (1.020 g, 2.41 mmol), KHMDS (0.418 g, 2.28 mmol) and 2-chlorobenzaldehyde (0.26 ml, 0.32 g, 2.3 mmol) by procedure B. The neat crude product was treated with oxalyl chloride (0.19 ml, 0.29 g, 2.3 mmol) to give the product as a clear oil that crystallised on standing (0.53 g, 95 %).

^{1}H NMR (500 MHz, $CDCl_3$) of alkene after oxalyl chloride treatment (*Z/E* ratio = 94:6):

Assigned to *E*-isomer: δ 7.28 (td, $J = 7.6, 1.2$, 2H), 7.24-7.19 (m, 2H) [34, 35].

Assigned to *Z*-isomer: δ 7.37 (dt, $J = 4.6, 1.8$, 2H), 7.16-7.09 (m, 2H), 7.01 (m, 2H), 6.99-6.93 (m, 2H), 6.85 (s, 2H, *H*C=C*H*) [35].

The phosphorus-containing residue from the filtration was dissolved in dry THF (5 ml) and cooled to −78 °C under a nitrogen atmosphere. A solution of $LiAlH_4$ in dry Et_2O (1 mol L^{-1}, 2.0 ml, contains 2.0 mmol) was added drop-wise to the mixture while stirring. This was then allowed to warm to room temperature while stirring over 30 min. The reaction flask was then cooled to 0 °C, and ethyl acetate

(10 ml) was added to quench any remaining $LiAlH_4$, followed by saturated NH_4Cl solution. The phases were separated and the aqueous layer was washed twice with ethyl acetate (10 ml each). The combined organic phases were dried over $MgSO_4$, filtered, and concentrated *in vacuo*. NMR of the crude product showed the crude sample of triphenylphosphine to contain 15 % phosphine oxide (derived from hydrolysis of residual phosphonium salt). Recrystallisation from methanol gave white crystals of triphenylphosphine (0.30 g, 50 %).

15. 2,2′-difluorostilbene [84–86]. Table 3.1 entry 15.

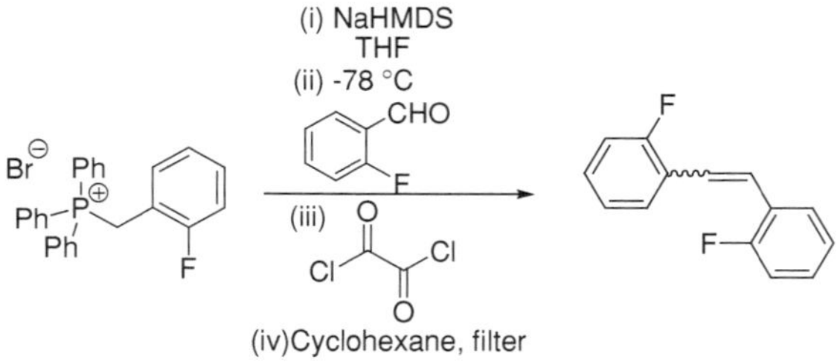
(i) NaHMDS
THF
(ii) -78 °C
CHO
F
(iii)
O
Cl
Cl
O
(iv)Cyclohexane, filter
Br
Ph
Ph
P
Ph
F
F
F

From 2-fluorobenzylmethyldiphenylphosphonium bromide (290 mg, 0.640 mmol), KHMDS (125 mg, 0.63 mmol) and 2-fluorobenzaldehyde (0.067 ml, 0.79 g, 0.64 mmol) by procedure B. The crude product was treated with oxalyl chloride (0.06 ml, 0.09 g, 0.7 mmol) to give the product as a clear oil that crystallised on standing (0.13 g, 93 %).

^{1}H NMR (300 MHz, $CDCl_3$) of alkene mixture:

No signals could be assigned to the *E*-isomer [84].

Assigned to *Z*-isomer: δ 7.07-6.96 (m, 2H), 6.96-6.85 (m, 2H), 6.76 (s, 2H, *H*C=C*H*) [86].

^{19}F NMR (282 MHz, $CDCl_3$) of crude product:

Assigned to *E*-isomer: δ −117.9-−118.0 (m).

Assigned to *Z*-isomer: δ −115.0-−115.1 (m).

Integrations indicate a *Z/E* ratio of 94:6.

16. 4-phenylbut-3-en-2-one [55]. Table 3.1 entry 16.

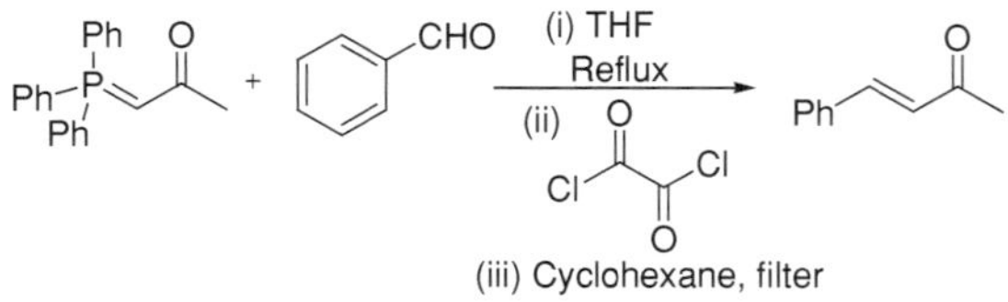
Ph
O
Ph
P
Ph
+
CHO
(i) THF
Reflux
(ii)
O
Cl
Cl
O
(iii) Cyclohexane, filter
O
Ph

To a solution of acetonylidenetriphenylphosphorane (872 mg, 2.74 mmol) in dry THF at 20 °C was added benzaldehyde (0.28 ml, 0.29 g, 2.7 mmol) dropwise. The reaction was refluxed (oil bath heated to 70 °C) while stirring overnight. The THF solvent was then removed *in vacuo*. The crude product was treated with oxalyl chloride (0.20 ml, 0.30 g, 2.4 mmol) to give the product as an oil (0.35 g, 88 %). ^{1}H NMR (300 MHz, $CDCl_3$) of mixture of enones.

Assigned to *E*-isomer: δ 7.60-7.36 (m, 6H, aromatic H & H-4), 6.72 (d, J = 16.3, 1H, H-3), 2.38 (s, 3H, CH_3) [55].

Assigned to *Z*-isomer: δ 6.90 (1H, d, ^3J = 12.7 Hz, H-4), 6.20 (1H, d, ^3J = 12.7 Hz, H-3), 2.14 (3H, s, CH_3) [55].

Integrations indicate *Z/E* ratio of 3:97.

17. 1-cyclopentyl-3-methylbut-1-ene. Table 3.1 entry 17.

(i) NaHMDS Et$_2$O; (ii) cyclopentane-CHO, -20 °C; (iii) oxalyl chloride (ClCOCOCl); (iv) Pentane, filter

For this reaction, procedure B was followed with the following modifications: Diethyl ether was used in place of THF, the reaction was carried out at −20 °C rather than −78 °C, and pentane was used in place of cyclohexane. No reaction of this ylide and aldehyde occurs below approximately −20 °C as judged by a lack of decolourisation of the ylide solution by addition of aldehyde at temperatures lower than this. From (*iso*-butyl)triphenylphosphonium bromide (250 mg, 0.63 mmol), NaHMDS (115 mg, 0.63 mmol) and cyclopentanecarboxaldehyde (0.067 ml, 62 mg, 0.63 mmol) by procedure B. The crude product was treated with oxalyl chloride (0.06 ml, 0.09 g, 0.7 mmol) to give the product as an oil (0.28 g, contains small impurities). The crude product obtained in this way contained only *Z*-alkene and no phosphorus or aldehyde, but decomposed if left to stand. The stereochemistry of the single isomer of alkene produced could not be determined by NMR due to the pseudosymmetry of the alkene. However, based on the frequently observed *Z*-selectivity observed in Li-salt free Wittig reactions of alkylidenetriphenylphosphoranes with secondary aldehydes, this alkene is tentatively assigned to be of *Z*-configuration.

^{1}H NMR (300 MHz, $CDCl_3$) δ 5.20-5.06 (m, 2H, *H*C=C*H*), 2.77-2.55 (m, 2H, C*H*Me$_2$ & *c*-C_5H_9 H-1), 1.83-1.51 (m, 6H, *c*-C_5H_9 ring hydrogens), 1.30-1.12 (m, 2H, *c*-C_5H_9 ring hydrogens), 0.95 (dd, J = 6.7, 6H, C*H*$_3$).

^{13}C NMR (75 MHz, $CDCl_3$) δ 136.2 & 132.9 (C = C), 38.3 (*c*-C_5H_9 C-1), 34.0 & 25.4 (*c*-C_5H_9 C-2 & C-3), 26.8 (*C*HMe$_2$), 23.5 (*C*H$_3$).

18. 1-(2-bromophenyl)-3-methylbut-1-ene [80] Table 3.1 entry 18.

(i) KHMDS THF
(ii) -78 °C
CHO
Br
(iii) oxalyl chloride
(iv)Cyclohexane, filter

From (*iso*-butyl)triphenylphosphonium bromide (555 mg, 1.39 mmol), KHMDS (308 mg, 1.54 mmol) and 2-bromobenzaldehyde (0.16 ml, 0.25 g, 1.4 mmol) by procedure B. The crude product was treated with oxalyl chloride (0.13 ml, 0.20 g, 1.54 mmol) to give the phosphorus-free product as a clear oil (0.28 g, 90 %). ^{1}H NMR (300 MHz, $CDCl_3$) of purified alkene:

Assigned to *E*-isomer: δ 6.67 (d, J = 15.8 Hz, 1H), 6.13 (dd, J = 15.8, 6.8 Hz, 1H), 1.11 (d, J = 6.7 Hz, 6H) [80].

Assigned to *Z*-isomer: δ 6.31 (d, J = 11.4 Hz, 1H), 5.64-5.51 (app t, 1H), 1.01 (d, J = 6.6 Hz, 6H) [80].

Integrations indicate a *Z/E* ratio of 82:18. Also present was a multiplet with signals of the *E* and *Z*-isomers overlapping (integration relative to 1H of *Z*-isomer): δ 2.72-2.48 (m, 1.2H, *E* and *Z* C*H*Me_2).

The phosphorus-containing residue left after filtration of the 1-(2-bromophenyl)-3-methylbut-1-ene was dissolved in dry THF (5 ml) and cooled to −78 °C under a nitrogen atmosphere. A solution of $LiAlH_4$ in dry Et_2O (1 mol L^{-1}, 1.7 ml, contains 1.7 mmol) was added drop-wise to the mixture while stirring. This was then allowed to warm to room temperature while stirring over 30 min. The reaction flask was then cooled to 0 °C, and ethyl acetate (10 ml) was added to quench any remaining $LiAlH_4$. The contents of the reaction flask were filtered to remove insoluble salts (especially phosphonium salt). The residue in the flask and in the filter funnel was washed with further ethyl acetate, and the ethyl acetate washings were combined and washed with NH_4Cl solution (5 ml), dried over $MgSO_4$, filtered, and concentrated *in vacuo*. White crystals of triphenylphosphine resulted (0.29 g, 80 %, contains ca. 5 % triphenylphosphine oxide).

^{1}H NMR (300 MHz, $CDCl_3$) δ 7.34-7.24 (m).

^{31}P NMR (121 MHz, $CDCl_3$) δ −5.3 (lit. −4.8) [87].

References

1. Shriver DF, Drezdzon MA (1986) The manipulation of air sensitive compounds, 2nd edn. McGraw-Hill, New York
2. Trippett S, Walker DM (1926) J Chem Soc 1266. Only MP (242-244 °C) and elemental analysis data are given
3. Muller G, Abicht HP, Waldkircher M, Lachmann J, Lutz M, Winkler M (2001) J Organomet Chem 622:121 (Characterisation of phosphonium iodide salt)
4. Cvengros J, Toma S, Marque S, Loupy A (2004) Can J Chem 82:1365
5. Fleming I, Loreto MA, Wallace IHM (1986) J Chem Soc Perkin Trans 1:349
6. Shieh R.-L, Lin R.-L, Hwang J.-J, Jwo J.-J (1998) J Chin Chem Soc (Taipei, Taiwan) 45:517 (Characterisation of phosphonium chloride salt)
7. Mitsumori T, Koga N, Iwamura H (1994) J Phys Org Chem 7:43
8. Staab HA, Gunthert P (1977) Chem Ber 110:619. Only MP data (265–266 °C) and elemental analysis data (Calculated C 53.7 %, H 3.79 %, P 5.54 %) are quoted
9. Cresp TM, Giles RGF, Sargent MV, Cresp TM, Giles RGF, Sargent MV, Brown C, Smith DO'N (1974) J Chem Soc, Perkin Trans 1:2435. Only MP (242–244 °C) and elemental analysis data are given
10. Yamataka H, Nagareda K, Ando K, Hanafusa T (1992) J Org Chem 57:2867
11. Uziel J, Riegal N, Aka B, Figuère P, Jugé S (1997) Tetrahedron Lett 38:3405
12. Moghaddam FM, Farimani MM (2010) Tetrahedron Lett 51:540
13. Based on method presented in Miyano M, Stealey MA (1975) J Org Chem 40:2840
14. Bell TW, Sondheimer F (1981) J Org Chem 46:217
15. Palaček J, Kvíčala J, Paleta O (2002) J Fluorine Chem 113:177
16. Dabkowski W, Ozarek A, Olejniczak S, Cypryk M, Chojnowski J, Michalski J (2009) Chem Eur J 15:1747
17. Gannett PM, Nagel DL, Reilly PJ, Lawson T, Sharpe J, Toth B (1064) J Org Chem 1988:53
18. Nesmeyanov NA, Rebrova OA, Mikul'shina VV, Petrovsky PV, Robas VI, Reutov OA (1976) J Organomet Chem 110:49
19. Cornforth J, Cornforth RH, Gray R (1982) J Chem Soc, Perkin Trans 1:2289
20. Stephan M (MS), Sterk D, Modec B, Mohar B (2007) J Org Chem 72:8010
21. Vedejs E, Marth CF, Ruggeri R (1988) Iodide salt characterisation. J Am Chem Soc 110:3940
22. Vedejs E, Marth CF, Ruggeri R (1988) J Am Chem Soc 110:3940
23. Dunne EC, Coyne EJ, Crowley PB, Gilheany DG (2002) Tetrahedron Lett 43:2449
24. Yeo Q, Kinney EP, Zheng C (2004) Org Lett 6:2997
25. Li J, Hua R, Liu T (2010) J Org Chem 75:2966
26. Peng Z-Y, Ma F-F, Zhu L-V, Xie X-M, Zhang Z (2009) J Org Chem 74:6855
27. Petrova J, Momchilova S, Kirilov M (1985) Phosphorus Sulfur 24:243 (Z isomer only)
28. Kabalka GW, Li NS, Tejedor D, Malladi RR, Trotman S (1999) J Org Chem 64:3157
29. Dong D-J, Li HH, Tian SK (2010) J Am Chem Soc 132:5018
30. de Meijere A, Song ZZ, Lansky A, Hyuda S, Rauch K, Noltemeyer M, Koenig B, Knieriem B (1998) Eur J Org Chem 2289
31. Kupchen SM, Wormser HC (1965) J Org Chem 30:3792
32. Arela RK, Leadbeater NE (2005) J Org Chem 70:1786
33. Cella R, Stefana HA (2006) Tetrahedron 62:5656
34. Wyatt P, Warren S, McPartlin M, Woodruffe T (2001) J Chem Soc, Perkin Trans 1:279
35. Engman L (1984) J Org Chem 49:3559
36. Wallace TW, Wardell I, Li KD, Leeming P, Redhouse AD, Challand SR (1995) J Chem Soc, Perkin Trans 1:2293
37. Terfort A, Brunner H (1996) J Chem Soc, Perkin Trans 1:1467
38. Rauniyar V, Zhai H, Hall DG (2008) J Am Chem Soc 130:8481 (*E*-isomer only)

39. Bellinger GF, Campbell WE, Giles RGF (1982) J Chem Soc Perkin Trans 1:2819 (*Z*-isomer only)
40. Mintas M, Jakopcic K, Klasnic L (1977) Z Naturforsch B 32b:181
41. Kim IS, Dong GR, Jung YH (2007) J Org Chem 72:5424
42. Park CP, Dong-Pyo K (2010) J Am Chem Soc 132:10102
43. Ueda S, Okada T, Nagasawa H (2010) Chem Commun 46:2462
44. Wu Y, Zhao J, Chen J, Pan C, Li L, Zhang H (2009) Org Lett 11:597
45. Ando K (1999) J Org Chem 64:8406
46. Jia C, Lu W, Oyamada J, Kitamura T, Matsuda K, Irie M, Fujiwara Y (2000) J Am Chem Soc 122:7252
47. Chinatareddy VR, Ellern A, Verkade JG (2010) J Org Chem 75:7166
48. Ruan J, Li X, Saidi O, Xiao J (2008) J Am Chem Soc 130:2424
49. Ohe K, Takahashi H, Uemura S, Sugita N (1987) J Org Chem 52:4859
50. Blakemore PR, Ho DKH, Mieke Nap W (2005) Org Biomol Chem 3:1365
51. Minin PL, Walton JC (2004) Org Biomol Chem 2:2471
52. Poeylaut-Palena AA, Testero SA, Mata EG (2008) J Org Chem 73:2024
53. Mutter R, Campbell IB, Martin de la Nava EM, Merritt AT, Wills M (2001) J Org Chem 66:3284
54. Bull SD, Davies SG, Delgado-Ballester S, Kelly PM, Kotchie LJ, Gianotti M, Laderas M, Smith AD (2001) J Chem Soc, Perkin Trans 1:3112 (*E*-isomer only)
55. Bellassoued M, Majidi A (1993) J Org Chem 58:2517
56. Yamamoto K, Watanabe M, Ideta K, Mataka S, Thiemann T (2005) Z Naturforsch 60b:1299 (*E*-isomer only)
57. Ballerini E, Minuti L, Piermatti, O (2010) J Org Chem 75:4251 (*E*-isomer only)
58. Vedejs E, Fleck T, Hara S (1987) J Org Chem 52:4637
59. Vedejs E, Fleck T (1989) J Am Chem Soc 111:5861
60. Davioud-Charvet E, McLeish MJ, Veine DM, Giegel D, Arscott LD, Andricopulo AD, Becker K, Mueller S, Schirmer RH, Williams Jr CH, Kenyon GL (2003) Biochemistry 42:13319 (*E*-isomer only)
61. Lin YM, Li Z, Casarotto V, Ehrmantraut J, Nguyen AN (2007) Tetrahedron Lett 48:5531 (*E*-isomer only)
62. Following procedure from: Paterson I, Gardner NM, Guzmán E, Wright AE (2009) Bioorg Med Chem 17:2282
63. Jiao L, Yuan C, Yu Z-X (2008) J Am Chem Soc 130:4421
64. Touchard FP (2004) Procedure from. Tetrahedron Lett 45:5519
65. Still WC, Gennari C (1983) Tetrahedron Lett 24:4405
66. Yamamoto K, Watanabe M, Ideta K, Mataka S, Thiemann T (2005) Z Naturforsch B 60:1299 (*E*-isomer only)
67. Ram RN, Manoj TP (2008) J Org Chem 73:5633
68. Henichart J-P, Houssin R, Bernier J-L (1986) J Het Chem 23:1531
69. Tanaka K, Takeaki S (2005) Org Lett 7:3561
70. House OH, Giese RW, Kronberger K, Kaplan JP, Simeone JF (1970) J Am Chem Soc 92:2800
71. Blake AJ, Harding M, Sharp JT (1994) J Chem Soc, Perkin Trans 1:3149 (*E*-isomer only)
72. Colberg JC, Rane A, Vaquer J, Soderquist JA (1993) J Am Chem Soc 115:6065
73. Rim C, Son DY (2003) Org Lett 5:2443
74. Akgun E, Glinski MB, Dhawan KL, Durst T (1981) J Org Chem 46:2730
75. Padwa A, Rieker WF, Rosenthal RJ (1985) J Am Chem Soc 107:1710
76. Maryanoff BE, Reitz AB, Mutter MS, Inners RR, Almond HR Jr, Whittle RR, Olofson RA (1986) J Am Chem Soc 108:7664
77. Huang C, Tang X, Fu H, Jiang Y, Zhao Y (2006) J Org Chem 71:5020
78. Han L-B, Kambe N, Ogiwa A, Ryu I, Sonada N (1993) Organometallics 12:473
79. Afarinkia K, Binch HM, Modi C (1998) Tetrahedron Lett 39:7419
80. Gan Y, Spencer TA (2006) J Org Chem 71:5870

81. Gurjar MK, Khaladkar TP, Borhade RG, Murugan A (2003) Tetrahedron Lett 44:5183
82. Krueger EB, Hopkins TP, Keaney MT, Walter MA, Boldi AM (2002) J Comb Chem 4:229
83. Tronchet JMJ, Gentile B (1979) Helv Chim Acta 62:2091
84. Baati R, Mioskowski C, Baram D, Karche R, Falck JR (2008) Org Lett 8:2949
85. Vikic-Topic D, Meic Z (1986) J Mol Struct 142:371
86. Coyne EJ (1997) Arynes, ylides and chiral epoxides from substituted stilbenes. Ph.D. Thesis, UCD, Belfield, Dublin 4, pp 222–223
87. Ramnial T, Taylor SA, Bender ML, Gorodetsky B, Lee PTK, Dickie DA, McCollum BM, Pye CC, Walsby CJ, Clyburne JAC (2008) J Org Chem 73:801

Printed by Publishers' Graphics LLC
BT20130210.19.19.97